# ANATOMY AND PHYSIOLOGY

# ANATOMY AND PHYSIOLOGY

## John Raynor
Borough of Manhattan Community College,
City University of New York

**HARPER & ROW, PUBLISHERS**
New York  Hagerstown  San Francisco  London

Sponsoring Editor: Joe Ingram/William M. Burgower
Project Editor: H. Detgen
Designer: T. R. Funderburk
Production Supervisor: Kewal K. Sharma
Photo Researcher: Myra Schachne
Compositor: Progressive Typographers, Inc.
Printer: The Murray Printing Company
Binder: Halliday Lithograph Corporation
Art Studio: Robert Frank; Eric G. Hieber Associates Inc.

**Anatomy and Physiology**

Copyright © 1977 by John Raynor

All rights reserved. Printed in the United States of America. No part of this book may be used or reproduced in any manner whatsoever without written permission except in the case of brief quotations embodied in critical articles and reviews. For information address Harper & Row, Publishers, Inc., 10 East 53rd Street, New York, N.Y. 10022.

Library of Congress Cataloging in Publication Data

Raynor, John, 1942–
   Anatomy and physiology.

  Includes index.
  1. Human physiology.  2. Anatomy, Human.
I. Title. [DNLM: 1. Anatomy. 2. Physiology.
QS4 R276a]
QP34.5.R38      612      76-1942
ISBN 0-06-045339-7

# Contents

Preface   xiii

1. Introduction   1
2. Chemistry   19
3. The Cell   39
4. Tissues and Membranes   70
5. The Skeletal System   85
6. The Muscular System   118
7. The Nervous System   154
8. Senses and Motor Pathways   184
9. The Endocrine System   211
10. Circulation: Introduction to the Circulatory System and the Blood   236
11. The Heart   251
12. The Blood Vessels and Lymphatics   266
13. Blood Pressure, Blood Flow, and Capillary Exchange   293
14. The Respiratory System   308
15. The Digestive System   330
16. Energy Metabolism   363
17. The Urinary System   383
18. The Reproductive System   410

Glossary   437
Index   443

# Expanded Contents

Preface xiii
1. INTRODUCTION 1
    General Principles 2
    Levels of Organization 4
    The Organ Systems 5
    Body Regions 12
    Terms Describing Anatomical Location 14
    Planes of the Body 14
    Body Cavities 16
    Learning New Terms 16
    Objectives 17
2. CHEMISTRY 19
    Atoms 20
    Molecules 24
    States of Matter 25
    Water 26
    Acids and Bases 28
    Carbon Compounds 29
    Carbohydrates 30
    Lipids 32
    Proteins 34
    Chemical Reactions 35
    Objectives 38
3. THE CELL 39
    General Cell Structure 40

Movement of Substances Across the Cell
  Membrane   42
Cellular Energy Metabolism   51
Protein Synthesis   60
Mitosis   66
Objectives   68

4. TISSUES AND MEMBRANES   70
   Epithelial Tissue   71
   Connective Tissue   73
   Muscle Tissue   78
   Nerve Tissue   79
   Membranes   79
   The Skin   80
   Objectives   84

5. THE SKELETAL SYSTEM   85
   Bone Tissue   87
   Cartilage Tissue   88
   Bone Structure   89
   Bone Growth and Repair   91
   Deposition and Release of Bone Calcium   92
   The Skeletal Bones   93
   Types of Movement   115
   Objectives   116

6. THE MUSCULAR SYSTEM   118
   Introduction   119
   Structure of a Skeletal Muscle Cell   121
   Mechanism of Contraction   123
   Energy for Muscular Contraction   129
   Force of Contraction   131
   The Skeletal Muscles   133
   Objectives   152

7. THE NERVOUS SYSTEM   154
   Introduction   155
   Divisions of the Nervous System   156
   Cells of the Nervous System   157
   Mechanism of Neural Activity   159
   Spinal Reflex   162
   Structure of the Spinal Cord   164
   Spinal Nerves   166
   The Brain   170
   Cranial Nerves   178
   Protection of the Central Nervous System   179
   Objectives   182

8. SENSES AND MOTOR PATHWAYS   184
   Touch and Pressure   185
   Temperature and Pain   188

Proprioception 189
Taste 190
Smell 191
The Eye and Vision 192
The Ear and Hearing 201
Balance 206
Motor Pathways to Skeletal Muscle 206
Autonomic Nervous System 207
Objectives 210

9. THE ENDOCRINE SYSTEM 211
Introduction 212
Pituitary Gland (Hypophysis) 215
Thyroid Gland 222
Parathyroids 227
Adrenal Glands 229
Pancreas 233
Other Structures with Endocrine Activity 234
Objectives 234

10. CIRCULATION: INTRODUCTION TO THE CIRCULATORY SYSTEM AND THE BLOOD 236
Introduction 237
Blood 238
White Blood Cells and Body Defense 242
Hemostasis 245
Blood Types and Blood Transfusions 247
Objectives 250

11. THE HEART 251
Introduction 252
Chambers of the Heart 254
Heart Valves 256
Contraction of the Heart 257
Cardiac Cycle 260
Heart Sounds 262
Cardiac Output 262
Objectives 265

12. THE BLOOD VESSELS AND LYMPHATICS 266
Arteries 267
Capillaries 276
Veins 277
Lymphatic System 287
Spleen 291
Thymus 291
Objectives 292

13. BLOOD PRESSURE, BLOOD FLOW, AND CAPILLARY EXCHANGE 293
Introduction 294

Maintenance of Stable Arterial Pressure   295
Capillary Blood Flow   301
Capillary Exchange   302
Hemorrhage   304
Objectives   307

14. THE RESPIRATORY SYSTEM   308
Respiratory Tract   309
Lungs   314
Mechanism of Respiration   316
Respiratory Volumes   322
Composition of Air   322
Exchange and Transport of Gases   323
Control of Respiration   326
Objectives   328

15. THE DIGESTIVE SYSTEM   330
Digestive Tube Structure   331
Peristalsis   333
Peritoneum   333
Nutrients   334
The Digestive Tract   338
Liver   347
Pancreas   351
Defecation   360
Objectives   361

16. ENERGY METABOLISM   363
Conversion of Food Energy to ATP   364
Storage of Energy   367
Usage of Stored Energy   369
Control of Metabolism   371
Diabetes Mellitus   376
Energy Balance   378
Regulation of Body Temperature   379
Objectives   381

17. THE URINARY SYSTEM   383
Functions of the Urinary System   384
Anatomy of the Urinary System   386
Basic Processes of Urine Formation   391
Control of Excretion   393
Fluid and Electrolyte Balance   395
Acid-Base Balance   401
Micturition   407
Diseases of the Kidney   407
Artificial Kidney   408
Objectives   409

18. THE REPRODUCTIVE SYSTEM   410
Male Reproductive Tract   411

Male Accessory Glands  414
Male Reproductive Hormones  415
Spermatogenesis  416
Seminal Fluid  418
Delivery of Sperm to the Female  418
Male Puberty  419
Female Reproductive Tract  419
Menstrual Cycle  423
Pregnancy  427
Lactation  433
Female Puberty  433
Female Menopause  433
Sex Determination  434
Objectives  434

Glossary  437
Index  443

# Preface

This text is designed to present the basic concepts of human anatomy and physiology for students planning a career in one of the health professions. Training programs in all of the health professions are based on the fact that a knowledge of the normal structures and functions of the human body is required to understand medical problems and how they can be treated. In this text I have tried to include and emphasize material that the student will apply to later professional courses.

Our present understanding of human anatomy and physiology is much broader than can be presented in an introductory text of this type. In being selective I have tried to emphasize the basic physiological processes and the reasons these processes take place. It is hoped that this information will provide a framework into which later knowledge can be placed.

Each chapter of the text includes behavioral objectives. These objectives are meant to serve two purposes. They are meant to delineate information the student should retain upon completion of each chapter and also to serve as a way for the student to test himself or herself on the subject matter.

The excellent illustrations were supplied by Robert Frank and by Eric G. Hieber Associates. Every attempt has been made to make the illustrations clear and readable. In addition to the anatomical illustrations, there are a large number of schematic illustrations which are meant to clarify physiological processes. Flowcharts are also used to sum up sequences of events involved in physiological processes. All of these illustrations are integrated with the written material in the text.

I would like to thank Dr. Ronald Slavin for his many helpful suggestions as to the type of material relevant to an introductory anatomy and physiology course. Finally, I would like to thank Joe Ingram, Holly Detgen, T. R. Funderburk, and Howard S. Leiderman of Harper & Row for all of the time and effort they have put into making this book a reality.

J.R.

# 1 Introduction

GENERAL PRINCIPLES
LEVELS OF ORGANIZATION
THE ORGAN SYSTEMS
BODY REGIONS
TERMS DESCRIBING ANATOMICAL
   LOCATION
PLANES OF THE BODY
BODY CAVITIES
LEARNING NEW TERMS

Human **anatomy** is the study of the structure of the human body, and human **physiology** is the study of how the human body functions. Because the structure of the body is interrelated with the way in which it functions it is reasonable to study anatomy and physiology as one subject.

Although the study of anatomy and physiology requires learning a large amount of detailed information, most of the information is relatively straightforward and should not be tremendously difficult to understand. In general the human body functions in a reasonable and logical way and most of the detailed information can be fit into an overall pattern. We are going to begin our description of human anatomy and physiology by looking at certain general principles that hopefully will serve as a framework for learning the details of anatomy and physiology.

## GENERAL PRINCIPLES

**1. From an anatomical and physiological point of view survival is the object of living.**

The structure and function of the human body are directed toward one end: survival. This includes survival of each individual part of the body, such as the liver or kidney, the survival of the body as a whole, and the survival of human beings as a form of life via reproduction. In looking at any particular aspect of structure or function you should always try to see in what way it aids in survival. For example, when it is hot out we sweat. Is there any point in sweating? Does it aid in survival? The evaporation of water from any object uses up heat and cools the object. When it is hot out we sweat; the evaporation of the sweat from our skin cools our body. If the body became too hot it would no longer function properly.

**2. Cells are the basic functional unit of the body.**

The human body is made up of trillions of tiny structures called **cells.** Cells are the basic working unit of the body. In each of these cells an enormous number of activities take place. Some of these activities serve to keep the cell itself alive while other activities contribute to the survival of the entire body. Therefore if one were to look at a particular cell in the pancreas, many of the activities taking place in that cell would be directed toward keeping itself alive. However, the cells of the pancreas also manufacture chemicals that are secreted into the digestive tract where they aid in the breakdown of food. Through the production of these digestive chemicals the cells of the pancreas contribute to the survival of the entire body.

In Chapter 3 we will discuss in detail those aspects of cell structure and function that are common to all cells. However, throughout the book you should keep in mind the fact that any activity that takes place in the body is really the result of activities that take place in the cells. Thus the pumping of blood by the heart is due to the action of heart cells, and the formation of urine by the kidney is a result of activities that take place in kidney cells.

### 3. The environment surrounding cells must be kept stable for cells to survive.

Cells are exceedingly delicate and can only function in a very stable environment. The immediate environment of all the cells of the body is the surrounding fluid, known as the **extracellular fluid.** The temperature, pressure, acidity, salt content, etc., of this fluid determine the conditions under which the cells function. It is from the extracellular fluid that the cells must obtain their nutrients and into which the cells must deposit their wastes. This extracellular fluid is referred to as the **internal environment,** the environment within the body. This is to distinguish it from the **external environment,** the environment outside the body. Whereas the condition of the external environment can vary over a reasonably large range, the conditions of the internal environment must be kept quite stable. For example, a change in external temperature from 98.6°F to 60°F would not affect the activities of most cells. On the other hand, a lowering of the internal temperature from its normal value of 98.6°F to 60°F would greatly slow down all cellular activity.

The idea that the internal environment must be kept constant was first developed by the nineteenth-century French physiologist Claude Bernard, and was later elaborated upon by the American physiologist W. B. Cannon. Cannon introduced the word **homeostasis** to mean a stable internal environment.

### 4. Most specialized body functions are directed toward maintaining homeostasis.

Survival depends on maintaining a stable internal environment, homeostasis. We have already stated that within each cell two types of activities take place. Some activities serve to keep the cell itself alive. Other activities are specialized activities of that particular type of cell. All of the specialized activities of cells act either directly or indirectly to maintain homeostasis. For example, one of the specialized functions of liver cells is to help maintain a stable amount of sugar in the extracellular fluid. If the level of sugar in the extracellular fluid begins to fall, the liver cells produce additional sugar which can be added to the extracellular fluid. If the amount of sugar in the extracellular fluid begins to get too high the liver cells remove sugar from the fluid. By adding sugar when there is not enough in the extracellular fluid and by removing sugar when there is too much, the liver helps to maintain a stable level of sugar in the internal environment.

**5. Body functions must be controlled to meet the needs of different situations.**

Although the internal environment must be kept very stable, the external world in which we live is continuously changing. The temperature, the amount and type of food available, and the need for physical activity are but a few of the factors that can change from day to day and from place to place. In order to maintain a stable internal environment we must be able to adjust to a changing external environment. For example, our internal temperature stays close to 98.6°F whether the external temperature is 105°F or 45°F. We accomplish this, in part, by sweating when it is hot and by shivering when it is cold. In order to have the appropriate response to a particular situation, body function must be controlled. Thus something must stimulate the sweat glands to secrete large amounts of sweat when it is hot and small amounts of sweat when it is cold. Likewise something must stimulate the rhythmic muscle tremors of shivering when it is cold but not when it is hot. To a very large extent the various body functions are controlled by the nerves and hormones. Thus nerves and hormones act to stimulate the sweat glands to secrete sweat when it is hot and to stimulate the muscle cells to tremor when it is cold.

# LEVELS OF ORGANIZATION

In order to carry out their specialized activities the cells of the body are grouped together into larger structures. A **tissue** consists of a group of similar cells along with the material between the cells, which are organized to carry out a particular function. There are four major types of tissues: epithelial, connective, nervous, and muscular, each of which will be discussed in Chapter 4.

Different types of tissues are combined into larger functional units known as **organs.** An organ is defined as a group of tissues working together to perform a particular function. The heart, for instance, is an organ made up of epithelial tissue which protects it, muscle tissue which is responsible for the actual contractions, nervous tissue which controls it, and connective tissue which holds the other tissues together.

Finally a number of different organs may act together to perform a particular function. Such a collection of organs is known as an **organ system.** The lungs and the tubes that connect the lungs with the outside air form the respiratory system, which is responsible for the exchange of oxygen and carbon dioxide between the internal and external environments.

In summary there are four basic levels of organization within the body: the individual cells, groups of similar cells organized into tissues, groups of tissues organized into organs, and groups of organs organized into organ systems.

## THE ORGAN SYSTEMS

Survival depends on maintaining homeostasis, a stable internal environment in which the cells can live and carry out their functions. At this point we are going to describe briefly each of the major organ systems and at least some of the ways in which they contribute to homeostasis. There are really two reasons for doing this: one is to give you an idea of the basic physiological needs of the body and the other is to give you an idea of at least some of the functions of the major internal organs. An overall idea of what goes on in the body should make it easier to learn the detailed information about the various organ systems given in Chapters 5 through 18.

Survival depends on obtaining from the outside world the raw materials that the cells require to carry out their functions. These raw materials include: molecules that can provide the energy necessary to do the work of the cells; molecules that can become part of the structure of the body; vitamins, which aid the cells in carrying out certain chemical reactions; minerals; and water. With the exception of oxygen, all of the substances taken into the body from the external world enter through the **digestive system.** The digestive system must then break down the food (digest it) into particles small enough to enter the bloodstream. Those substances in the food that for one reason or another cannot be absorbed into the blood are eliminated from the body as feces.

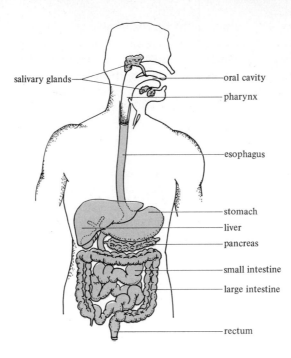

**Figure 1-1** The digestive system.

Basically the digestive system consists of a tube that runs from the mouth to the anus, as illustrated in Figure 1-1. This tube, the **digestive tract,** consists of the **oral cavity** where food enters the body; the **pharynx** (throat) located at the back of the oral cavity; the **esophagus** which is a long tube connecting the pharynx with the stomach; the **stomach** which stores food and partially digests it; the **small intestine** in which most of the food is digested and absorbed into the blood; and the **large intestine** and **rectum** in which the unabsorbed food is stored until its elimination from the body. In addition to the digestive tract, the digestive system contains certain structures which secrete chemicals that aid in digestion: the salivary glands in the mouth and the liver and pancreas in the abdomen.

The only substance needed from the outside world which does not enter the body through the digestive system is oxygen, a gas used in the process by which the energy in food is released for use by the cells. The exchange of gases between the external environment and the internal environment is carried out by the **respiratory system.** The respiratory system brings oxygen into the body and eliminates carbon dioxide, a potentially poisonous gas produced by the combination of oxygen and food.

The major structures of the respiratory system are shown in Figure 1-2. Gases enter or leave the body through either the nasal cavity or the oral cavity. These two cavities are joined at the pharynx, which has one opening into the esophagus through which

**7** The organ systems

food passes on its way to the stomach and one opening into the **larynx** through which air passes on its way to the lungs. The larynx, which contains the two vocal cords used in speaking, guards against the entry of food into the lungs. An opening from the larynx leads into a tube, the **trachea,** which branches into two **bronchi,** one to each lung. Within the lungs the bronchi branch into smaller and smaller tubes which finally end in tiny sacs called alveoli. At the alveoli oxygen enters the blood to be carried to the cells, and carbon dioxide, which has been removed from the cells, enters the respiratory tract to be eliminated from the body. Thus the respiratory tract consists of a set of tubes through which oxygen passes as it moves from the mouth or nose, where it enters the body, to the alveoli, where it enters the blood; carbon dioxide passes through these same tubes as it moves from the alveoli, where it leaves the blood, to the mouth or nose, where it leaves the body. Some of these tubes—the small branching bronchi—as well as the tiny air sacs where gases enter or leave the blood—the alveoli—are located within the lungs.

Food from the digestive tract and oxygen from the respiratory tract must be transported to cells throughout the body. The **circulatory system** functions to transport material from one part of the body to another much as a subway system transports people from one part of a city to another. The circulatory system is composed of the **blood,** which is the transporting fluid that moves throughout the body, the **blood vessels,** which are tubes through which the blood flows, and the **heart,** which pumps the blood through the blood vessels.

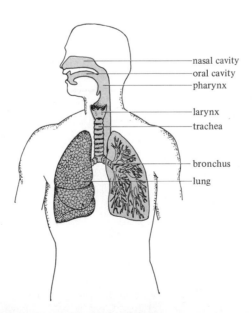

**Figure 1-2** The respiratory system.

**8** INTRODUCTION

Just as the circulatory system must transport food and oxygen to the cells it must also transport wastes such as carbon dioxide away from the cells. All of the wastes produced by the cells must eventually be eliminated from the body. Because it is a gas, carbon dioxide is eliminated by the respiratory system. Most wastes, however, are eliminated by the **urinary system.** In addition to eliminating wastes, the urinary system plays a role in regulating the amount of salt and water contained within the body.

The **kidneys** are the major functional units of the urinary system, which is illustrated in Figure 1-3. Waste materials are removed from the blood as it flows through the kidneys and are used to form the **urine,** a fluid that is eliminated from the body. Once the urine is formed in the kidneys it flows through the **ureters** to the **bladder** where it is stored until its elimination from the body. The tube that carries urine out of the body from the bladder is called the **urethra.**

The **skeletomuscular system** provides the framework to which the various internal organs such as the heart, stomach, liver, and kidney are connected. This framework is designed in such a way as to protect the internal organs and to provide a reasonable amount of movement. This system is often divided into two systems: the **skeletal system,** shown in Figure 1-4a, which consists of the bones that form the solid framework, and the **muscular system,** shown in Figure 1-4b, which consists of the skeletal muscles that can move the bones.

Reproduction, although not essential for an individual's survival, is necessary for the survival of human beings as a form of life. The function of the **reproductive system** is to create new human beings and to provide for their nourishment and protection until they develop to a point at which they are capable of sustaining independent life. This involves the formation of sperm in the male

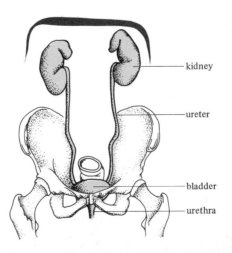

**Figure 1-3** The urinary system.

**9** The organ systems

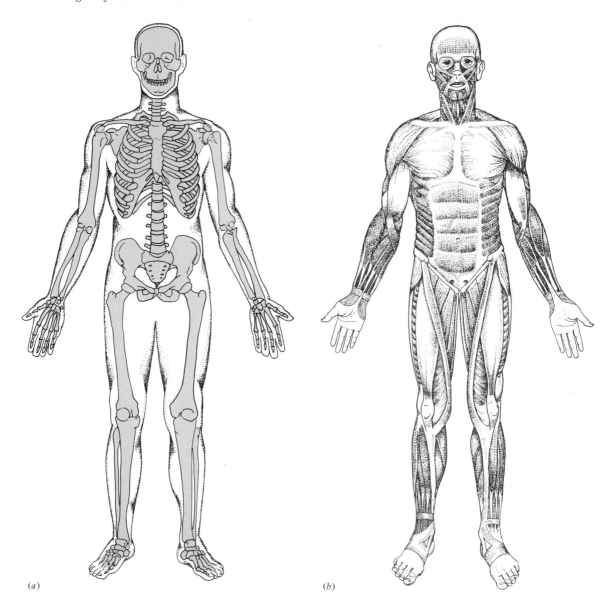

(a)    (b)

**Figure 1-4** The skeletomuscular system: (a) the skeletal system and (b) the muscular system.

body and eggs in the female body, the fertilization of the egg by the sperm, the growth and development of the fertilized egg in the female body for nine months, the delivery of the fetus, and the nourishment of the newborn baby. Obviously the male and female reproductive systems have different roles in this process and are designed differently.

The male reproductive system is illustrated in Figure 1-5. The **testes,** which are located outside of the abdomen in the **scrotal sac,**

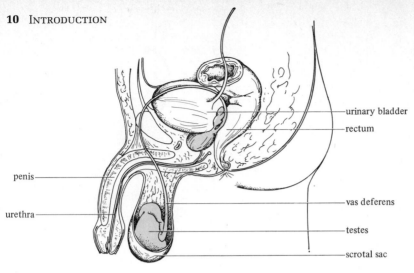

Figure 1-5 The male reproductive system.

produce the sperm as well as the male reproductive hormone testosterone. Once produced in the testes, sperm pass through the **vas deferens** and **ejaculatory duct** to the urethra. The **urethra** is a tube that passes through the penis. Both sperm and urine are transported out of the body through this tube.

The female reproductive system is more complicated than the male system in that it must provide for the growth and development of the eggs once they are fertilized. Figure 1-6 illustrates the major parts of the female reproductive system. The **ovaries** are somewhat analogous to the male testes, they produce the eggs and the female reproductive hormones estrogen and progesterone. The **oviducts,** or **fallopian tubes,** connect the ovaries with the uterus, where the fertilized eggs develop. The **vagina** is a tube that connects the uterus with the outside of the body. Notice that in the

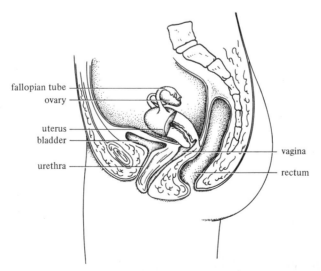

Figure 1-6 The female reproductive system.

## 11 The organ systems

female body the reproductive system and the urinary system are completely separate, whereas in the male they join at the urethra. The mammary glands, which provide nourishment for the baby after it is born, also form a part of the female system.

Each of the systems we have discussed so far is controlled to function in a way that is appropriate to the prevailing conditions. The liver and pancreas should secrete digestive chemicals into the digestive tract when it contains food but not when it is empty; more air should be moved in and out of the lungs when a person is running than when he is sleeping; the kidneys should form more urine after a person drinks a six-pack of beer than beforehand. The job of regulating the other organ systems is the function of the **endocrine system** and the **nervous system.**

The endocrine system regulates body function through the action of chemicals called **hormones.** The hormones are manufactured and secreted into the blood by a group of structures, the **endocrine glands,** which are located at various parts of the body as illustrated in Figure 1-7. Each of the endocrine glands is designed

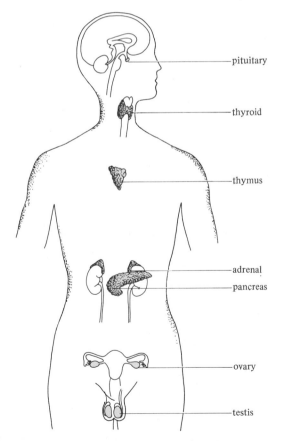

**Figure 1-7** The endocrine system.

for the production of one or more particular hormones. Once the hormones are secreted into the blood they are transported to the particular cells whose function they regulate. For example, when a baby sucks on its mother's nipple the pituitary gland releases a hormone called oxytocin into the blood. The oxytocin is transported by the blood to the breast where it stimulates the secretion of milk. In other words oxytocin regulates the secretion of milk by the breast so that milk is secreted only when the baby is nursing.

The nervous system regulates body function through the action of neurons, a network of cells that extends throughout the body. The nervous system is exceedingly complex, and the ways in which it works are not very well understood at the present time. From the simplest point of view one could say that the nervous system carries out two basic types of activities: it detects changes in the internal and external environment, and it directs the appropriate response to these changes. For example, if you touch something hot with your hand the nervous system detects the heat and directs the contraction of the muscle, which pulls your hand away from the hot object.

Anatomically the nervous system consists of **receptors** that detect changes in the environment; the **central nervous system,** composed of the brain and spinal cord, which directs the response to these changes; and the **peripheral nervous system,** which connects the central nervous system with the receptors and with the muscles and glands which actually produce the response.

## BODY REGIONS

The various surface regions of the body have specific anatomical names. Figure 1-8 gives both the anatomical and common names of the major body regions. For example, the forehead is the frontal region, the eye is the orbital region, the mouth is the buccal region, etc. Learning the names of these surface regions at this point will aid in learning the names of the underlying structures later. For instance from Figure 1-8 you can see that the femoral region is the upper leg. The bone of the upper leg is called the femur, one of the muscles of the upper leg is the quadriceps femoris, the blood vessel carrying blood to the upper leg is the femoral artery, the blood vessel carrying blood away from the upper leg is the femoral vein, and one of the nerves of the upper leg is the femoral nerve.

**13** Body regions

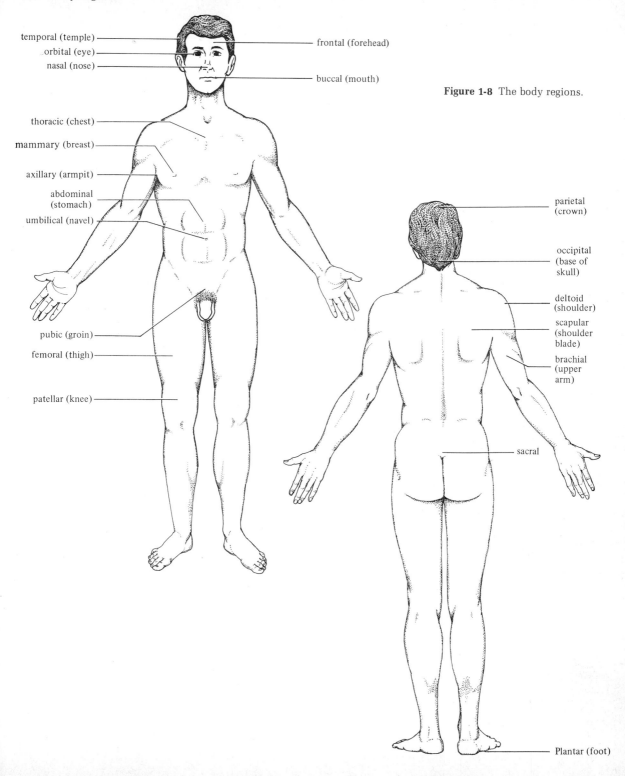

**Figure 1-8** The body regions.

# TERMS DESCRIBING ANATOMICAL LOCATION

The relative locations of the various parts of the body are described by a special set of anatomical terms listed in Table 1-1. These terms apply when the body is in the anatomical position. In this position the person would be standing erect, facing forward, with the arms at the sides and the palms forward.

**Table 1.1** Terms Describing Anatomical Location

| Term | Definition | Example |
|---|---|---|
| Superior (cranial) | Toward the head | The nose is superior to the mouth. |
| Inferior (caudal) | Toward the feet | The waist is inferior to the shoulder. |
| Anterior (ventral) | Toward the front | The nose is on the anterior (ventral) side of the face. |
| Posterior (dorsal) | Toward the back | The backbone is posterior (dorsal) to the ribs. |
| Medial | Toward the midline | The navel is medial to the hips. |
| Lateral | Toward the side | The eyes are lateral to the nose. |
| Internal | Toward the inside | The heart is internal to the ribs. |
| External | Toward the outside | The skin is external to the muscles. |
| Proximal | Toward the point of attachment or origin | The elbow is proximal to the hand. |
| Distal | Further from the point of attachment or origin | The hand is distal to the elbow. |
| Central | Toward the anatomical center (also refers to the principal part) | The central canal is a space in the center of the spinal cord. |
| Peripheral | Toward the outside or surface of the body (also refers to extensions of the principal part) | Peripheral nerves are extensions of the central nervous system. |
| Visceral | Related to the internal organs | The visceral layer of the pleura covers the lungs. |
| Parietal | Related to the body walls | The parietal layer of the pleura lines the inside wall of the chest. |

In using these terms you should bear in mind the fact that they refer to the relative location of body parts. The nose is superior to the mouth, but it is inferior to the eyes. The wrist is distal to the elbow, but it is proximal to the fingers.

# PLANES OF THE BODY

The body is a three-dimensional structure, having height, width, and depth. Unfortunately, for the most part you will have to learn about this three-dimensional structure through the combined use of the flat two-dimensional drawings used in this textbook and by your instructor. These two-dimensional drawings are made by cutting a slice, called a section, through the body or through one of its parts and by drawing what is exposed. In looking at a drawing it is important to know from what angle the drawing is made. Three

## 15  Planes of the body

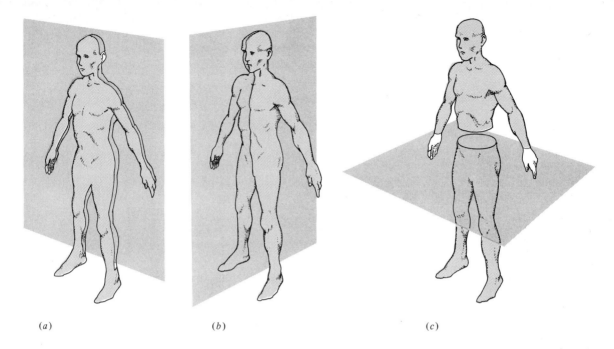

**Figure 1-9**  The planes of the body: (a) frontal, (b) sagittal, and (c) transverse.

fundamental planes are used in cutting sections through the body. These are shown in Figure 1-9.

A **frontal** or **coronal** plane, illustrated in Figure 1-9a, passes lengthwise through the body to divide the front from the back. A frontal section shows a structure from the front. A **sagittal plane,** illustrated in Figure 1-9b, passes lengthwise through the body dividing it into left and right sides. A sagittal section shows a structure from the side. A **transverse plane,** illustrated in Figure 1-9c, cuts across the body to divide the top from the bottom.

Figure 1-10 shows frontal and sagittal views of the brain. The brain clearly looks different from the two different angles.

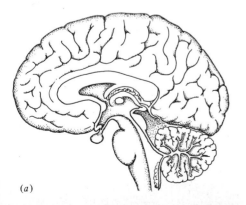
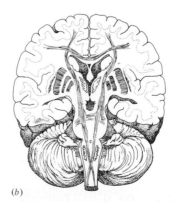

**Figure 1-10**  The brain as seen in (a) sagittal and (b) frontal sections.

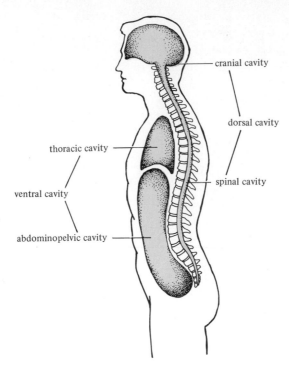

**Figure 1-11** The body cavities.

## BODY CAVITIES

The inside of the body is divided into two major cavities, the **ventral cavity** and the **dorsal cavity.** Each of these is in turn divided into two portions, as illustrated in Figure 1-11.

The ventral cavity is divided into a **thoracic portion** and an **abdominopelvic portion** by the main muscle of respiration, the diaphragm. The dorsal cavity is divided into a **cranial portion,** contained within the skull, and a **spinal portion,** contained within the backbone. The thoracic portion of the ventral cavity contains the heart and lungs. The stomach, intestine, liver, pancreas, kidney, bladder, and female reproductive organs are all located in the abdominopelvic portion of the ventral cavity. Within the dorsal cavity the brain is found in the cranial portion and the spinal cord is found in the spinal portion.

## LEARNING NEW TERMS

Learning anatomy and physiology involves learning a whole new vocabulary. This can be made easier if you see those features that are common to many of the words you will learn. A large number of the new words are made up of certain prefixes, roots, and suffixes derived from Greek and Latin. A root is a basic word that can be modified by certain letters before it (prefixes) and certain letters after it (suffixes). For example, *glyco-* is a root word for sugar. *Hyper-* is a prefix meaning above normal and *-emia* is a suffix meaning in the blood. Hyperglycemia means high blood sugar. If the prefix *hyper-* (above) is replaced by the prefix *hypo-* (below) a new word, hypoglycemia is derived meaning low blood sugar. Likewise *hepata-* is the root word for liver and *-itis* is a suffix meaning inflammation. Therefore hepatitis refers to an inflammation of the liver. Some of the most common prefixes, roots, and suffixes can be found in the Glossary.

## OBJECTIVES FOR THE STUDY OF THE INTRODUCTION

At the end of this unit you should be able to:

1. Define the words anatomy and physiology.
2. Name the basic functional unit of the body.
3. Explain what is meant by the internal environment.
4. Explain the concept of homeostasis.
5. Explain the need for control over body function.
6. Name the four levels of organization in the body and describe each.
7. State three functions of the digestive system.
8. Name in order the structures of the digestive tract which food passes through as it moves from the mouth to the anus.
9. State two functions of the respiratory system.
10. Name in order the structures of the respiratory tract through which air passes during inspiration.
11. State the main function of the circulatory system.
12. Name the three components of the circulatory system.
13. State the two functions of the urinary system.
14. Name the structures through which urine flows as it passes from the kidney to the outside of the body.
15. State a function of the skeletal system and of the muscular system.
16. Name the male organ that produces sperm and the tubes through which sperm flow out of the body.
17. Name the female organ that produces eggs, the organ in which the fetus develops, and the tube that connects these two organs.
18. Name the two systems that control body function.
19. Name the two structures that compose the central nervous system.
20. State the location of the following body regions: frontal, temporal, parietal, occipital, nasal, orbital, buccal, cervical, thoracic, abdominal, mammary, umbilical, scapular, lumbar, saeral, deltoid, axillary, brachial, pubic, pelvic, femoral, patellar, and plantar.

21. Define the following terms and be able to use them to describe the relative location of various parts of the body: superior, inferior, anterior, posterior, ventral, dorsal, medial, lateral, internal, external, proximal, distal, central, peripheral, visceral, parietal.
22. State the view of the body obtained from each of the following sections: frontal, sagittal, and transverse.
23. Name the two major internal cavities and the two subdivisions of each.
24. Define the terms root, prefix, and suffix.

# 2 Chemistry

ATOMS
MOLECULES
STATES OF MATTER
WATER
ACIDS AND BASES
CARBON COMPOUNDS
CARBOHYDRATES
LIPIDS
PROTEINS
CHEMICAL REACTIONS

A living human being may be seen as a collection of chemical substances organized in a particular pattern. The properties of the chemicals that make up the human body and the laws that govern their interactions will determine the manner in which the body functions. In this chapter we are going to look at some basic chemistry as it applies to the human body. This information is intended to help you in understanding the ways in which the body normally functions and the types of things that happen to it as a result of various diseases.

## ATOMS

All matter is composed of certain fundamental units called **atoms.** These atoms are exceedingly small, about **0.000000004** inches in diameter. This means that even in a small amount of any substance there are a tremendous number of atoms. One ounce of salt, for instance, contains 300,000,000,000,000,000,000,000 atoms. The structure of an atom is illustrated in Figure 2-1.

The **nucleus,** or center, of the atom is made up of two identical-sized particles called **protons** and **neutrons.** At a relatively great distance from the nucleus are much smaller particles called **electrons,** which move in orbit around the nucleus. Protons, neutrons, and electrons are distinguished not only by their size and

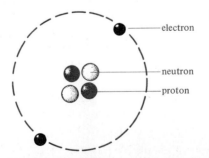

**Figure 2-1** Structure of an atom.

**21** Atoms

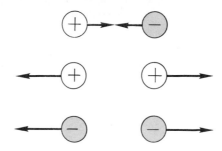

**Figure 2-2** Interactions of charged particles.

location but also by their electrical charge. Protons have a positive charge, neutrons have no charge, and electrons have a negative charge.

The electrical charge of a particle is a measurement of its ability to attract or repel other electrically charged particles. Substances can have either a positive or negative charge or they can be neutral. Figure 2-2 illustrates the interactions between substances with different types of electrical charges. Particles with the same type of charge repel each other whereas particles with opposite charges attract each other. Thus two positively charged particles repel each other, two negatively charged particles repel each other, and a positively charged particle and a negatively charged particle attract each other. Neutral particles are neither attracted nor repelled by either type of charged particle.

Substances made up of identical atoms are called **elements.** So far 105 different elements have been discovered, each of which differs from the other in the number of protons contained in the nucleus. The number of protons contained in the nucleus of a particular element is known as its **atomic number.** The number of electrons orbiting around the nucleus is equal to the number of protons in the nucleus; the atoms are neutral. The number of neutrons can be less than, equal to, or greater than the number of protons.

Because protons and neutrons have identical weights and electrons are approximately 2000 times lighter, most of the weight of an atom is contained in its nucleus. The sum of the number of protons and the number of neutrons in an atom of a particular element is known as its **atomic weight.** Oxygen has eight protons and eight neutrons and therefore has an atomic weight of 16. Sodium has 11 protons and 12 neutrons and an atomic weight of 23.

The various elements are classified in order of increasing atomic number in the periodic table, illustrated in Figure 2-3. Each element is represented by a **symbol** consisting of one or two letters. For example the first element, hydrogen, represented by the symbol H, has an atomic number of 1 and an atomic weight of 1. Potassium, on the other hand, represented by the symbol K, has an atomic number of 19 and an atomic weight of 39. This tells you that a potassium atom weighs 39 times as much as a hydrogen atom. The periodic table is also arranged in such a way that elements with similar properties are grouped under each other. Strontium, atomic number 38, is listed just under calcium, atomic number 20

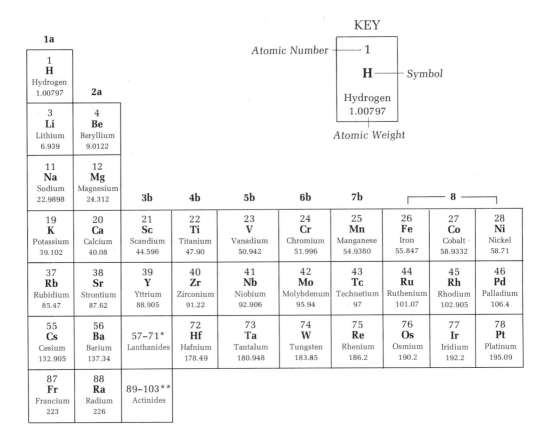

**Figure 2-3** Periodic table of the elements.

in the 2a column. It is not surprising that strontium and calcium have similar properties. An unfortunate result of this is that radioactive strontium produced by atomic fallout can be taken up by bone in place of calcium. Once in the bone it destroys the bone marrow and can cause cancer.

|  |  |  |  |  |  |  | 0 |
|---|---|---|---|---|---|---|---|
|  |  |  |  |  |  |  | 2<br>**He**<br>Helium<br>4.0026 |
|  |  | 3a | 4a | 5a | 6a | 7a |  |
|  |  | 5<br>**B**<br>Boron<br>10.811 | 6<br>**C**<br>Carbon<br>12.01115 | 7<br>**N**<br>Nitrogen<br>14.0067 | 8<br>**O**<br>Oxygen<br>15.9994 | 9<br>**F**<br>Fluorine<br>18.9984 | 10<br>**Ne**<br>Neon<br>20.183 |
| 1b | 2b | 13<br>**Al**<br>Aluminum<br>26.9815 | 14<br>**Si**<br>Silicon<br>28.086 | 15<br>**P**<br>Phosphorus<br>30.9738 | 16<br>**S**<br>Sulfur<br>32.064 | 17<br>**Cl**<br>Chlorine<br>35.453 | 18<br>**Ar**<br>Argon<br>39.948 |
| 29<br>**Cu**<br>Copper<br>63.546 | 30<br>**Zn**<br>Zinc<br>65.37 | 31<br>**Ga**<br>Gallium<br>69.72 | 32<br>**Ge**<br>Germanium<br>72.59 | 33<br>**As**<br>Arsenic<br>74.9216 | 34<br>**Se**<br>Selenium<br>78.96 | 35<br>**Br**<br>Bromine<br>79.904 | 36<br>**Kr**<br>Krypton<br>83.80 |
| 47<br>**Ag**<br>Silver<br>107.868 | 48<br>**Cd**<br>Cadmium<br>112.40 | 49<br>**In**<br>Indium<br>114.82 | 50<br>**Sn**<br>Tin<br>118.69 | 51<br>**Sb**<br>Antimony<br>121.75 | 52<br>**Te**<br>Tellurium<br>127.60 | 53<br>**I**<br>Iodine<br>126.9044 | 54<br>**Xe**<br>Xenon<br>131.30 |
| 79<br>**Au**<br>Gold<br>196.967 | 80<br>**Hg**<br>Mercury<br>200.59 | 81<br>**Tl**<br>Thallium<br>204.37 | 82<br>**Pb**<br>Lead<br>207.19 | 83<br>**Bi**<br>Bismuth<br>208.980 | 84<br>**Po**<br>Polonium<br>210 | 85<br>**At**<br>Astatine<br>210 | 86<br>**Rn**<br>Radon<br>222 |

| 65<br>**Tb**<br>Terbium<br>158.924 | 66<br>**Dy**<br>Dysprosium<br>162.50 | 67<br>**Ho**<br>Holmium<br>164.930 | 68<br>**Er**<br>Erbium<br>167.26 | 69<br>**Tm**<br>Thulium<br>168.934 | 70<br>**Yb**<br>Ytterbium<br>173.04 | 71<br>**Lu**<br>Lutetium<br>174.97 |
|---|---|---|---|---|---|---|
| 97<br>**Bk**<br>Berkelium<br>247 | 98<br>**Cf**<br>Californium<br>251 | 99<br>**Es**<br>Einsteinium<br>254 | 100<br>**Fm**<br>Fermium<br>257 | 101<br>**Md**<br>Mendelevium<br>256 | 102<br>**No**<br>Nobelium<br>254 | 103<br>**Lw**<br>Lawrencium<br>257 |

Not all of the elements that occur in nature play an important role in the makeup of the body. Four elements, carbon, hydrogen, nitrogen, and oxygen, make up 96% of the body weight. In addition, calcium, phosphorus, potassium, sulfur, sodium, magnesium, chlorine, iron, and iodine play important roles in body chemistry.

## MOLECULES

Most of the atoms in the body are combined into more complex structures called **molecules.** A molecule consists of two or more atoms held together by electrical forces known as **chemical bonds.** The composition of a molecule is given by its **formula,** which indicates the number of each type of atom that is present in the molecule. Figure 2-4 illustrates a few physiologically important molecules and their formulas. Water, which makes up 60% of the body weight, consists of two hydrogen atoms combined with one oxygen atom. Its formula is $H_2O$. Oxygen ($O_2$), which we need to burn food, consists simply of two oxygen atoms bound together. Carbon dioxide ($CO_2$) consists of two oxygen atoms bound to one carbon atom.

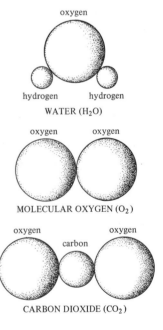

Figure 2-4 Chemical formulas of water, oxygen, and carbon dioxide.

Not all of the molecules in the body are so small. Insulin, the hormone that regulates the use of sugar, has the formula $C_{257}H_{380}N_{64}O_{76}S_6$. This means there are 257 carbon atoms, 380 hydrogen atoms, 64 nitrogen atoms, 76 oxygen atoms, and 6 sulfur atoms in one molecule of insulin. The hemoglobin molecule, which is responsible for the transport of oxygen in the blood, is even much larger than insulin. In reading the formula for any molecule the letters tell you what types of elements are present in the molecule and the numbers after and below the letters tell you the number of atoms of that element contained in one molecule. If there is no number after and below a letter it means there is only one atom of that element in the molecule.

There are two basic types of bonds that bind the atoms in a molecule. **Ionic bonds** are formed when electrons move from one atom to another, as illustrated in Figure 2-5. Because of the particular arrangement of its orbiting electrons, chlorine, like the other elements in its group on the periodic table, has a strong tendency to attract electrons from other atoms. Sodium and the other elements in its group have a tendency to lose electrons. When a sodium atom and a chlorine atom approach each other, the chlorine atom pulls away one of the sodium electrons. Remember, an atom normally has the same number of positive protons as negative electrons and is neutral. Once a chlorine atom acquires an extra electron from sodium it has more negative electrons than positive protons and thus has a charge of −1. Likewise the sodium atom that has lost an

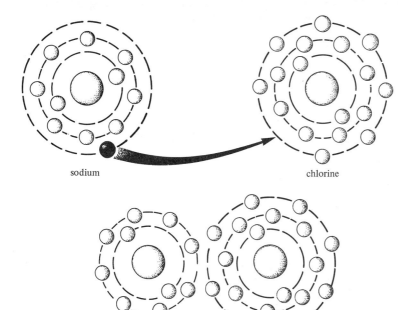

**Figure 2-5** Formation of an ionic bond.

electron is left with more positive protons than negative electrons and a charge of +1. Atoms or molecules that acquire an electrical charge are called **ions.** The movement of electrons from sodium atoms to chlorine atoms lends to the formation of $Na^+$ and $Cl^-$ ions. The mutual attraction between positive and negative charges forms the ionic bond holding the $Na^+$ and $Cl^-$ ions together into molecules of NaCl.

The second type of chemical bond holding the atoms in a molecule together is called a **covalent bond.** A covalent bond is formed by the sharing of electrons between the atoms in a molecule. This is illustrated in Figure 2-6. When two atoms of hydrogen come in contact they form a molecule of hydrogen, $H_2$. In this molecule each hydrogen electron orbits around both hydrogen nuclei, as indicated by the arrows. This sharing of electrons between the two hydrogen atoms forms the covalent bond that holds the atoms together.

## STATES OF MATTER

Molecules do not exist in isolation; rather, each molecule exists in the midst of many other molecules. The form taken by the molecules as a group depends upon the force of attraction between the molecules and upon the temperature. All molecules are continuously moving as a result of heat energy. The higher the temperature

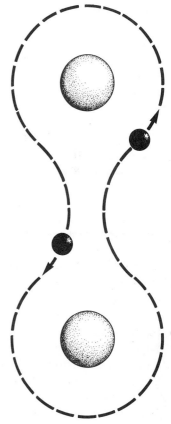

**Figure 2-6** Covalent bond in a hydrogen molecule.

the faster a molecule will move. On the other hand, molecules can be bound to one another in the same way as the atoms are bound in the molecule. The binding of one molecule to another interferes with its ability to move freely. Thus the attraction between molecules tends to hold them in place whereas heat energy tends to make molecules move about.

Three states of matter are distinguished depending on how freely molecules can move. In a **solid** the molecules are bound quite strongly to each other. Each molecule vibrates, but it is held in a fixed position and cannot move relative to the other molecules. Thus a solid is characterized by having a fixed shape. When a solid is heated it eventually turns into a **liquid.** The additional heat allows the molecules to move somewhat independently of each other. Thus a liquid can flow and change shape, although there is still enough attraction between the molecules of a liquid to maintain a continuous mass. If a liquid is heated it eventually turns into a **gas.** In a gas the molecules move almost completely independently of each other. The motion of a gas molecule is changed only if it collides with another gas molecule. The human body is a complex mixture of solids, liquids, and gases. Some structures, such as bones and teeth, are almost entirely solid. The inside of cells and the fluid that surrounds them is a liquid. The air we take into our lungs is a gas.

## WATER

Water is the medium in which all of the physiological activities necessary for life take place. This is just as true for forms of life such as human beings, which live on land, as it is for forms of life that live in the lakes, rivers, and oceans. Life on land is possible only because terrestrial plants and animals can incorporate large amounts of water within their bodies. This water forms the major component both of the extracellular fluid which surrounds the cell and of the interior of the cell itself. For a typical person 60–70% of the body weight is water. This means that a 150-pound person contains about 100 pounds of water.

Water has many properties that account for its tremendous importance within the human body. For one thing it takes a relatively large amount of heat to change the temperature of water. This helps maintain a constant body temperature in the face of a changing external temperature. Physiologically the most important property of water is that at body temperature it is a liquid in which many substances can be easily dissolved. When a substance is dissolved its individual molecules are separated and are surrounded by the molecules of the substance in which it is dissolved, as illustrated in

# 27 Water

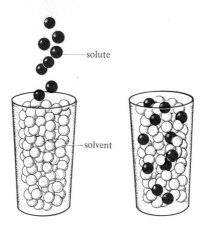

**Figure 2-7** Dissolving of a solute in a solvent.

Figure 2-7. A substance in which other substances can be dissolved is called a **solvent**. The substances that are dissolved are called **solutes**.

The reason that water is such a good solvent has to do with the distribution of electrical charge in the water molecule, illustrated in Figure 2-8. Each of the hydrogen atoms in a water molecule is bound to the oxygen atom by a covalent bond, the sharing of electrons. However, the oxygen atom has a somewhat stronger attraction for the shared electrons than the hydrogen atoms have. Because of this the shared electrons spend more time orbiting the oxygen atom than they do orbiting the hydrogen atom, giving the oxygen part of the water molecule a slightly negative charge and the hydrogen part of the water molecule a slightly positive charge. A molecule in which there is a separation of charge is referred to as a **polar molecule,** and a covalent bond in which there is a separation of charge is called a polar covalent bond. Water is a polar molecule and the covalent bonds between the oxygen and hydrogen of the water molecule are called polar covalent bonds.

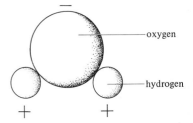

**Figure 2-8** Distribution of electrical charge in a water molecule.

Substances are dissolved in water as a result of electrical interactions between the water molecule and the solute being dissolved. Figure 2-9 illustrates this by showing what happens when sodium chloride (NaCl), table salt, is placed in water. The positive sodium ions are attracted to the negative oxygen parts of water molecules and the negative chloride ions are attracted to the positive hydrogen parts of water molecules. The water molecules pull the sodium and chloride ions apart thereby dissolving them. Ions, such as sodium and chloride, when dissolved in water are called **electrolytes.** The term electrolyte refers to the fact that the ions have an electrical charge.

Because water is the most abundant molecule in the body and is also such a good solvent most of the other molecules in the body are dissolved within the water molecules. This means that the

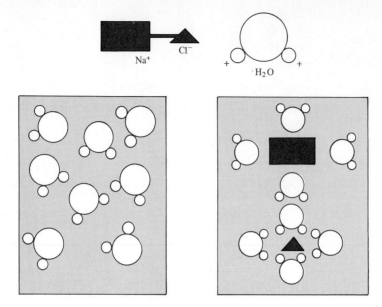

Figure 2-9 Sodium chloride dissolving in water.

chemical interactions between these other molecules take place within the water in which they are dissolved.

The amount of any substance dissolved in a specific volume of water is known as its **concentration.** The concentration of a substance is really a measurement of how close together or densely packed its dissolved molecules are within a solution.

## ACIDS AND BASES

Of the various electrolytes dissolved in the body water, one of the most important is the hydrogen ion, $H^+$. The concentration of $H^+$, in the body fluids, depends upon the relative amounts of acid and base present. An **acid** is a substance that releases hydrogen ion, as illustrated by hydrochloric acid and carbonic acid in Figure 2-10a. A **base** is a substance that combines with hydrogen ions, as illustrated by hydroxide and bicarbonate in Figure 2-10b. The addi-

| | |
|---|---|
| $HCl \longrightarrow H^+ + Cl^-$ <br> hydrochloric acid | $OH^- + H^+ \longrightarrow H_2O$ <br> hydroxide ion |
| $H_2CO_3 \longrightarrow H^+ + HCO_3^-$ <br> carbonic acid | $HCO_3^- + H^+ \longrightarrow H_2CO_3$ <br> bicarbonate ion |
| (a) | (b) |

Figure 2-10 (a) The release of hydrogen ion by an acid and (b) the combination of hydrogen ion with a base.

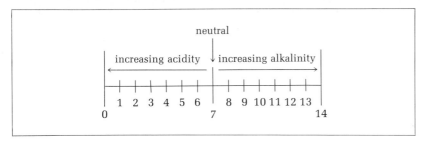

**Figure 2-11** pH scale.

tion of acids to the body fluids raises their $H^+$ concentration whereas the addition of bases to the body fluid lowers their $H^+$ concentration.

The $H^+$ concentration is usually expressed in terms of a special scale of numbers called the pH scale, illustrated in Figure 2-11. A pH of 7 indicates a neutral solution—the acid and base present balance each other. A pH of less than 7 indicates an acid solution—the acid present is more powerful than the base, whereas a pH of greater than 7 indicates a basic (alkaline) solution—the base present is more powerful than the acid. The lower the pH of a solution, the more hydrogen ion there is present. A solution with a pH of 2 has a higher $H^+$ concentration than a solution with a pH of 4. Conversely, the higher the pH, the less $H^+$ is present. A solution with a pH of 10 contains less $H^+$ than a solution with a pH of 8. The pH of a solution goes down when its $H^+$ concentration increases and the pH goes up when the $H^+$ concentration decreases.

Normally the pH of the body fluids is in the range 7.35–7.45; they are slightly basic. Because any change in pH, either an increase or decrease in $H^+$ concentration, can interfere with normal cellular function, the pH must be very carefully regulated. The regulation of the acid-base balance of the body, so that a pH of 7.35–7.45 is maintained, will be discussed in detail in Chapter 17.

## CARBON COMPOUNDS

Within the body water the chemical interactions of molecules which contain the element carbon are the key to both the structure of the body and its various physiological activities. The fact that carbon atoms can be used as a framework on which a variety of large molecules can be built makes them so important in all living systems. The building of large molecules out of carbon atoms depends on the ability of carbon to form four covalent bonds and thus to combine with four other atoms. The atoms that carbon most frequently combines with are hydrogen, oxygen, nitrogen, and other carbon atoms.

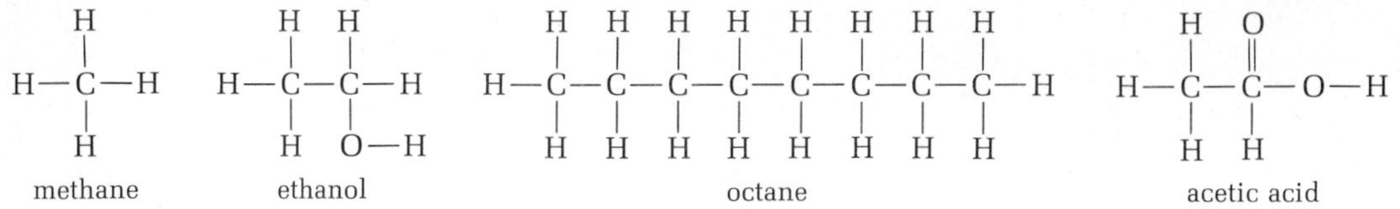

```
      H              H  H           H  H  H  H  H  H  H  H           H  O
      |              |  |           |  |  |  |  |  |  |  |           |  ‖
   H—C—H          H—C—C—H       H—C—C—C—C—C—C—C—C—H             H—C—C—O—H
      |              |  |           |  |  |  |  |  |  |  |           |  |
      H              H  O—H         H  H  H  H  H  H  H  H           H  H
   methane         ethanol                    octane                 acetic acid
```

Figure 2-12 shows a few examples of relatively small molecules that contain carbon. Methane is one of the gases released from the digestive tract when a person lets air, ethanol is the alcohol a person may drink, octane is one of the gases a person may put into his car, and acetic acid is the vinegar a person may sprinkle on his salad. In looking at these four compounds you should notice a few things. One is that each carbon atom forms four bonds. In the case of acetic acid two of the bonds connect the same oxygen atom; this is called a **double bond.** Second, the fact that carbon can combine with other carbons permits the formation of chains of carbon atoms as for instance in octane, which has a chain of eight carbon atoms. Finally these four compounds, which are all formed by the combination of carbon with either hydrogen or hydrogen and oxygen, are exceedingly different substances. As a result of the ability of carbon to combine with a number of other elements and to form long chains, a great variety of large carbon molecules can be formed. Large carbon-containing molecules play a role in every aspect of anatomy and physiology. At this point we are going to look at three fundamental types of molecules that contain carbon: carbohydrates, lipids, and proteins. Other types of carbon-containing molecules will be discussed as they become relevant throughout the book.

**Figure 2-12** Examples of carbon compounds.

## CARBOHYDRATES

The **carbohydrates** (sugars) are the basic source of energy for cellular activity. Within the cells they can be burned (combined with oxygen), and the energy that is released can be used to perform cellular work. In addition to serving as a source of cellular energy the carbohydrates can be converted to other types of molecules which form part of the structure of the cell or which carry out physiological functions.

The carbohydrates are formed from carbon, hydrogen, and oxygen in the proportion of two hydrogens and one oxygen for each carbon ($C_nH_{2n}O_n$). The simplest sugars are called **monosaccharides,** the most important of which are **glucose, fructose,** and **galactose,** shown in Figure 2-13. All three of these monosaccharides are six-carbon compounds with a ring-type structure. Of the three, glucose

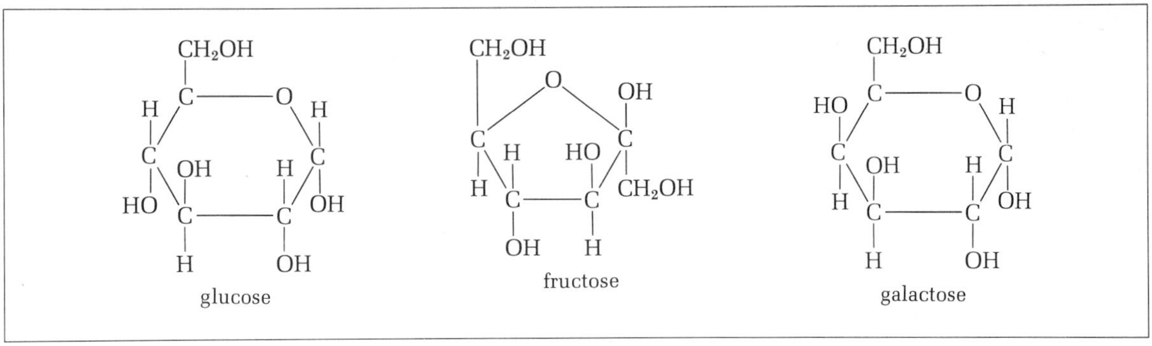

# 31 Carbohydrates

[Structural formulas of glucose, fructose, and galactose shown in ring form]

**Figure 2-13** Monosaccharides.

is by far the most important. As it is the primary sugar transported by the blood to the cells, glucose is often referred to as blood sugar. The other two monosaccharides, fructose and galactose, are converted to glucose before they are used by the cells. The process by which the cells obtain energy from glucose will be discussed in detail in Chapter 3, and the formation of glucose as well as the conversion of glucose into other molecules will be discussed in Chapter 16.

Simple sugars can be linked together to form chains of sugar molecules. A chain composed of two simple sugar molecules is called a **disaccharide,** two of which are sucrose and lactose. **Sucrose** is common table sugar, an extract of beet or sugar cane. As eating habits have changed over the last 100 years sucrose has formed an increasingly larger proportion of the diet to the point where the average American consumed 175 pounds of sugar in 1972. Some physicians feel that the huge amounts of sucrose we consume may be one of the causes of the high incidence of diabetes and heart disease in the population. **Lactose** is the sugar found in milk. As we will discuss in Chapter 15 sucrose and lactose are normally broken down in the digestive tract into monosaccharides which are then absorbed into the blood. Sucrose and lactose themselves cannot be absorbed into the blood. Some infants are born with an inability to break down lactose in their digestive tract into its two monosaccharides. Because milk is the major food in the diet of an infant this inability to use lactose can leave a child exceedingly undernourished, a problem compounded by the fact that undigested lactose causes severe diarrhea.

Long chains of simple sugars are called **polysaccharides.** Physiologically the three most important polysaccharides are glycogen, starch, and cellulose, all of which are composed of glucose molecules linked together. **Glycogen** is the form in which animals, including ourselves, store glucose in the body for later use. The formation and use of glycogen will be discussed in detail in Chapter 16. **Starch** and **cellulose** are plant polysaccharides. Starch is the form in which plants store glucose; cellulose is a tough fibrous material which helps hold together the plant structure.

## LIPIDS

The lipids are a group of carbon-containing compounds which are characterized by the fact that they do not dissolve readily in water. Like the carbohydrates the lipids are composed of carbon, hydrogen, and oxygen; however, the lipids contain a much smaller proportion of oxygen atoms than do the carbohydrates. The three most important groups of lipids are the fats, the phospholipids, and the steroids.

**Fats** are the primary form in which our body stores energy. If we eat more food than we need the excess food, rather than being eliminated, is stored as fat. This stored fat can then be broken down and used by the cells at a later time when our food intake is inadequate. Fat also plays a role in insulating our body against temperature changes and in protecting our internal organs.

Fats are composed of two types of subunits, fatty acids and glycerol, as illustrated in Figure 2-14. A **fatty acid** is composed of a straight chain of carbon atoms, usually 10 to 18 carbons long, to

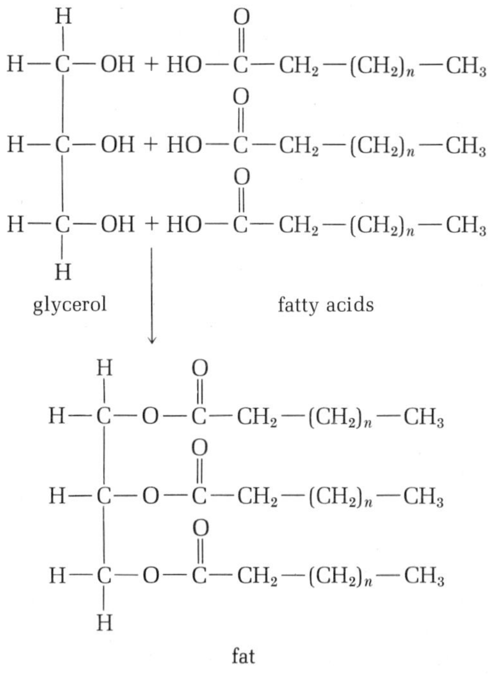

**Figure 2-14** Fat.

$$\begin{array}{c}
\text{H} \\
| \\
\text{H}-\text{C}-\text{OH} + \text{HO}-\overset{\overset{\text{O}}{\|}}{\text{C}}-\text{CH}_2-(\text{CH}_2)_n-\text{CH}_3 \\
| \\
\text{H}-\text{C}-\text{OH} + \text{HO}-\overset{\overset{\text{O}}{\|}}{\text{C}}-\text{CH}_2-(\text{CH}_2)_n-\text{CH}_3 \\
| \\
\text{H}-\text{C}-\text{OH} + \text{HO}-\overset{\overset{\text{O}}{\|}}{\text{C}}-\text{CH}_2-(\text{CH}_2)_n-\text{CH}_3 \\
| \\
\text{H}
\end{array}$$

glycerol       fatty acids

$$\downarrow$$

$$\begin{array}{c}
\text{H} \\
| \\
\text{H}-\text{C}-\text{O}-\overset{\overset{\text{O}}{\|}}{\text{C}}-\text{CH}_2-(\text{CH}_2)_n-\text{CH}_3 \\
| \\
\text{H}-\text{C}-\text{O}-\overset{\overset{\text{O}}{\|}}{\text{C}}-\text{CH}_2-(\text{CH}_2)_n-\text{CH}_3 \\
| \\
\text{H}-\text{C}-\text{O}-\overset{\overset{\text{O}}{\|}}{\text{C}}-\text{CH}_2-(\text{CH}_2)_n-\text{CH}_3 \\
| \\
\text{H}
\end{array}$$

fat

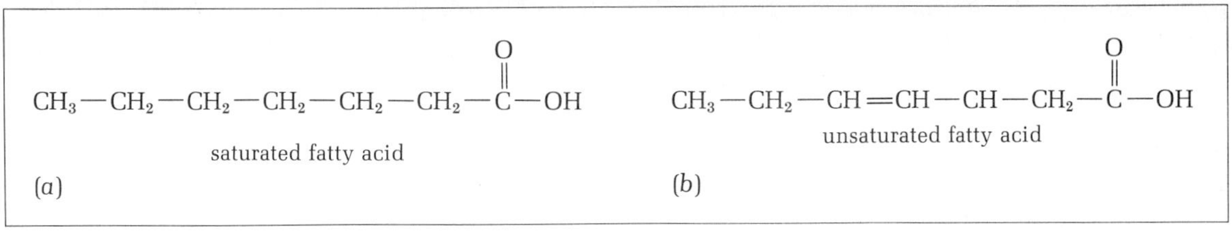

$$CH_3-CH_2-CH_2-CH_2-CH_2-CH_2-\overset{\overset{O}{\|}}{C}-OH$$

saturated fatty acid

(a)

$$CH_3-CH_2-CH=CH-CH-CH_2-\overset{\overset{O}{\|}}{C}-OH$$

unsaturated fatty acid

(b)

**Figure 2-15** (a) Saturated fat and (b) unsaturated fat.

which hydrogen atoms and an acid group are attached, whereas **glycerol** is a three-carbon sugar. The combination of one fatty acid with glycerol forms a **monoglyceride,** two fatty acids with glycerol a **diglyceride,** and three fatty acids with glycerol a **triglyceride.**

Plants can also store food as fat. However, the fat of plants is somewhat different from the fat of animals, as illustrated in Figure 2-15. Many of the plant fats have double bonds between some of the carbons; these are called **unsaturated fats;** if there are many double bonds they are called **polyunsaturated fats.** Animal fat, in contrast, is composed primarily of **saturated fat,** a type of fat that contains no double bonds between the carbons.

A great deal of research in recent years has indicated that excess animal fat in the diet may lead to cardiovascular disease. This is because the liver converts saturated fat into cholesterol, a substance that seems to play a role in the formation of deposits on the walls of blood vessels. These deposits can block the flow of blood leading to tissue damage or even death. The liver cannot convert unsaturated fat into cholesterol; therefore it appears that vegetable fat represents less of a health hazard than does animal fat.

**Phospholipids** are structurally quite similar to fats except that a group containing phosphorus and sometimes nitrogen is attached to one of the carbons of glycerol. The phospholipids are an important part of the membrane surrounding the cells.

**Steroids** are lipids that are composed of four interlocking carbon rings to which other groups are attached, as illustrated in Figure 2-16 by the male reproductive hormone testosterone and the female reproductive hormone estrogen. Notice the similarity of the

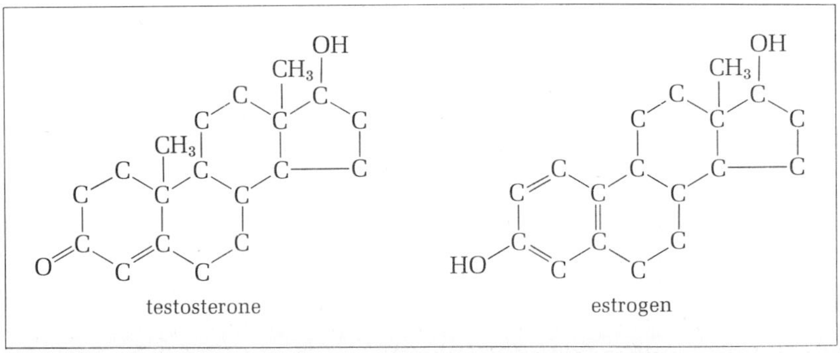

**Figure 2-16** Examples of steroids.

structures of these steroids despite the fact that their physiological activities are quite different. Testosterone is responsible for male secondary sexual characteristics such as a deep voice, hair growth on the face, and narrow hips whereas estrogen is responsible for female secondary sexual characteristics such as a high voice, smooth skin, and wide hips. This illustrates the fact that a slight variation in the chemical structure of a molecule can lead to a tremendous difference in the way in which it acts.

## PROTEINS

**Proteins** comprise over 50% of the organic material in the body and play an essential role in almost every aspect of anatomy and physiology. Proteins are the building blocks from which the cells are formed, and they are the core of the material that holds the various cells together. Proteins regulate the chemical activity inside of cells, act as hormones, are responsible for the contractions of muscles, fight infection, and carry out many other specialized functions too numerous to mention here. We will be dealing with proteins in every chapter of the book.

Proteins are composed of a chain of smaller subunits called **amino acids**. As illustrated in Figure 2-17, amino acids are composed of carbon, hydrogen, oxygen, and nitrogen atoms. They are characterized by containing an acid group, COOH; an amino group, $NH_2$; a hydrogen atom; and a fourth group known as a side chain — all of which are attached to the same carbon. Some side chains contain sulfur atoms. There are approximately 20 amino acids which differ by the side chain they contain. Figure 2-18 shows a few of the amino acids and their particular side chain. It should be emphasized again that all of the amino acids contain the acid group and the amino group. Only the side chains are different.

Amino acids are linked together into a chain to form a protein through the combination of the amino group of one amino acid with the acid group of another, as illustrated in Figure 2-19. The bond that holds amino acids together is called a **peptide bond**. No-

**Figure 2-17** General structure of an amino acid.

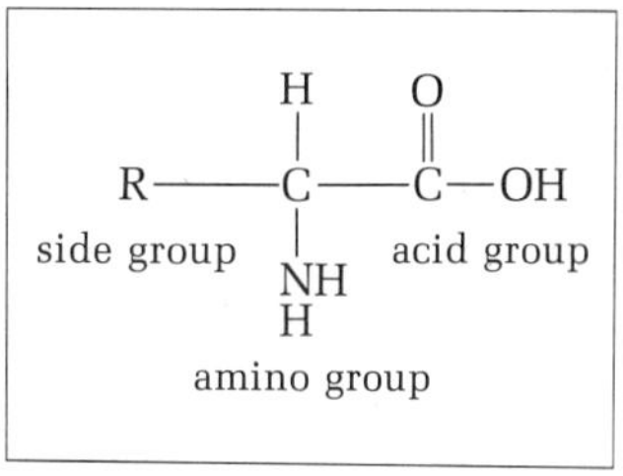

**Figure 2-18** Examples of specific amino acids.

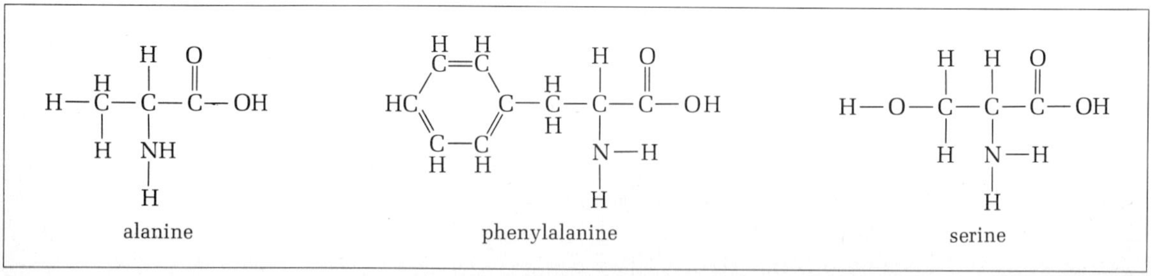

alanine

phenylalanine

serine

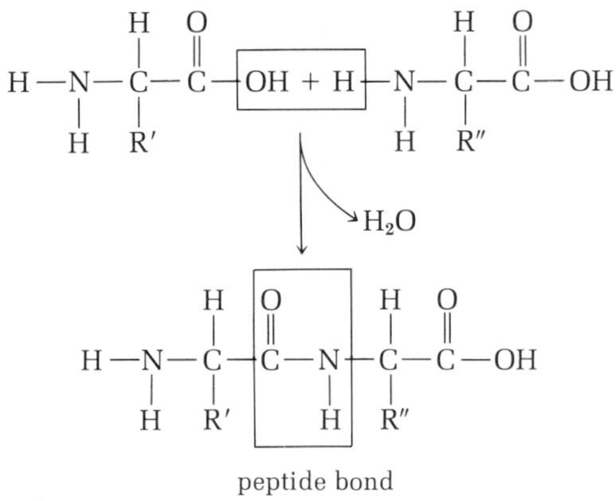

**Figure 2-19** Formation of a peptide bond.

tice that each peptide bond results in the formation of a molecule of water, an OH coming from the acid group and an H from the amino group. Proteins are characterized by the particular order of amino acids and by the length of the chain. The length of protein chains varies enormously—some contain only a few amino acids whereas others contain as many as 50,000.

We have discussed some of the important types of chemicals found within the body. Table 2-1 shows the relative abundance of these chemicals.

**Table 2-1** Relative Abundance of Major Chemicals Within the Body

| Type of Molecule | Body Weight (%) |
|---|---|
| Water | 60 |
| Protein | 17 |
| Lipid | 15 |
| Electrolytes | 5 |
| Carbohydrate | 1 |
| Others | 2 |

## CHEMICAL REACTIONS

The various molecules contained within the body are continuously involved in chemical reactions. These reactions can be divided into two basic types: **catabolic** or degradative reactions that involve the breakdown of larger molecules into smaller molecules and **anabolic** or synthetic reactions that involve the combination of smaller molecules into larger molecules. The term **metabolism** refers to the sum of all the chemical reactions that take place.

Chemical reactions are usually represented by chemical equations that indicate in the form of formulas the substances which are reacting and the products which are formed. For example, the chemical equation

$$C_6H_{12}O_6 + 6O_2 \longrightarrow 6H_2O + 6CO_2$$

indicates that glucose and oxygen react to form water and carbon dioxide. The number in front of the formula for each molecule indicates the relative number of molecules of that type involved in the reaction. In the above example one glucose molecule combines with six oxygen molecules to form six water molecules and six carbon dioxide molecules.

In order for any chemical reaction to take place the reacting substances must come in contact with each other in such a way as to break old chemical bonds and form new ones. This usually requires that a particular part of one molecule come in contact with a particular part of another molecule, as illustrated schematically in Figure 2-20. Molecules that do not "fit" together will not react. Even if molecules are able to react a number of different factors will affect the speed or rate at which they react. Concentration, temperature, and catalysts are three of the factors that affect the rate.

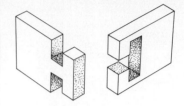

**Figure 2-20** Reaction sites.

The rate at which a reaction takes place will depend, among other things, on the amount of reacting substance that is present, as illustrated in Figure 2-21. An increase in the concentration of the reacting substances will increase the frequency with which they come in contact and react.

**Figure 2-21** Effect of concentration on reaction rate.

Temperature also affects the rate at which a chemical reaction will take place. As the temperature of the reacting substances is increased they move about faster, which increases the chances of their coming in contact and reacting. Because the body temperature is normally kept quite constant, temperature normally does not play a role in altering the speed of chemical reactions in the body. However, the increased temperature caused by a fever can greatly increase the rate of metabolism of the body. Conversely, during certain types of surgery the body temperature is lowered to slow down the rate of metabolism.

The most important factor affecting the speed at which chemical reactions take place in the body is the presence of **enzymes.** Enzymes are proteins that act as catalysts; they speed up chemical reactions without undergoing any change themselves.

The exact mechanisms by which enzymes accelerate chemical reactions are quite complicated and by no means completely understood. However, it is generally felt that enzymes act by orienting the reacting substances in such a way as to allow their reacting portions to come in contact. This is illustrated in Figure 2-22. Each of the reacting substances combines with the enzyme. Once the substances combine with the enzyme they are in a position to combine with each other. After the reacting substances combine with each other they separate from the enzyme and the enzyme can be used again. In order to orient the molecules that are reacting properly the enzyme must be specific for the particular type of reaction that is taking place.

Most of the chemical reactions that occur in the body require enzymes to act at physiologically significant rates. For example, starch placed in a glass of water remains intact whereas starch taken into the digestive tract is rapidly broken down by enzymes

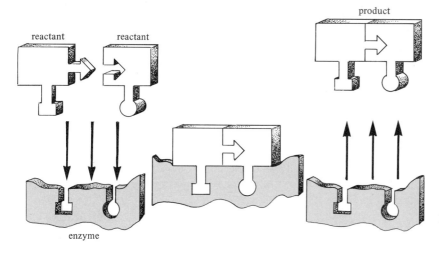

**Figure 2-22** Proposed mechanism of enzyme action.

into individual glucose molecules. Basically every chemical reaction that takes place in the body requires an enzyme and thus the pattern of chemical activity in each cell is determined by the particular enzymes present in that cell. In the next chapter we will discuss how cells manufacture enzymes and what determines which particular enzymes a cell manufactures.

Many enzymes can only function in the presence of certain accessory factors called **coenzymes.** As the name implies, a coenzyme aids an enzyme in speeding up a chemical reaction. Some coenzymes, particularly small metal ions such as copper and magnesium, seem to act by aiding the combination of the reacting substance with the enzyme. Other coenzymes seem to act by removing a portion of the reacting substance and passing it on to another molecule. The coenzymes that act in this second way are small organic molecules which the body is usually not able to manufacture itself. These coenzymes are the **vitamins** which must be included in our diet. If there is an inadequate amount of a particular vitamin in our diet the chemical reaction in which it acts as a coenzyme will be hindered. On the other hand an excess of vitamins is of no particular value because the rates of chemical reactions are limited by the amounts of reacting substances and enzyme available. We will discuss the particular process in which each of the most important vitamins is involved as it comes up in the book.

## OBJECTIVES FOR THE STUDY OF CHEMISTRY

At the end of this unit you should be able to:

1. Name the three types of particles which are found in the atom.
2. Name which two particles are found in the nucleus.
3. State the electrical charge of protons, neutrons, and electrons.
4. State which type of charges attract each other and which type repel each other.
5. Name the four most abundant elements in the body.
6. Explain how a molecule differs from an atom.
7. Distinguish between a chemical symbol and a chemical formula.
8. Distinguish between an ionic bond and a covalent bond.
9. Define the term ion.
10. Distinguish between a solid, a liquid, and a gas.
11. Distinguish between a solute and a solvent.
12. Define the term concentration.
13. Define the terms acid and base.
14. State the relationship between changes in hydrogen ion concentration and changes in pH.
15. State whether a solution is acidic or basic once the pH is known.
16. State the normal pH of the body fluids.
17. Name the three types of atoms that compose carbohydrates.
18. Distinguish between monosaccharides, disaccharides, and polysaccharides.
19. State whether each of the following is a monosaccharide, disaccharide, or polysaccharide: glucose, fructose, galactose, sucrose, lactose, glycogen, starch, and cellulose.
20. State the main characteristic of lipids.
21. Name the two subunits of a fat.
22. Distinguish between a monoglyceride, diglyceride, and triglyceride.
23. Distinguish between saturated and unsaturated fats.
24. Name the subunits from which proteins are made.
25. Name the four types of atoms found in all amino acids.
26. Name the bond which holds amino acids together.
27. Distinguish between catabolic and anabolic chemical reactions.
28. Define the term metabolism.
29. Name three factors that influence the rate at which a chemical reaction will take place.
30. Define the terms enzyme and coenzyme.

# 3 The Cell

GENERAL CELL STRUCTURE
MOVEMENT OF SUBSTANCES ACROSS
  THE CELL MEMBRANE
CELLULAR ENERGY METABOLISM
PROTEIN SYNTHESIS
MITOSIS

In the previous chapter we discussed some of the different types of chemical compounds that make up the human body. If these compounds were simply placed in a large container, most of the activities we associate with life would not take place; yet within the human body these same compounds perform all of the functions essential to life. The explanation for this is that when these compounds are mixed together in a container they are disorganized—they are not arranged in any particular pattern—whereas in the human body they are arranged into very highly organized structures and their interactions are carefully controlled. The cell is the basic organizational unit into which chemical compounds are arranged within the human body.

Although there are a great many different types of cells within the body, each adapted to carry out a particular function, there are certain features common to all of these cells. The features of structure and function which are common to all cells will be the subject of this chapter. Later chapters will discuss the specific features of structure and function which are characteristic of particular types of cells.

## GENERAL CELL STRUCTURE

Figure 3-1 is a schematic drawing of a cell, illustrating the basic organelles which are found in most cells. The inside of the cell is separated from the extracellular fluid which surrounds the cell by a thin barrier, the cellular membrane. Because this barrier must be crossed by any substance entering or leaving the cell it serves to regulate the flow of material into and out of the cell.

The large object at the center of the cell, the **nucleus,** serves as the control center of the cell. Located within the nucleus of each human cell (with the exception of sperm cells and egg cells) are 46 **chromosomes,** the long thread-like structures which contain the **genes.** These genes which are the hereditary units passed from gen-

# 41 General cell structure

eration to generation control every activity that takes place within the cell. As will be discussed in a subsequent section of this chapter the genes control cellular activity as a result of their control over protein production in the cell. The nucleus is surrounded by a membrane, the **nuclear membrane,** which controls the flow of material into and out of the nucleus. The nuclear membrane is a double membrane composed of two membranes similar in structure to the cellular membrane, with a space in between. The large object inside the nucleus is called the **nucleolus** and seems to play a role in translating the instructions contained within the genes into the actual production of proteins.

The entire cell area between the nucleus and the cell membrane is collectively referred to as the **cytoplasm.** Located within the cytoplasm are a number of different organelles: the centrioles, the ribosomes, the endoplasmic reticulum, the Golgi apparatus, the mitochondria, and the lysosomes.

Scattered throughout the cytoplasm are thousands of tiny spherical structures called **ribosomes.** The ribosomes are the site of protein synthesis within the cell. It is at the ribosomes that amino acids are linked together into proteins. Some of the ribosomes float freely within the cytoplasm; others are attached to the **endoplasmic reticulum,** an elaborate system of membranes within the cytoplasm outside the nucleus. It appears that the free-floating ribosomes make proteins that will be used within the cell itself, whereas the

**Figure 3-1** Parts of a typical cell.

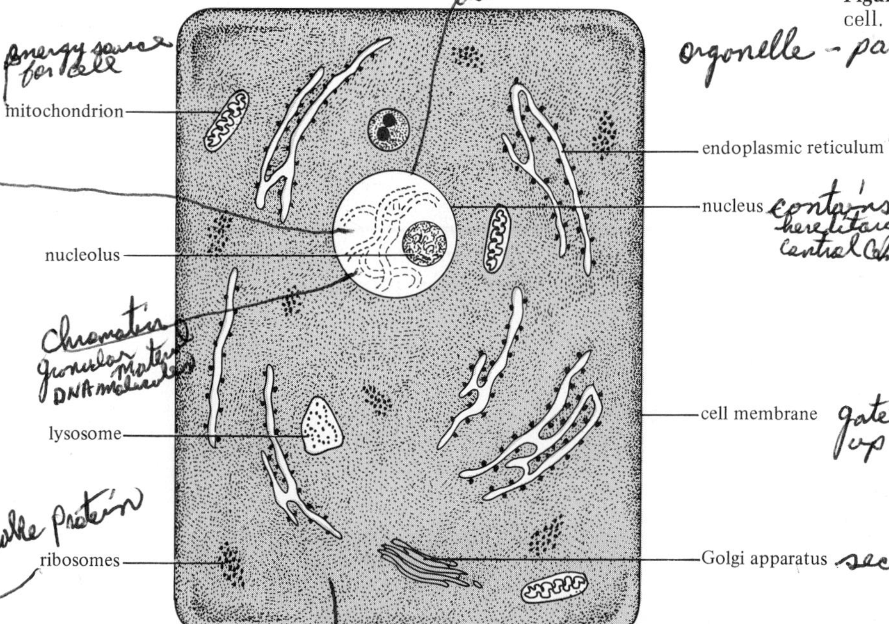

ribosomes attached to the endoplasmic reticulum make proteins that are secreted from the cell.

The endoplasmic reticulum forms the walls of an extensive series of canals which serve as a channel for the transport of materials from one part of the cell to another. Since the endoplasmic reticulum is connected to the nuclear membrane, substances can move quite readily from the nucleus into the canals formed by the endoplasmic reticulum and then throughout the cytoplasm. Endoplasmic reticulum that is lined with ribosomes is referred to as **rough endoplasmic reticulum;** when no ribosomes are present it is called **smooth endoplasmic reticulum.** As we have mentioned, ribosomes that are attached to the endoplasmic reticulum manufacture proteins that are secreted by the cells. Smooth endoplasmic reticulum probably contains the enzymes needed for the manufacture of lipids within the cell.

Located just next to the nucleus are two structures, oriented at right angles to each other. These are the **centrioles,** which play an essential role in pulling the chromosomes apart during cell division, a process that will be discussed later in the chapter.

Once proteins that are destined to be secreted are manufactured at the ribosomes along the endoplasmic reticulum, they are transported through the canals of the endoplasmic reticulum to another set of membranes, the **Golgi apparatus.** This special structure consists of a series of membrane-lined sacs arranged in parallel. The protein which is to be secreted is concentrated in the Golgi apparatus and is often combined with carbohydrate.

The **mitochondria** are small organelles distributed throughout the cytoplasm. They are composed of an outer smooth membrane and an inner membrane which contains many folds known as **cristae.** These folds contain most of the enzymes which are used in the conversion of food energy into a form of energy which the cells can use to do cellular work.

The final organelles we are going to mention are the **lysosomes,** small membrane-enclosed sacs that contain powerful enzymes capable of breaking down large molecules such as proteins, lipids, and polysaccharides. As these enzymes are capable of breaking down the cell itself they are normally contained within the lysosomes. Under certain circumstances, to be discussed in later chapters, these enzymes are released and used to digest large molecules which are taken into cells.

## MOVEMENT OF SUBSTANCES ACROSS THE CELL MEMBRANE

The cellular membrane acts to regulate the movement of materials between the inside of the cell and the extracellular fluid. It is able

**43** Movement of substances across the cell membrane

to do this because it is a semipermeable barrier—certain substances can cross the membrane and others cannot. In order for any substance to enter or leave the cell there must be some force causing the substance to move and it must be able to pass through the cell membrane. Therefore in examining the movement of substances into and out of cells we have two factors to consider: the forces that cause molecules to move and the ways in which the structure of the cellular membrane affects what type of substances can pass through it. We are going to look at the structure of the cellular membrane first.

At the present time there is a great deal of controversy as to the exact structure of the cellular membrane, illustrated in Figure 3-2. It is known that the membrane is exceedingly thin, approximately 75 Angstroms or 1/300,000th of an inch, and that it is composed of lipid and protein. There is disagreement, however, as to the manner in which the lipid and protein are arranged. The most widely held view is that the membrane is composed of an inner and outer layer of protein with a double layer of lipid in the middle. It is believed that the middle lipid layer is composed of phospholipids and steroids oriented perpendicularly to the protein molecules of the inner and outer layers. More recent research, however, has indicated that there are probably portions of the membrane with lipid at the surface (not sandwiched between protein) and that some of the proteins penetrate the width of the membrane.

In order to explain the movement of certain substances in and out of the cell, physiologists have postulated that the membrane contains pores approximately 7 Angstroms wide, the width of two water molecules. Since this is too small to be seen even with an electron microscope, no one has yet been able to prove that they exist.

In order for any substance to cross the membrane it must be able to pass through the lipid and protein portion of the membrane or it must be able to pass through one of the pores. In our discussion of lipids in the previous chapter we stated that lipids are

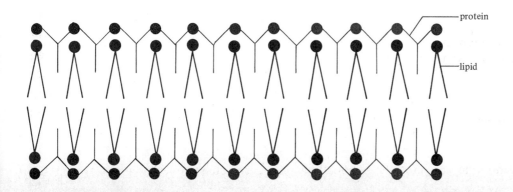

**Figure 3-2** Bilayer structure of the cellular membrane.

defined as molecules that do not mix readily with water. This is because lipid molecules do not have a positive and negative portion, they are not polar, and therefore are not electrically attracted by the polar water molecules. In general, molecules that dissolve readily in a polar substance such as water do not dissolve readily in a nonpolar substance such as lipid, and vice versa. This is important to mention because the membrane is composed to a large extent of lipid, so that those substances that cannot pass through the pores must be dissolved in lipid if they are to cross the membrane. Because the pores are relatively narrow only very small molecules, such as water, can cross the membrane by passing through the pores. Molecules that are too large to pass through the pores and that cannot be dissolved in the lipid layer will not cross the membrane. We have looked at the structure of the membrane and the ways in which this structure affects the movement of substances through the membrane. We must now turn to a discussion of the forces that cause molecules to move in the first place.

All molecules are continuously in motion as a result of heat energy. The higher the temperature, the faster the molecules will move. Each molecule in a solution, such as exists in the body, will move in one direction until it collides with another molecule and bounces off, moving in a different direction, as illustrated in Figure 3-3. Because the molecules in a solution are quite close together, they are continuously bumping into each other and changing the direction of their movement. The result of this is that at any moment there are molecules moving in all directions, with no more molecules moving in one direction than in another. Figure 3-4 is a schematic illustration of how this random motion of molecules can lead to the movement of substances into or out of cells.

Imagine a situation in which two solutions are separated by a membrane and there is a substance that can cross the membrane

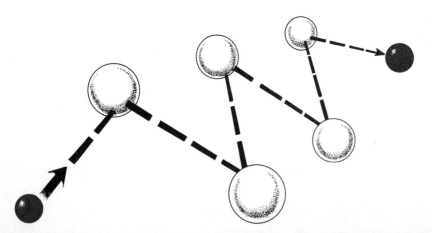

**Figure 3-3** Movement of a molecule.

## 45 Movement of substances across the cell membrane

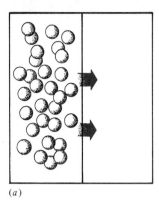
(a)

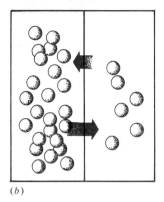
(b)

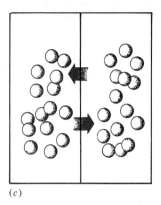
(c)

**Figure 3-4** Diffusion across a membrane.

present in the solution on one side of the membrane, but not in the solution on the other side, as illustrated in Figure 3-4a. Since the molecules of this substance are in continuous motion, some of the molecules will move in the direction of the membrane and cross into the second solution. Once this happens some of the molecules of the substance now in the second solution will begin to move back into the first solution, while other molecules will continue to move from the first solution into the second; that is, molecules are now moving across the membrane in both directions, as illustrated in Figure 3-4b. However, as long as there are more molecules of the substance in solution 1 than there are in solution 2, the rate of movement from 1 to 2 will be greater than the rate of movement from 2 back to 1. This means that the number of molecules of the substance in solution 2 will be increasing and the number of molecules of the substance in 1 will be decreasing. Eventually the concentration of the substance will be the same in both containers, as illustrated in Figure 3-4c. At this point the rate of movement from 2 to 1 will be the same as the rate of movement from 1 to 2, and the concentration of the substance in both solutions will become stable. This situation is referred to as an equilibrium. But look at what happened. We started out with the substance present only in solution 1 and we ended up with the substances present at equal concentrations in both solutions. The substance moved from an area of higher concentration to an area of lower concentration, a process known as **diffusion.** Diffusion is one of the processes by which a substance can move into or out of cells. If there is a difference in the concentration of a substance inside and outside of the cell, and if the substance can pass through the cell membrane, it will diffuse from the area of higher concentration to the area of lower concentration. For example, oxygen is continuously being used up in the cell so that the concentration of oxygen is less within cells than it is in the extracellular fluid. Because oxygen can pass through the cellular membrane it is continuously diffusing from an area of high concentration, the extracellular fluid, to an area of low concentration, the inside of the cell.

The movement of water from an area of high water concentration through a semipermeable membrane to an area of low water concentration is referred to as **osmosis.** Osmosis is simply a special name for the diffusion of water.

Normally the term concentration is used to describe the amount of solute dissolved in a particular volume of water. The more solute that is dissolved, the higher its concentration. However, as the concentration of solute in a solution is increased, the concentration of water in the solution is decreased, as illustrated in Figure 3-5.

Within any volume of solution there is only room for a certain number of molecules. If solute molecules are added to a solution they will push aside some of the water molecules and occupy the space formerly occupied by these water molecules. This decreases the water concentration of the solution. The water concentration of any solution depends upon the number of solute particles that are dissolved in the solution independent of the type of solute. The number of solute particles depends on the concentration of the solute and if the solute breaks up into ions, the number of ions formed by each molecule. The total number of solute particles dissolved in a liter of solution is known as its **osmolar concentration.** The higher the osmolar concentration of a solution, the lower its water concentration.

We have said that water moves by osmosis from an area of high water concentration to an area of low water concentration.

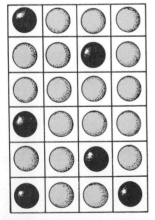
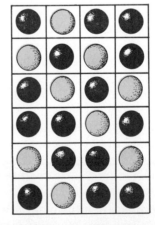

Figure 3-5 Relationship between solute and water concentration.

**47** Movement of substances across the cell membrane

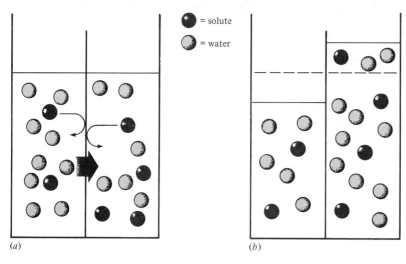

Figure 3-6 Osmosis.

This is the same thing as saying that water moves from an area of low osmolar concentration to an area of high osmolar concentration, as indicated in Figure 3-6.

Figure 3-6a illustrates a solution with an osmolar concentration of 1 osmole/liter separated from a solution with an osmolar concentration of 2 osmoles/liter by a semipermeable membrane, one that is permeable to water but not to the dissolved solute. Because the solute concentration is higher in the solution on the right than in the solution on the left the water concentration is higher in the solution on the left than in the solution on the right. Water will move by osmosis from an area of high water concentration to an area of low water concentration, from the solution on the left into the solution on the right. This movement of water from left to right will increase the volume of the solution on the right and decrease the volume of the solution on the left, as illustrated in Figure 3-6b. Another way of looking at this situation is to say that water is moving from the dilute solution into the concentrated solution until both solutions have the same concentration of dissolved particles.

The increase in volume brought about by the movement of water into a solution by osmosis leads to an increase in the pressure of the solution. This increase in pressure caused by osmosis is referred to as the **osmotic pressure** of the solution. The **potential osmotic pressure** of a solution is the pressure that would develop in the solution if it were separated from pure water by a semipermeable membrane. The significance of potential osmotic pressure is that it is a measurement of the tendency of water to move into a particular solution. The higher the potential osmotic pressure of a solution, the more likely it is that water will move into the solution. Water moves from a solution of lower potential osmotic pressure to an area of higher potential osmotic pressure. The potential osmotic pressure of any solution depends on its osmolar concentration, the number of dissolved particles of solute it contains. As the number of dissolved particles in a solution increases, its potential osmotic pressure increases. This increases the tendency of water to move into the solution.

Water is the main component of both the extracellular fluid which surrounds cells and the intracellular fluid within cells; water can also cross the cellular membrane; therefore it is essential that the concentration of water be the same inside and outside the cells. If the concentration of water becomes higher outside the cells, water will move into them by osmosis and they will swell. On the other hand if the water concentration is higher inside the cells water will move out of the cells and they will shrink.

A solution in which normal body cells will neither swell nor shrink is called an **isotonic** solution. A solution in which cells will shrink is called a **hypertonic solution.** A solution in which cells will swell is called a **hypotonic solution.** Figure 3-7 illustrates the effects of isotonic, hypertonic, and hypotonic solutions upon cell size.

A hypertonic solution has a higher potential osmotic pressure than normal cells. Water will move out of the cells into the solution and the cells will shrink. An isotonic solution has the same potential osmotic pressure as the cells so that there is no net movement of water in either direction. A hypotonic solution has a lower po-

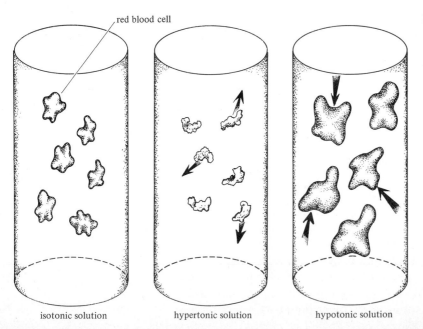

**Figure 3-7** Effects of isotonic, hypertonic, and hypotonic solutions on cell size.

tential osmotic pressure than normal cells which means that it has a higher water concentration. Water will thus move from a hypotonic solution into cells, causing them to swell. One application of all this is that solutions that are injected into the body should always be isotonic or they may cause damage to cells.

We have discussed osmosis at such length because it is fundamental to many physiological processes. We will use the concept of osmosis in our discussion of digestion, urine formation, circulation, and water balance.

In both the cases of diffusion and osmosis the membrane simply serves as a passive barrier. Again, if there is a concentration difference of a particular substance between the inside and the outside of the cell, and if the substance can pass through the membrane, it will move from an area of higher concentration to an area of lower concentration. However, the membrane is not always merely a passive structure; rather it is actively involved in transporting many substances in and out of the cell.

For example, many substances, such as sugars and amino acids, which are too big to pass through the pores and which do not dissolve readily in lipid still manage to diffuse in or out of the cell quite rapidly. In order to explain this phenomenon physiologists have postulated that the membrane contains certain molecules that act as **carriers,** that is, they help other molecules across the membrane. This is illustrated in Figure 3-8. The idea is that a molecule, for example, glucose, which cannot cross the membrane by itself, will combine with a carrier at one side of the membrane and that the carrier along with the molecule will then diffuse across the membrane. Once across the membrane the molecule that is being transported will be released. The carrier is then free to take another molecule of the substance back across the membrane in the opposite direction. The carrier is somewhat analogous to a ferryboat that takes cars back and forth across a river. Diffusion in which a substance is helped across the membrane is referred to as **facilitated diffusion.**

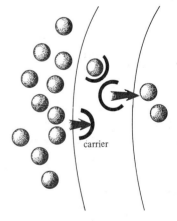

**Figure 3-8** Facilitated diffusion.

A more complex type of membrane involvement in the movement of substances in and out of cells is a process known as **active transport.** In facilitated diffusion substances move from a higher concentration to a lower concentration. The membrane only helps them across. Active transport on the other hand involves the movement of a substance from an area of lower concentration to an area of higher concentration. This is the reverse of the direction in which substances would naturally move by diffusion, thereby requiring the use of energy by the cell. The difference between diffusion and active transport is analogous to the difference between coasting down a hill on a bicycle and pedaling a bicycle to the top of a hill. The active transport of material across cell membranes requires a great deal of energy—some biologists estimate this to be as much as 20% of all the energy used in cells.

Although many theories have been proposed to explain the mechanism by which active transport takes place, our understanding of the process is still quite limited. Most of the theories suggest that active transport requires some sort of carrier, as illustrated in Figure 3-9.

The substance being actively transported, indicated by the dark circles, might combine with a carrier at one surface of the membrane. According to this theory the substance along with the carrier then diffuses to the other side of the membrane where the transported substance is released. At this point energy supplied by the cell is used to change the shape of the carrier. The carrier then diffuses back across the membrane either alone or in combination with a substance being actively transported in the opposite direction. Once the carrier has moved back across the membrane it returns to its original shape and the cycle is repeated. It should be emphasized again that this is but one of many theories and it will take a great deal more research to understand this process fully.

Active transport is the means by which the body maintains a much higher concentration of $K^+$ inside the cells than in the extracellular fluid, and maintains a much higher concentration of $Na^+$ in the extracellular fluid than inside of cells. The active transport of $Na^+$ and $K^+$ across the cell membrane is illustrated in Figure 3-10.

Some type of active transport mechanism, often referred to as a $Na^+$ pump, continuously moves $K^+$ into the cell and $Na^+$ out of the cell. This builds up the intracellular concentration of $K^+$ and

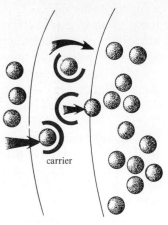

**Figure 3-9** Active transport.

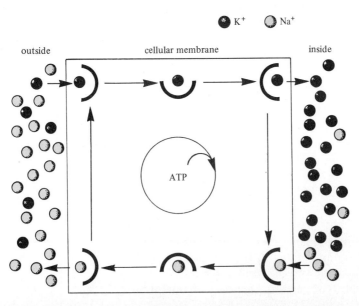

**Figure 3-10** Active transport of sodium and potassium across the cellular membrane.

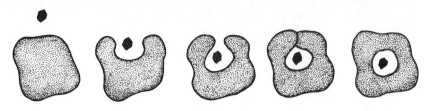

Figure 3-11 Phagocytosis.

the extracellular concentration of Na$^+$. The cellular membrane is relatively impermeable to the diffusion of Na$^+$ and K$^+$ so that they do not normally diffuse back down their concentration gradients very rapidly. Any sodium that does diffuse into the cell is actively transported back into the extracellular fluid and likewise K$^+$ that diffuses out of the cell is actively transported back into the cell. This separation of Na$^+$ and K$^+$ will assume great significance when we discuss the mechanism of nerve impulses, muscle contraction, and the control of acid–base balance. Active transport is also an essential process in the absorption of food from the digestive tract and in the formation of urine by the kidney.

Another mechanism in which the membrane participates in the transport of material in or out of the cell is a process known as **pinocytosis** or **phagocytosis.** This process is illustrated in Figure 3-11.

A portion of the membrane surrounds the particle that is to be transported into the cell and forms a small pocket or vesicle in which the particle is contained. The vesicle then moves into the interior of the cell where its walls are eventually broken down, thereby causing it to release the particle into the cytoplasm. When the particle being taken into the cell is small or liquid the process is referred to as pinocytosis; when the particle is large or solid the process is called phagocytosis.

## CELLULAR ENERGY METABOLISM

Energy is defined as the ability to do work, that is, the ability to change things in some way. Energy that is being stored is referred to as **potential energy,** whereas energy in the form of a change taking place is referred to as **kinetic energy.** A ball at the top of a hill has potential energy because of its position; when the ball rolls down the hill this potential energy is converted to the kinetic energy of the moving ball.

There are many different forms of energy: coal, heat, light, and sound are but a few of the many forms in which energy can occur. Although energy can be neither created nor destroyed, it can under the appropriate circumstances be converted from one form to another. For example, a power plant converts the energy stored in coal into electrical energy, which is then conducted through wires to your house. Within your house this electrical energy can be converted into light energy by a light bulb, heat energy by a toaster, or sound energy by a radio.

Just as a radio requires energy to do its work, the production of sound waves, a cell requires energy to do its work. Active transport, the synthesis of large molecules, and the contraction of muscles are all examples of work performed by cells. In general, energy is available to cells in the form of glucose, amino acid, and fat molecules which are absorbed from the digestive tract and carried to the cells by the blood. The job of the cell then is to convert the energy stored in these molecules into the kinetic energy of work being performed such as active transport or muscular contraction.

Glucose, amino acids, and fats all contain potential energy in the form of the chemical bonds which hold their various atoms together. Each chemical bond contains an amount of energy equal to the amount of energy required to form the bond in the first place. By a series of chemical reactions cells can break apart these chemical bonds, thereby causing their stored energy to be released. If this energy were released in an uncontrolled way all of the released energy would be converted to heat and be unavailable for the performance of cellular work. In order to avoid this the cells break the molecules apart in a series of small steps and use the energy which

GLYCOLYSIS

## 53 Cellular energy metabolism

is released to form a molecule, **adenosine triphosphate,** which can be used directly to do cellular work. In other words the cells first convert the energy stored in glucose, amino acid, or fat molecules into energy stored in adenosine triphosphate (ATP) and then they use the energy of ATP to do cellular work. The purpose of this is to minimize the amount of stored energy that is converted to heat and to maximize the amount of stored energy that can be used to do cellular work. At this point we are going to examine the basic processes by which cells convert food energy into ATP energy. We will discuss the ways in which cells use ATP to do cellular work as they become relevant throughout the book.

The exact mechanisms by which cells break down glucose, amino acids, and fats are complicated and involve a great many chemical reactions, each of which requires a separate enzyme. We are not going to describe each of the individual reactions but only those steps of the process that either highlight the basic mechanism of what is taking place or that will become important in later parts of the book.

The breakdown of glucose can be divided into two basic processes: **glycolysis** and the **Krebs (citric acid) cycle.** Glycolysis involves the conversion of glucose, a six-carbon compound, into two molecules of **pyruvic acid,** a three-carbon compound. Figure 3-12 shows the chemical reactions by which this takes place. We are not showing you these reactions in order for you to memorize

**Figure 3-12** Chemical reactions of glycolysis.

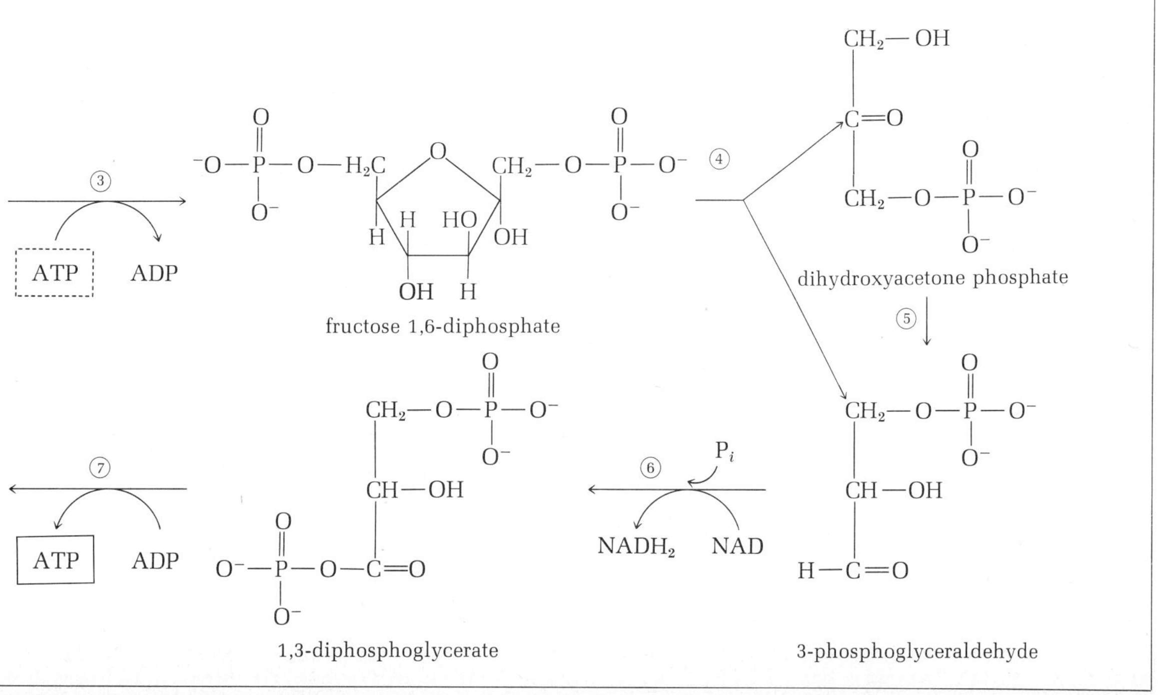

them, but only to give you an appreciation of the fact that the conversion involves many chemical reactions, each of which involves a relatively small chemical change. The overall process of glycolysis is summarized in Figure 3-13 where only the carbon atoms of glucose and pyruvic acid are shown. During glycolysis two molecules of ATP are formed; this represents only a small amount of the energy stored in glucose.

Once pyruvic acid has been formed the breakdown of glucose is completed by a series of reactions known as the Krebs (citric acid) cycle. These reactions are summarized in Figure 3-14 and illustrated in detail in Figure 3-15. The first step involves the removal of carbon dioxide from pyruvic acid and the combination of the remaining two-carbon compound, the acetyl group, with a substance known as coenzyme A. Coenzyme A is formed from pantothenic acid, a B vitamin, and is a carrier for the acetyl group in a number of different reactions. The compound formed from the combination of the acetyl group and coenzyme A is called **acetyl-CoA**. Coenzyme A transfers the acetyl group formed from pyruvic acid to a four-carbon compound already present in the cell, leading

**Figure 3-13** Summary of glycolysis showing only carbon atoms.

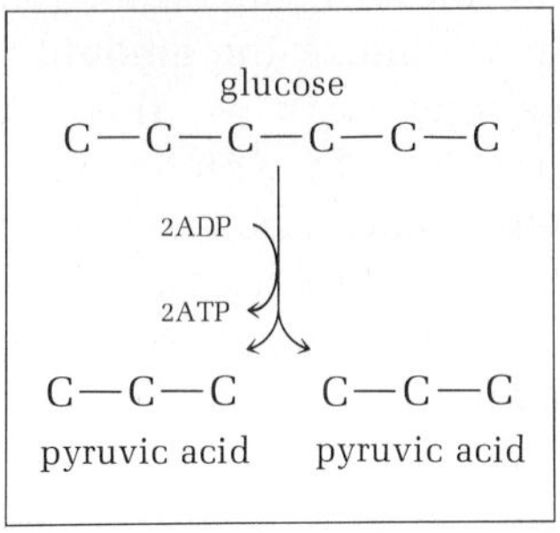

**Figure 3-14** Summary of the Krebs cycle.

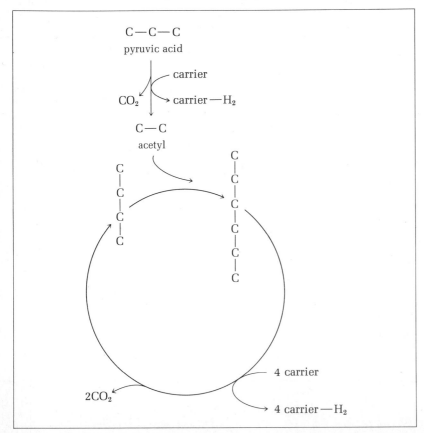

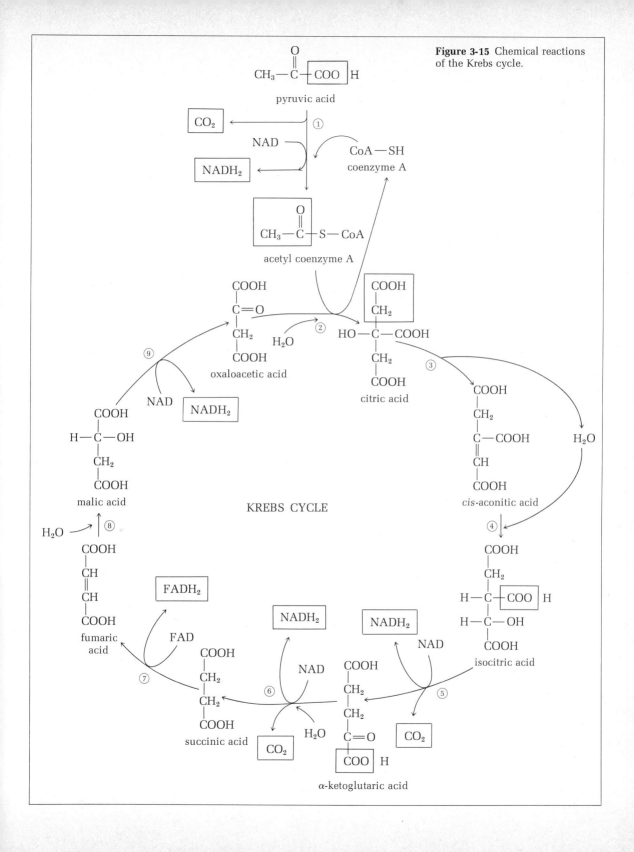

**Figure 3-15** Chemical reactions of the Krebs cycle.

to the formation of a six-carbon compound called **citric acid.** The six-carbon citric acid is then converted to carbon dioxide and a five-carbon compound, which is in turn converted to carbon dioxide and a four-carbon compound. Finally this four-carbon compound is converted into the four-carbon compound that combined with acetyl-CoA at the start of this series of reactions. In looking at this process it is clear why these reactions are referred to as a cycle; the four-carbon compound which combines with acetyl-CoA at the beginning of the series of reactions is reformed at the end of the series of reactions. It can then combine with another molecule of acetyl-CoA and begin a new cycle.

As a result of the Krebs cycle the carbon and oxygen atoms of pyruvic acid are converted into carbon dioxide. Because the pyruvic acid was originally formed from glucose, what is really happening is that the carbon and oxygen atoms of glucose are being converted to carbon dioxide.

By this point you are probably wondering about the whole point of the Krebs cycle: how is it used to form ATP? The key to the formation of most of the ATP that can be derived from glucose lies in the hydrogen atoms of glucose. We have already shown that when a glucose molecule is broken down its carbon and oxygen atoms are converted to carbon dioxide. The hydrogen atoms obtained from the breakdown of glucose are combined with oxygen from the air we breathe to form water, a process that releases a considerable amount of energy. This released energy is used to combine adenosine diphosphate (ADP) with phosphate to form the high-energy ATP. The process by which hydrogen and oxygen are combined into water with the resultant formation of ATP is referred to as **oxidative phosphorylation,** oxidative meaning that the process involves oxygen and phosphorylation referring to the fact that a phosphate group is added to another molecule.

During one of the steps of glycolysis and five of the steps of the Krebs cycle a pair of hydrogen atoms are removed and combined with either nicotinamide adenine dinucleotide (NAD), a coenzyme formed from the vitamin niacine, or flavine adenine dinucleotide (FAD), a coenzyme formed from the vitamin riboflavin. The NAD and FAD then transport the hydrogens to a portion of the mitochondrial membrane, which contains a special group of proteins known as cytochromes. The job of the cytochromes is to combine the hydrogen with oxygen in such a way as to maximize the conversion of the released energy into the formation of ATP. The reactions of oxidative phosphorylation can be summarized as follows:

$$O_2 + ADP + P + NAD \cdot H_2 + FAD \cdot H_2 \longrightarrow ATP + NAD + FAD + H_2O + heat$$

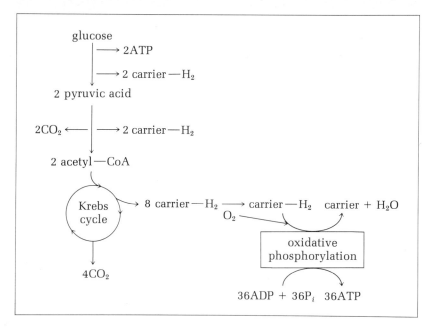

Figure 3-16 Formation of ATP from glucose.

The point of all of this is that it is really the reactions of oxidative phosphorylation which lead to the formation of most of the cell's ATP. Of the 36 new ATPs that can be formed from the breakdown of glucose, 32 are formed by oxidative phosphorylation and four are formed directly in one of the breakdown steps.

The breakdown of glucose and the conversion of its energy into the energy of ATP are summarized in Figure 3-16. The carbon and oxygen atoms of glucose are converted to carbon dioxide; the hydrogen atoms are combined with the oxygen we obtain by breathing to form water. Energy released by this process is used to combine ADP and phosphate into the high-energy compound ATP. We can represent the entire process by the equation

$$C_6H_{12}O_6 + ADP + P \longrightarrow CO_2 + H_2O + ATP + heat$$

ATP formed by this process is used as a source of energy for the various types of work performed by the cell: carbon dioxide, water, and heat are by-products of the reaction. Carbon dioxide is expired into the outside air, water is eventually lost through urination or sweating, and heat helps to maintain the body temperature.

The manner in which cells convert the energy contained in fats or amino acids into ATP is similar in many respects to the manner in which they convert the energy of glucose into ATP. Fats and amino acids are first converted to molecules which can enter the Krebs cycle; then by means of the reactions of the Krebs cycle and oxidative phosphorylation, hydrogen is removed and combined

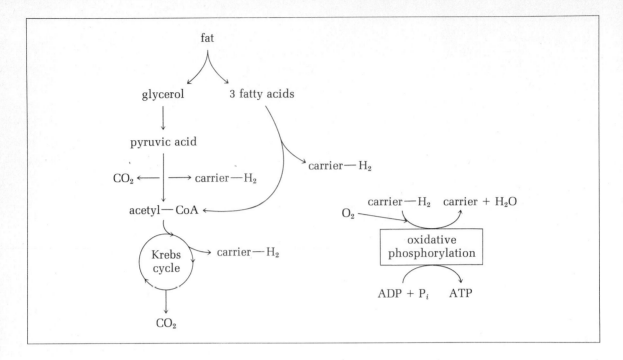

**Figure 3-17** Formation of ATP from fat.

with oxygen to form water with the released energy being used to convert ADP and phosphate into ATP. Figure 3-17 summarizes the conversion of the energy of fat into ATP.

Fat is first broken down into its two components: glycerol and fatty acids. Glycerol is then converted into one of the compounds formed during the glycolytic conversion of glucose to pyruvic acid. Once formed this compound can be converted into pyruvic acid, which can enter the Krebs cycle. Fatty acids are converted to acetyl-CoA, which can directly enter the Krebs cycle. During the conversion of fatty acids to acetyl-CoA a considerable number of hydrogens are removed. The hydrogens removed during the Krebs cycle as well as the hydrogens removed during the conversion of fatty acids to acetyl-CoA are combined with oxygen by oxidative phosphorylation with the resultant formation of ATP.

Each of the amino acids has a somewhat different structure, and the exact mechanism by which each is broken down is different. However, each of the amino acids can be converted by one series of reactions or another into a compound that can enter the Krebs cycle. This process is summarized in Figure 3-18.

Amino acid molecules differ from glucose and fat molecules in that they contain nitrogen atoms as well as carbon, hydrogen, and oxygen atoms. In order for amino acids to be converted into molecules that can enter the Krebs cycle the nitrogen atoms must first be removed. Once the nitrogen is removed the remaining mole-

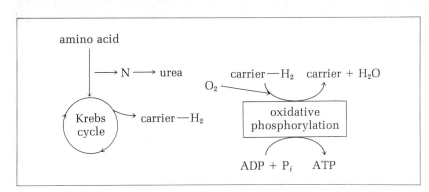

**Figure 3-18** Formation of ATP from amino acid.

cule, composed of carbon, hydrogen, and oxygen atoms, is fed into the Krebs cycle, and once again the hydrogen atoms are removed and combined with oxygen by oxidative phosphorylation with the resultant formation of water and ATP. The nitrogen that is removed in the breakdown of amino acids is usually converted into **urea,** and then eliminated from the body as part of the urine.

Figure 3-19 reviews the basic steps in the conversion of the

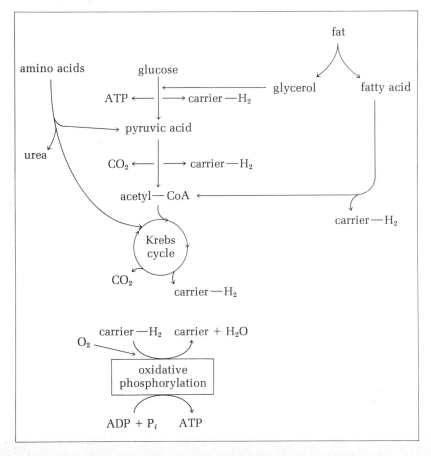

**Figure 3-19** Summary of the formation of ATP from glucose, fats, and amino acids.

energy contained in glucose, fats, and amino acids into the energy of ATP. Each of these molecules is first converted into a molecule that can enter the Krebs cycle and then by the reactions of the Krebs cycle and oxidative phosphorylation into carbon dioxide and water. The energy which is released is used to combine ADP and phosphate into the high-energy compound ATP, a compound which serves as a source of energy for all of the work performed by the cell.

## PROTEIN SYNTHESIS

Proteins have two crucial functions within cells: they are the enzymes that control which chemical reactions take place within the cell and they form an essential part of the physical structure of the various cell organelles such as the cellular membrane, the endoplasmic reticulum, and the mitochondria. Even the nonprotein components of cellular structure are dependent on the protein content of the cell because they are manufactured by chemical reactions that require enzymes (proteins). This means that the physiological activity and physical structure of each cell depend on the particular proteins that are manufactured (synthesized) by that cell. It is the genes, the hereditary units passed from generation to generation and from cell to cell, that determine which proteins cells are able to synthesize and consequently the nature of their physiological activity and physical structure. At this point we are going to discuss the manner in which genes direct the production of proteins; in Chapter 18 we will discuss the patterns in which genes are passed from generation to generation.

Proteins are composed of chains of amino acids. Each individual type of protein is characterized by which of the 20 different amino acids it contains and the particular order in which these amino acids are arranged. This is illustrated in Figure 3-20. Imagine the circle, square, triangle, and cylinder as representing four dif-

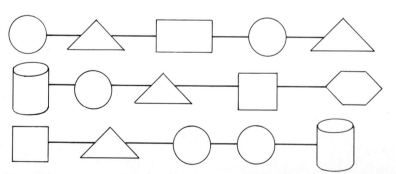

**Figure 3-20** Schematic representation of how different proteins can be made from the same amino acids.

**61** Protein synthesis

ferent amino acids that could be used to make proteins. The various chains would then represent some of the proteins that could be made with these amino acids. Notice that these proteins differ with regard to the particular amino acids they contain, the order in which the amino acids are arranged, and the length of the chain. As the cell has 20 different amino acids to work with and can form proteins that vary in length from a few amino acids to thousands of amino acids, an astronomical number of different proteins could conceivably be produced by the cell. The genes contain a code that determines which of these proteins the cell will actually manufacture.

Genes are composed of a type of molecule called deoxyribonucleic acid (DNA), the largest molecule found within the cell. Each DNA molecule is composed of a chain of smaller subunits called nucleotides which consist of three molecular parts: a phosphate group, a pentose (five-carbon) sugar, and a nitrogenous base, as indicated in Figure 3-21.

Each strand of DNA consists of a chain of four nitrogen bases linked together by sugar and phosphate molecules. There are four nitrogen bases: adenine, guanine, cytosine, and thymine. It is the particular sequence in which the nitrogen bases occur in the DNA molecule that determines the sequence in which the cell will combine amino acids into a protein. Before discussing how the code contained in the DNA molecule is used to produce a protein we must first discuss the structure of DNA a little further.

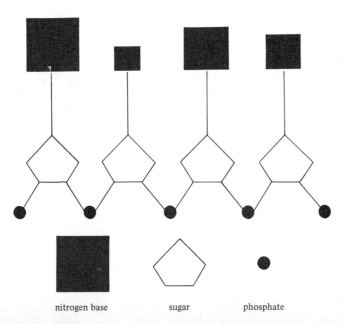

**Figure 3-21** Combination of nucleotides into a chain of DNA.

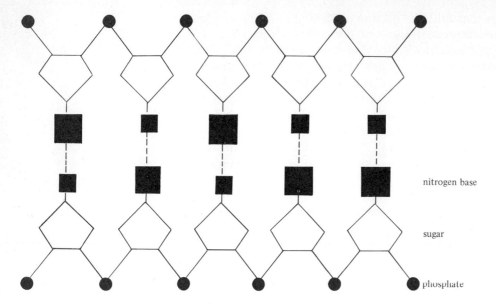

A DNA molecule is actually composed of two strands of nucleotides, as indicated in Figure 3-22. The two strands are loosely held together by chemical bonds between the nitrogen bases of one strand and the nitrogen bases of the other strand. However, the nitrogen bases must combine in a particular pattern, adenine must combine with thymine, and cytosine must combine with guanine. The two strands of DNA, held together by **base pairing** between adenine and thymine as well as between cytosine and guanine, are wound into a coil known as a double helix, illustrated in Figure 3-23.

We have already stated that it is the order of nitrogen bases in a strand of DNA which codes for the order of amino acids in the protein. More specifically, three nitrogen bases form the code for one amino acid as illustrated in Figure 3-24. Suppose the nitrogen-base sequence that codes for the amino acid histidine is cytosine-adenine-thymine (from now on we will represent a nitrogen base by the first letter of its name), the nitrogen-base sequence that codes for the amino acid tyrosine is TAC, and the nitrogen-base sequence that codes for the amino acid arginine is AGA. If the genetic DNA had the nitrogen-base sequence CATTACAGA the protein that is being coded for would have the amino acid sequence histadine-tyrosine-arginine. On the other hand if the DNA had the nitrogen-base sequence CATAGATAC, the protein should have the amino acid sequence histidine-arginine-tyrosine. The point of all this is that the order of nitrogen bases in the DNA of a gene codes for the order in which amino acids should be arranged into a protein with three nitrogen bases forming the code for each amino acid.

**Figure 3-22** Combination of two strands of DNA through pairing of nitrogen bases.

**Figure 3-23** Three-dimensional double-helix structure of DNA.

DNA does not actually combine amino acids into proteins; it only codes for the type of proteins the cell should manufacture. It is similar in this respect to an architect's blueprint of what a building should look like. Someone must still take the blueprint and use it to construct the building. In the case of the cell there has to be a mechanism of using the DNA code to actually build a protein. Furthermore because the DNA code is located on the chromosomes within the nucleus, and amino acids are combined into proteins at the ribosomes scattered throughout the cytoplasm, the cell must have a way of transferring the DNA code from the nucleus to the cytoplasm.

It is a compound called ribonucleic acid (RNA) which is responsible for transferring the DNA code from the genes to the ribosomes and then using this code to assemble amino acids in the proper sequence into a protein. RNA is similar to DNA in that it is composed of a chain of nucleotides, each consisting of a sugar, a phosphate group, and a nitrogen base. RNA differs from DNA, however, in three ways: it contains the sugar ribose instead of the sugar deoxyribose (this is why it is called RNA), it contains the nitrogen base uracil rather than the nitrogen base thymine, and it is single stranded whereas DNA is double stranded.

Figure 3-25 illustrates how a particular type of RNA, known as messenger RNA (mRNA), transfers the code for which type of protein is to be constructed from the DNA to the ribosomes. The first step in this process is the splitting apart of the two strands of DNA. One of the strands of DNA is then used to assemble a strand of mRNA with the sequence of nitrogen bases in the DNA molecule determining the sequence of nitrogen bases in the mRNA as a result of base pairing. Thus if the DNA has the nitrogen-base sequence

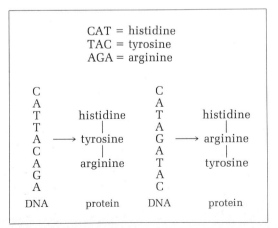

**Figure 3-24** Nitrogen base code of DNA.

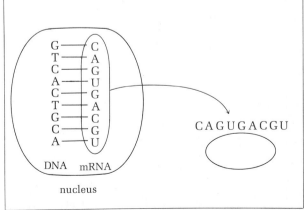

**Figure 3-25** Role of messenger RNA in protein synthesis.

GTCACTGCA, the mRNA will have the sequence CAGUGACGU; cytosine pairs with guanine and adenine pairs with thymine (as RNA contains uracil instead of thymine, uracil of RNA pairs with adenine of DNA). Another way of saying this is that the sequence of nitrogen bases in DNA serves as a mold or template from which only one particular sequence of nitrogen bases in RNA can be derived. This serves to transfer the code from the DNA to mRNA. Once formed the mRNA separates from the DNA and diffuses into the cytoplasm where it combines with one or more ribosomes.

Once the code has been transferred to the ribosomes, the amino acids must be brought to the ribosomes and lined up in the proper order. This is accomplished by another type of RNA known as transfer RNA (tRNA). There are 20 different types of tRNA, each of which contains the DNA code for a particular amino acid (with uracil replacing thymine in the code) and has the ability to combine with the amino acid for which it has the code. A molecule of tRNA will combine with the amino acid for which it is coded and will then diffuse to the ribosomes where it will combine by base pairing with the appropriate portion of the mRNA molecule, as indicated in Figure 3-26. This results in the amino acids being lined up in the required order. All that remains is for the amino acids to

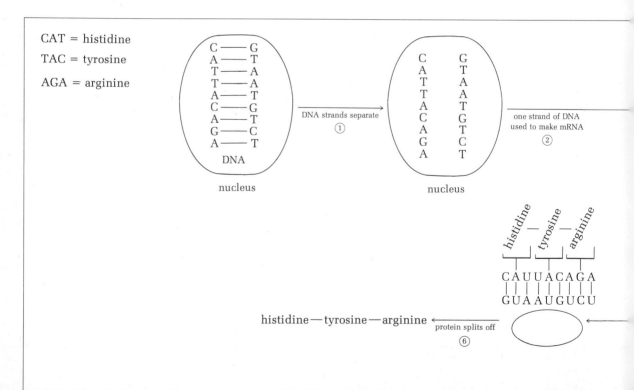

# 65 Protein synthesis

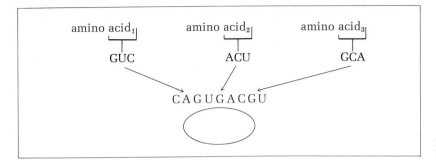

**Figure 3-26** Role of transfer RNA in protein synthesis.

be combined with each other into a protein and for the completed protein to break away from the ribosomes.

The entire process of protein synthesis is summarized in Figure 3-27. Genes, composed of the double-stranded DNA molecule, contain the codes for what types of proteins the cell should construct. The code is contained in the order of the nitrogen bases in the DNA molecule, with three consecutive nitrogen bases coding for each amino acid. In order to construct a protein the code must first be transferred from DNA to mRNA. This is accomplished by splitting the two strands of DNA and using one of the strands of

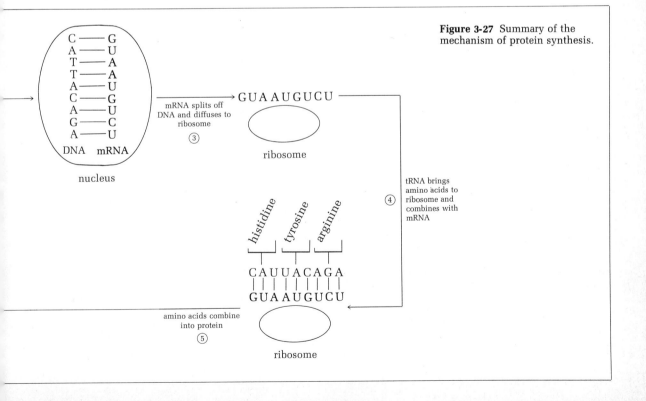

**Figure 3-27** Summary of the mechanism of protein synthesis.

DNA as a template for the construction of a strand of mRNA. The mRNA then diffuses out to the cytoplasm where it combines with ribosomes. Molecules of tRNA bring the amino acids to the ribosomes, and by base pairing with mRNA they line the amino acids up in the proper order. The amino acids then combine into a protein which separates from the ribosomes. This protein can then either become part of the physical structure of the cell, act as an enzyme, or be secreted out of the cell to function in some part of the body.

## MITOSIS

Most human cells have the ability to divide into two new cells which have the same genetic composition as the original cell, a process known as mitosis. It is this property of cells which is responsible for the growth of a single fertilized egg cell into an adult composed of many trillions of cells, and which enables the body to replace worn-out or damaged cells with new cells which function properly. Some types of cells divide quite rapidly whereas other types of cells divide slowly or not at all; for example, the cells of the bone marrow are continuously dividing to form new blood cells whereas adult nerve and muscle cells seem totally unable to divide. Even the same types of cells may divide at different rates under different circumstances. Skin cells divide much more rapidly at the site of a cut than in an intact portion. When cells begin to divide too rapidly a tumor is formed. A **tumor** is a growth of physiologically and anatomically unnecessary cells that unfortunately can use up the nutrients required by normal cells and distort the physical structure of the organ in which it is growing. In addition, tumor cells often detach themselves from the original growth and move to other parts of the body where they continue to divide, leading to the formation of new tumors.

At this point nobody really understands the forces that stimulate cells to divide or that stop cell division once the appropriate number of cells has been obtained. The actual process of mitosis, however, is fairly well understood.

The first step in mitosis involves duplication of all the genes of the dividing cell so that a complete set of genes is available for each of the two new cells that will be formed. This process is illustrated in Figure 3-28.

Genetic duplication is accomplished by a separation of the two strands of DNA and by the use of each of these strands as a template for the production of a new strand of DNA. The new strand of DNA is formed by base pairing in the same way that the strand of mRNA was formed during protein synthesis. Splitting of

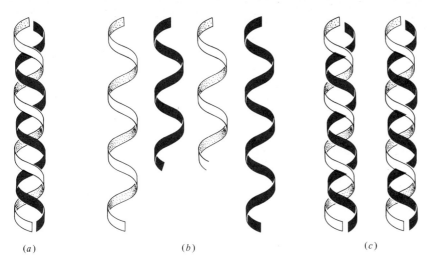

(a)  (b)  (c)

**Figure 3-28** Duplication of DNA.

the DNA molecule and the formation of a new strand of DNA along each of the original strands results in the formation of two new DNA molecules, each of which is identical to the other and to the original molecule of DNA. In other words the original gene has been duplicated.

Once the genes have been duplicated the cell must divide in such a way that each of the new cells gets a complete set of genes. This process is illustrated in Figure 3-29. During the period in

**Figure 3-29** Mitosis. (Photos by Lester Bergman & Assoc.)

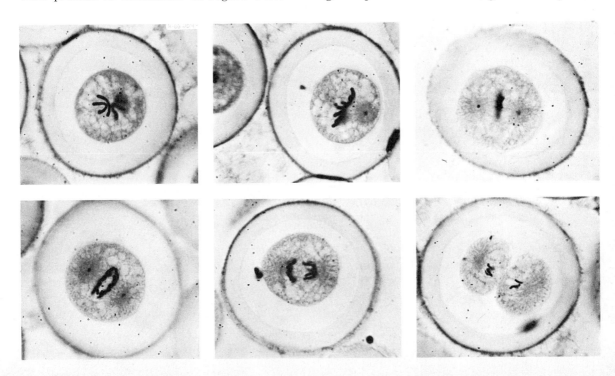

which the genes are duplicating, the various chromosomes are coiled around each other. This period is referred to as *interphase* to indicate that the cell is not in the process of actually dividing. Once the genes are duplicated the chromosomes begin to unwind and line up at the center of the nucleus while the nuclear membrane disappears. Duplication of the DNA during interphase has resulted in the formation of two identical chromosomes from each of the original chromosomes. The two identical chromosomes are bound together at a special region known as the **centromere.** While the duplicated chromosomes are lining up at the center of the nucleus, the tiny centrioles, located just outside the nucleus, are moving to opposite sides of the nucleus; fibers appear that connect the centrioles with each other and with the centromeres which hold the two strands of the chromosomes together. Once the double-stranded chromosomes are lined up at the center of the nucleus and are connected by fibers to the centrioles, the two strands break apart at the centromere. The fibers then shorten with the resultant movement of one chromosome of each pair toward one side of the cell and the other chromosome of the pair toward the other side. As the two sets of chromosomes are separated a new nuclear membrane forms around each set, and the entire cell splits into two new cells, each of which contains one set of chromosomes.

In summary, mitosis is cell division in which two new cells are formed which are identical in genetic composition to each other and to the original cell from which they are formed. This is accomplished by duplicating the genetic material, lining up the duplicated genetic material at the center of the nucleus, pulling the duplicated genetic material apart so that identical sets of genetic material move toward each side of the cell, and splitting the cell into two new cells.

## OBJECTIVES FOR THE STUDY OF THE CELL

At the end of this unit you should be able to:

1. State the function of the cellular membrane, nucleus, ribosomes, endoplasmic reticulum, centrioles, mitochrondria, lysosomes, and Golgi apparatus.
2. Name the structures in the nucleus that contain the genes.
3. Describe the process of diffusion.
4. Define the term osmosis.
5. Define the term osmolar concentration.
6. Distinguish between isotonic, hypertonic, and hypotonic solutions.
7. Distinguish between facilitated diffusion and active transport.
8. Describe the processes of pinocytosis and phagocytosis.
9. Define the word energy.
10. State the forms in which energy enters the cells.

11. Name the molecule that serves as the direct source of energy for cellular work.
12. Name the process by which glucose is converted to pyruvic acid.
13. Name the compound into which pyruvic acid must be converted in order to enter the Krebs cycle.
14. Name the process by which hydrogen and oxygen are combined in such a way as to form adenosine triphosphate and water.
15. Name the nitrogen waste product formed during the breakdown of amino acids.
16. Cite two important functions of proteins.
17. State the function of genes and the type of molecule of which they are composed.
18. Explain how genes code for proteins.
19. Explain the role of messenger RNA in protein synthesis.
20. Explain the role of transfer RNA in protein synthesis.
21. Define the term mitosis.
22. Explain the process of mitosis.

# 4
# Tissues and Membranes

EPITHELIAL TISSUE
CONNECTIVE TISSUE
MUSCLE TISSUE
NERVE TISSUE
MEMBRANES
THE SKIN

Although cells are the basic functional units that carry out the work necessary to maintain life, they do not conduct this work in isolation. All of the individual cells of the body are incorporated into larger organizational units, the simplest of which are the various types of tissues. A tissue is composed of a group of cells that are usually similar in structure and function and the intercellular material that holds these cells together. The cells and intercellular material of each type of tissue are organized in such a way as to carry out a particular function. The exact function of each type of tissue is related to the types of cells contained in the tissue, the way in which these cells are arranged, and the amount and type of intercellular material found between the cells.

Based on both structure and function, the various tissues are classified into four basic categories: (1) epithelial, (2) connective, (3) muscle, and (4) nerve. These major categories are in turn divided into subtypes which, in the case of connective tissue, represent tissues that vary considerably in both structure and function.

## EPITHELIAL TISSUE

**Epithelial tissue** consists of cells which are packed very closely together with almost no intercellular space between the cells. A small amount of intercellular material serves to hold the cells together. Because there is no room for blood vessels to pass through epithelial tissue, the epithelial cells must receive nutrients from and eliminate wastes into blood vessels that pass through adjacent tissues.

The fact that the cells of epithelial tissue are so closely packed enables this tissue to act as a protective barrier across which only certain substances can pass, as indicated in Figure 4-1. Many different substances can cross a tissue in which the cells are spaced widely apart by simply moving between the cells, whereas only those substances that can move through the cells or are small

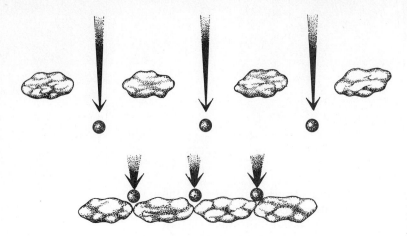

Figure 4-1 Protective barrier formed by epithelial tissue.

enough to squeeze between the cells can traverse epithelial tissue in which the cells are close together. The exact structure of each of the different types of epithelial tissue is related to its location and the types of substances that should or should not move across it.

Acting as a protective barrier across which only certain substances can travel, epithelial tissue lines the entire outer surface of the body, the inner walls of the various body cavities, and the inner walls of all the tubes within the body. Therefore the outer layer of the skin is epithelial tissue, the inner walls of the thoracic and abdominal cavities are epithelial tissue, and the inner linings of the respiratory tract, digestive tract, urogenital tract, and blood vessels are epithelial tissue.

An example of the protective function of epithelial tissue can be seen in the way in which it normally prevents bacteria from crossing any of the body surfaces and from gaining access to the underlying structures. In order to enter the body, bacteria must either travel through the skin or enter one of the external orifices such as the mouth, nose, anus, or urinary opening. Normally, bacteria cannot cross the skin unless it is cut, because they cannot move through the outer epithelial layer of the skin.

Bacteria that enter the body along with food cannot cross the epithelial tissue lining the digestive tract, and although some take up residence inside the digestive tract, most are ultimately expelled from the body through defecation. The bacteria that enter the respiratory tract along with the inspired air are prevented from passing into the underlying structures by the epithelial tissue lining the respiratory tract.

These bacteria are trapped in a mucus which is secreted by the epithelial lining; the mucus is periodically swallowed, and the bacteria, like those swallowed with food, are eliminated from the body by defecation. Likewise bacteria that enter through the anal or urogenital openings are usually trapped in the tubes they enter by the epithelial lining of these tubes. These bacteria are then eliminated from the body with the urine and feces or are simply destroyed within the tubes. The epithelial tissue is not the only weapon the body has to defend itself against agents of infection. If these agents cross the tissue, either because the tissue is physically injured or because the agents destroy it themselves, the body can defend itself by a number of other means. We will discuss these as they become relevant throughout the book.

In addition to its protective function, epithelial tissue performs the functions of absorption, secretion, and filtration. The epithelial tissue lining the digestive tract and kidney tubules is able to selectively absorb many substances from these tubes into the blood. Bacteria contained in the food cannot cross the digestive epithelial tissue, but sugars, fats, amino acids, salts, and water can move across the digestive epithelial tissue and enter the blood.

Epithelial tissue lining endocrine glands is adapted for the secretion of hormones into the blood, and epithelial tissue lining the digestive and respiratory tracts is designed for the secretion of a mucus that acts as a second protective layer, coating the epithelial tissue lining these tracts. Finally, the epithelial tissue lining the capillaries acts to filter the fluid that moves between the blood vessels and the intercellular spaces.

Depending on its exact location and the function it serves epithelial tissue can have any of a number of shapes, as illustrated in Figure 4-2. The various types of epithelial tissues are named according to the shape and arrangement of their cells. If the cells are flat, it is called **squamous epithelium;** if the cells are shaped like a cube, it is called **cuboidal epithelium;** and if the cells are tall and thin, it is called **columnar epithelium.** Epithelial tissue which is composed of only a single layer of cells is called **simple epithelium,** whereas epithelial tissue composed of several layers of cells is called **stratified epithelium.**

Stratified epithelium is generally found on surfaces of the body that should only be crossed by a few substances. The outer layer of the skin and the inner layer of the mouth are both composed of stratified squamous epithelium. Simple epithelium, on the other hand, is usually found on surfaces that many substances must be able to cross.

## CONNECTIVE TISSUE

**Connective tissue** is characterized by having cells that are spaced at some distance from one another with a relatively large amount of

(a)

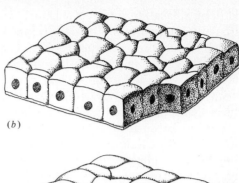
(b)

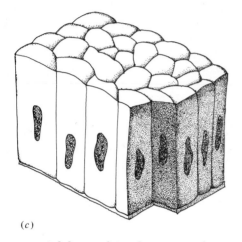
(c)

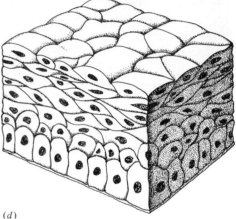
(d)

**Figure 4-2** Types of epithelial tissue: (a) simple squamous, (b) simple cuboidal, (c) simple columnar, and (d) stratified squamous.

material located in the spaces between the cells. The term, however, is somewhat of a catchall because tissues as dissimilar as blood and bone are both considered to be forms of connective tissue. Because of the diversity of the different tissues that fall into the general category of connective tissue, a variety of different systems are used for classifying them into subtypes.

We will use the one that divides connective tissue into four major subtypes: blood and lymph, connective tissue proper, cartilage, and bone. Blood and lymph will be discussed in Chapters 10 and 12, respectively, and bone and cartilage will be discussed in the next chapter, Chapter 5. In this chapter, we are going to look at connective tissue proper, which functions to hold the various parts of the body together.

Connective tissue proper is usually located just below the epithelium within an organ or as a sheet of tissue connecting one organ with another. Although its structure varies from one location in the body to another, connective tissue proper almost always contains a large number of fibers within its intercellular spaces. These fibers can be divided into three types: collagen fibers, elastin fibers, and reticular fibers.

**Collagen fibers** are composed of bundles of smaller fibers (often called fibrils) of the protein collagen. These fibers are quite tough, although somewhat flexible, and serve to strengthen the connective tissue which contains them. Collagen is the most abundant protein in the body. It constitutes close to 40% of the total body protein. Collagen fibers, once secreted, are retained in the intercellular spaces for the lifetime of a person, unless, of course, they are destroyed by physical injury or infection. However, these fibers undergo molecular changes over the years which may be responsible for some of the changes that take place during the aging process.

**Elastin fibers** are considerably thinner than collagen fibers and are composed of the protein elastin. As the name implies, these fibers are elastic: they can be stretched easily, but they will return to their original shape once the stretching force is removed. They are similar in this respect to a rubber band or a balloon. It is the elastic fibers contained within the connective tissue of the skin, for example, that enables the skin of a woman's abdomen to stretch to accommodate the growing fetus during pregnancy and then return almost to its original shape after delivery.

**Reticular fibers** are very small and as the name "reticular" indicates they are wound around each other into complex networks. These fibers are found in the connective tissue component of delicate structures, such as capillaries and small nerves, and at the point at which connective tissue is joined to other types of tissue.

In addition to the different types of fibers, there are a number of different kinds of cells that are found in connective tissue proper: the fibroblasts, macrophages, mast cells, adipose cells, and white blood cells.

The **fibroblasts** are responsible for manufacturing and secreting the various proteins which make up the intercellular fibers of connective tissue. As we discussed in the previous chapter, the protein production of a cell is controlled by its genes. In the fibroblasts, the genetic DNA directs the production of collagen, elastin, and reticular fibers. These proteins are probably assembled at ribosomes located along the endoplasmic reticulum of the fibroblasts and are then secreted by the Golgi apparatus into the intercellular spaces where they form fibers. Thus although the fibers are found outside the cells, the proteins that make up the fibers are produced within the cells.

Fibroblasts are particularly active during growth and after an injury, periods during which new tissue must be manufactured. During these periods, the fibroblasts begin to divide rapidly, enlarge, and greatly increase their rate of protein production. This provides the intercellular fibers necessary for the production of new connective tissue.

**Macrophages** are cells that are capable of removing undesirable material from the body by phagocytosis, the process by which large pieces of matter are engulfed into a cell and then digested by enzymes secreted from the cell's lysosomes. These cells are particularly common in the vicinity of blood vessels and they serve as a second line of defense against infectious agents that manage to cross the epithelium.

Macrophage cells are normally stationary. However, if there is any injury to the connective tissue, these cells will move, in much the same way as an amoeba moves, to the site of the injury; and by phagocytosis they will remove any debris caused by the injury. This clears the area in preparation for the formation of new tissue. Macrophage cells are also used to remove old blood cells from the circulation and to remove any tissue cells that are no longer functioning properly.

**Mast cells** are also particularly common in connective tissue located near blood vessels. They secrete two substances, heparin and histamine, which have very potent effects on the circulation. **Heparin** is an anticoagulant that prevents blood clotting. The exact physiological role of heparin secretion by the mast cells has not yet been determined, although it may be involved in preventing the formation of internal clots—thrombi—along the walls of blood vessels. A small amount of heparin is often added to syringes or test tubes that are to be filled with blood in order to prevent the blood from clotting. **Histamine** is released by mast cells when a tissue is injured. It acts to widen the blood vessels (vasodilation) and thus to increase blood flow to the injured area. This increased blood flow supplies the materials needed to repair the damage and to fight any agent that may have caused the damage. Histamine also increases the permeability of capillaries so that the material provided by the increased blood flow can move out of the blood vessels and to the site of the injury. The movement of fluid out of the blood vessels at the site of an injury leads to a local swelling, a process known as inflammation.

**Adipose cells** are connective tissue cells that are particularly adapted for the storage of fat. At a time when more energy (food) is taken into the body then is needed for the formation of ATP, the excess energy is stored as fat in the adipose cells. At a later time, when the need for ATP to do cellular work exceeds the amount that can be supplied by the food being eaten, fat is released from the adipose cells and is used as a source of ATP.

Many investigators now feel that the number of adipose cells a person has in his connective tissue is determined both genetically and by the diet eaten during the first few years of life. They also believe that to some degree adipose cells "demand" to be filled, that is, a person's appetite is related to the number of fat cells he has in his body. If this is the case, overstimulation of fat-cell development in a child by feeding a child too much may result in an

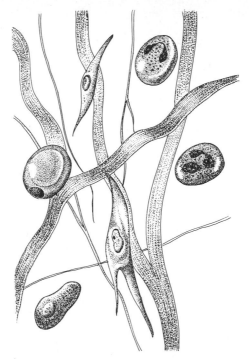

**Figure 4-3** Connective tissue proper.

adult who is predisposed to being overweight. Connective tissue with a large number of fat cells is often called **adipose tissue.**

In addition to fibroblasts, macrophages, mast cells, and adipose cells, connective tissue proper often contains a number of white blood cells. We will talk about the different types of white blood cells in Chapter 10.

The various cells and fibers of connective tissue are embedded in a ground substance (matrix) composed of water, salt, protein, carbohydrate, and lipid. This ground substance may be firm, jelly-like, or fluid, depending on the particular connective tissue.

Figure 4-3 illustrates the various types of cells and fibers found in connective tissue proper.

Connective tissue proper can be subdivided into two basic forms: loose connective tissue and dense connective tissue. **Loose connective tissue,** as the term indicates, is reasonably flexible and allows for a fair amount of movement between the structures it connects. This tissue is characterized by its relatively large amount of ground substance in which the various types of connective tissue cells and fibers are arranged in an irregular pattern. It is abundantly distributed throughout the body and serves a number of functions: it holds the various tissues of each organ together, it binds the different organs to each other, it helps suspend the internal organs in their proper position within the body cavities, and it forms a delicate protective layer around the small nerves and blood vessels located within the internal organs.

**Dense connective tissue** is characterized by an abundance of fibers and a scarcity of cells and ground substance, as illustrated in Figure 4-4. In some forms of dense connective tissue, the fibers are irregularly arranged as they are in loose connective tissue; however, they are much thicker and are woven into a tight net. This creates a very tough tissue that will resist a pulling force from any direction. Such dense connective tissue forms the dermis of the skin, the protective capsule of the internal organs, and the sheaths that surround bones and nerves. In other forms of dense connective tissue, the fibers are regularly arranged into thick, parallel bundles. This arrangement enables the tissue to withstand large pulling forces exerted in one direction. **Tendons,** which connect muscles to bones, and **ligaments,** which connect bones to other bones, are both forms of dense connective tissue in which the fibers are arranged in parallel.

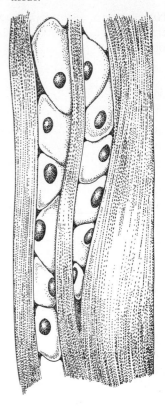

**Figure 4-4** Dense connective tissue.

## MUSCLE TISSUE

**Muscle tissue** is composed of cells that are capable of contracting. There are three different types of muscle found within the body: skeletal muscle, cardiac muscle, and smooth muscle. **Skeletal muscle,** as the name implies, is attached to the various bones of the skeleton and is responsible for body movement. This type of muscle can be voluntarily controlled. Skeletal muscle will be discussed in detail in Chapter 6. **Cardiac muscle** is the type of muscle tissue found within the heart. This muscle is structured in such a way as to enable the heart to contract as a coordinated unit. In Chapter 11 we will discuss the structure of cardiac muscle. **Smooth muscle** is contained within the walls of many of the body tubes, such as the blood vessels and the digestive tract. In general, smooth muscle serves either to propel material through the tube or to adjust the diameter of the tube. Smooth muscle will be discussed as it becomes relevant throughout the book. Figure 4-5 illustrates the various types of muscle tissue.

**Figure 4-5** Muscle tissue: skeletal, cardiac, and smooth.

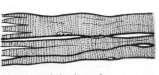

skeletal muscle

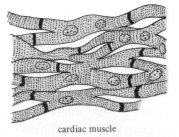

cardiac muscle

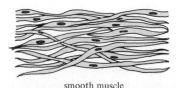

smooth muscle

## NERVE TISSUE

**Nerve tissue** is composed of cells capable of conducting electrical impulses and of certain types of supporting cells. This tissue will be discussed in detail in Chapter 7.

## MEMBRANES

Individual tissues, like individual cells, do not function in isolation; they are incorporated into larger organizational units. The membranes that cover the outer surface and many of the internal surfaces of the body are probably the simplest structures into which tissues are organized.

There are four basic types of membranes: mucous, serous, synovial, and cutaneous (skin), all of which are composed of a surface layer of epithelial tissue and a deeper layer of connective tissue.

**Mucous membranes** line all of the tracts (passageways) within the body that open to the exterior. Thus, mucous membranes form the lining of the digestive tract, respiratory tract, urinary tract, and reproductive tract. The epithelial layer of the mucous membrane is at the surface of the passageway, and the connective tissue layer is behind the epithelial layer. Mucous membranes, as the name implies, secrete **mucus,** a slimy substance composed of the protein polysaccharide **mucin,** which acts to lubricate the passages into which it is secreted. This secretion is carried out by specialized cells of the epithelial layer known as **goblet cells.**

All of the mucous membranes carry out two basic functions: protection and the secretion of mucus. In addition, the mucous membrane that lines the stomach and intestine is capable of absorbing substances into the blood. We will discuss individual mucous membranes in more detail when we examine the particular structures in which they are found.

Whereas mucous membranes line the passageways that open to the outside, **serous membranes** line the internal cavities which are closed to the outside. There are three principal serous membranes: the **pleura,** which lines the thoracic cavity; the **pericardium,** which forms the sac that surrounds the heart; and the **peritoneum,** which lines the abdominal cavity. Each of these serous membranes not only covers the walls of its cavity, but also folds back on itself and forms a covering around the internal organs contained within the cavity. The part of the membrane that lines the walls of the cavity is called the **parietal layer** and the part of the membrane that covers the organs is called the **visceral layer.** This is illustrated for the pleura in Figure 4-6. The parietal pleura lines

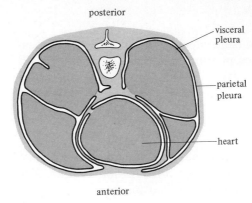

Figure 4-6 Two layers of the pleura.

the walls of the thoracic cavity and the visceral pleura covers the outer surface of the lungs. Epithelial tissue of the parietal pleura faces the epithelial tissue of the visceral pleura. Both of these layers of epithelial tissue secrete serosal fluid into the small space between the lungs and the walls of the thoracic cage. This fluid moistens the visceral and parietal pleura and allows them to slide smoothly over one another as the lungs move during respiration. In a similar manner, serosal fluid in the spaces between the visceral and parietal layers of other serosal membranes greatly reduces friction that might occur as various parts of the body move in relationship to each other.

**Synovial membranes** line joints, tendons, and other structures within the skeletal and muscular systems. We will consider them when we discuss these systems in the following two chapters.

## THE SKIN

The **cutaneous membrane,** or skin, is a complex structure that covers the entire outer surface of the body. Its essential functions include: protecting the body against injury and infection, reducing water loss from the body fluids, aiding in the regulation of body temperature, responding to touch and temperature stimuli from the external environment, and excreting certain waste products.

The skin protects the body by acting as a barrier. It prevents the millions of bacteria and viruses with which we come into daily contact from entering the body and damaging the internal structures and it resists mechanical forces which might otherwise tear at these internal structures. Because the skin is quite impermeable to water, it reduces the amount of water that evaporates out of the body fluids, in much the same way as the cover of a pot reduces the amount of water that evaporates out of the pot.

The various blood vessels of the skin help regulate the body temperature. If the body temperature begins to rise, blood flow to the skin is increased and some of the excess heat is lost by radiation into the external environment. On the other hand, if the body temperature begins to fall, blood flow to the skin is reduced so that body heat can be conserved. The sweat glands of the skin also aid in the regulation of body temperature. If the body temperature gets too high, the amount of sweat secreted by these glands is increased. Some of the excess body heat is used up in evaporating this sweat from the surface of the skin. Nerve endings in the skin are activated when the skin is touched and when the temperature of the external environment changes.

The skin, as illustrated in Figure 4-7, is composed of two basic layers: a thin outer layer, the epidermis; and a thick, inner layer, the dermis. A subcutaneous layer of loose connective tissue binds the skin to the structures that lie below it, but this layer is not considered part of the skin itself. The subcutaneous layer often contains many adipose cells and is referred to as the subcutaneous fat.

The **epidermis** is composed of stratified, squamous epithelial tissue which contains two types of cells: keratinizing cells and melanocytes.

**Keratinizing cells** produce the protein **keratin,** which makes the skin impermeable to the entry of micro-organisms from without and to the loss of water from within. These cells are produced by mitosis in the deepest layers of the epidermis and then are pushed gradually to the surface of the skin. As they move toward the surface, they become filled with keratin until keratin replaces all of the cell's cytoplasm and the cell dies. These cells eventually flake off

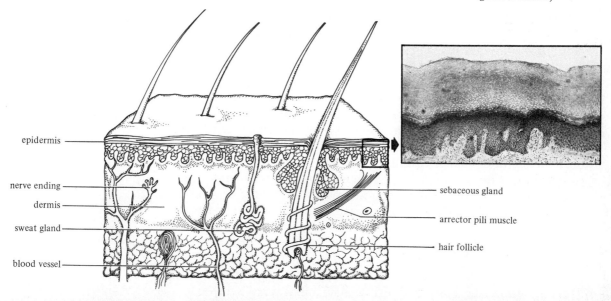

**Figure 4-7** The skin. (Photo by Lester Bergman & Assoc.)

and are replaced by new cells which have been pushed up from below.

The other type of cells in the epidermis, the **melanocytes,** are specialized for the production of the pigment **melanin,** which protects the body against the radiation of the sun. This pigment is partly responsible for the characteristic color of a person's skin. Melanocytes, like keratinizing cells, are produced in the deep layers of the epidermis. However, they do not migrate to the surface of the skin.

The number of melanocytes seems to be the same in people of all skin colors, but the ability of these melanocytes to produce melanin depends on a person's genetic makeup as well as a number of other factors. Sunlight stimulates the melanocytes to produce more melanin so that the longer one is exposed to the sun, the darker the skin becomes. In addition two of the hormones produced by the anterior pituitary gland—adrenocorticotropic hormone (ACTH) and melanocyte-stimulating hormone (MSH)—stimulate melanin production.

Skin color is determined not only by the amount of melanin it contains, but also by the amount and composition of the blood that flows through it. For example, high blood pressure, alcohol, and excess heat production all can cause the skin to take on a red or pink color by increasing the amount of blood flowing through it. On the other hand, after the loss of blood during a hemorrhage, a person's skin can become exceedingly pale owing to the tremendous decrease in the amount of blood flowing through it.

Under circumstances in which the blood does not contain enough oxygen, the skin can take on a bluish hue referred to as **cyanosis.** Generally the more pigment the skin contains, the harder it is to notice variations in skin color caused by changes in the blood flow through it. This must be taken into account in looking for signs of abnormal blood flow in people with darkly pigmented skin. In this case the lips and nails often offer a good site for observing any alterations in blood flow to the surface of the body.

The **dermis** of the skin is composed of dense connective tissues in which the intercellular collagen fibers are interlaced into a tightly woven net that gives a considerable degree of strength to the skin. All of the blood vessels and nerves that go to the skin are contained within the dermis as are the major parts of the hairs and glands which form the accessory structures of the skin.

The hairs of the skin are composed of two primary portions: the **root,** which is contained within the dermis, and the **shaft,** which extends through the epidermis and above the surface of the skin. Each hair arises as a result of cell division in the root. The newly formed cells grow upward and fill with keratin to form the shaft. The root of the hair and the tissues that surround it are referred to as the hair follicle. Small muscles, the **arrector pili muscles,** are attached to each hair follicle. When these muscles contract, as they do when a person is cold or frightened, the hairs stand on end. This action pulls up the small mounds of skin known as goose pimples.

Hairs are kept soft by an oily substance called **sebum,** which is secreted into the hair follicles by the **sebaceous glands.** Sebum seeps out of the hair follicle to the surface of the skin where it spreads out as a thin film of oil. This film helps keep the skin soft and pliable. The sebum that coats the skin also carries out a number of other functions: it reduces the rate at which the dead, keratinized cells of the epidermis flake off, it helps make the skin more impermeable to water, and it reduces heat loss from the body. If the sebaceous glands are somewhat underactive, the skin becomes dry and flakes off easily, whereas overactive sebaceous glands can cause the skin to become quite oily. **Acne** is a skin disease that results from a thickening of the hair follicles accompanied by oversecretion of the sebaceous glands. Upon exposure to air, which moves into the hair follicle, the sebum thickens and forms a discolored plug, a blackhead, which blocks the follicle outlet. The sudden increase in the level of sex hormones which occurs at puberty appears to stimulate acne formation.

**Sweat glands** are found in skin all over the body, but are most numerous in the palms of the hands, soles of the feet, and axillae (armpits). The base of a sweat gland, located in the dermis, is surrounded by nerve endings and capillaries. Sweat, composed primarily of water with a small amount of dissolved salts, is formed at the base of the gland and is then secreted through coiled ducts to the surface of the skin. The function of the sweat glands is to aid in the regulation of body temperature. When the body temperature begins to rise, the nerves at the base of the sweat glands stimulate the glands to increase their production and secretion of sweat. The evaporation of sweat from the surface of the skin uses up some of the excess body heat. This helps return the body temperature back to normal.

Normally, sweat evaporates from the skin as quickly as it is formed and we are unaware of sweat production. However, when the weather is hot and humid, the humidity of the air reduces the rate at which sweat evaporates from the surface of the skin, making it more difficult to keep the body temperature down. This is why the humidity, as much as the heat, determines how uncomfortable a person is on a hot day.

**Nails,** like hairs and the cells that make up the outer layer of the epidermis, are composed of dead epithelial cells that are filled with keratin. The cells are originally formed in the tissue below the half-moon-shaped white area (lanula), which is located at the base of the nail.

## OBJECTIVES FOR THE STUDY OF TISSUES AND MEMBRANES

At the end of this unit you should be able to:

1. Define the word tissue.
2. Name the four basic types of tissue.
3. Describe the structure of epithelial tissue.
4. State four functions of epithelial tissue.
5. Distinguish between squamous, cuboidal, and columnar epithelium.
6. Distinguish between simple and stratified epithelium.
7. State where epithelial tissue is found in the body.
8. Describe how the structure of connective tissue differs from that of epithelial tissue.
9. State where connective tissue proper is found in the body.
10. Name the three types of fibers found in the intercellular substance of connective tissue proper and state the major characteristics of each type.
11. Name the four types of cells found in connective tissue proper and state the major function of each.
12. Distinguish between loose and dense connective tissue.
13. State the major characteristics of nervous tissue and of muscular tissue.
14. Name the two types of tissue that comprise a membrane.
15. State the major locations in the body of mucous, serous, and synovial membranes.
16. Name at least four functions of the skin.
17. Describe the composition of the epidermis and of the dermis.
18. Distinguish between the function of keratinizing cells and of melanocytes.
19. Distinguish between the function of the sebaceous glands and of the sweat glands.

# 5
# The Skeletal System

BONE TISSUE
CARTILAGE TISSUE
BONE STRUCTURE
BONE GROWTH AND REPAIR
DEPOSITION AND RELEASE
 OF BONE CALCIUM
THE SKELETAL BONES
TYPES OF MOVEMENT

The skeletal system is composed of all the bones of the body. The hard and rigid structure of bone enables the skeletal system to act as a framework that supports the body and gives it shape. In this sense, the skeletal system is analogous to the steel girders which support and shape the structure of a large building.

The skeletal system also acts to protect many of the vital internal organs by forming the bony cases in which they are enclosed. Thus the bones of the skull surround and protect the brain and the rib cage surrounds and protects the heart and lungs. Even those internal organs that are not completely surrounded by bone are usually at least partially protected by bony structures.

The skeletal system is designed in such a way that it not only forms a supporting framework, but it is a supporting framework that can be moved. Contractions of the skeletal muscles, which are attached to the bones by tendons, pull at the bones and move them. The precise type of movement that takes place depends on the particular combination of muscles that pull at the bone and the flexibility of the joint to which the bone is attached.

Bones, unlike the steel girders that support a large building, are not simply inactive supporting structures. They are composed of metabolically active tissues which carry out a number of essential physiological activities. First, the various cells of a bone must carry out all of the metabolic activities necessary for the production, growth, maintenance, and repair of the bone tissue itself. The bones also play an essential role in maintaining the proper amount of calcium in the body fluids. When the level of calcium in the blood begins to fall, some of the calcium stored in the bones moves into the blood thus helping to restore the blood level of calcium back to normal. Finally, most of the blood cells are produced by the bone marrow, which is enclosed within certain bones.

In summary, the skeletal system carries out six major types of functions: it serves as a supporting framework; it protects many of the internal organs; it permits movement of the body; it produces, maintains, and repairs itself; it releases calcium when the body fluids become deficient in calcium; and it encloses the marrow, which produces the blood cells.

## BONE TISSUE

Bones, like other organs, are composed of a number of different types of tissues, but it is the osseous connective tissue, or bone tissue, that gives bones their characteristic properties. Bone tissue is typical of connective tissues in that it is composed of cells, intercellular fibers, and an intercellular ground substance. However, bone tissue is unique among all the tissues of the body in that its intercellular substance is calcified; it contains calcium phosphate and other calcium salts.

    This calcified intercellular substance of bone is referred to as the **bone matrix.** Although it is not yet completely understood as to how the calcium salts make bones firm, it is known that they act by binding intercellular collagen fibers to each other in a rigid pattern. Under very abnormal circumstances, tissues other than bone tissue become calcified, a process that greatly interferes with their ability to function properly.

    As shown in Figure 5-1, the bone matrix is arranged in layers known as **lamellae.** The bone cells, or **osteocytes,** are located

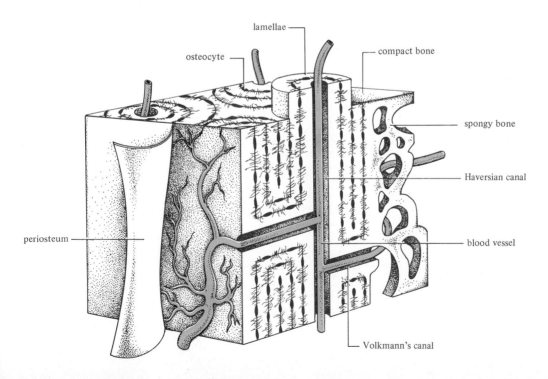

**Figure 5-1** Bone matrix.

between the layers of bone matrix. The function of these osteocytes is to maintain the structure of the bone matrix. Since nutrients and wastes cannot diffuse through the calcified bone matrix, there is a system of canals that runs through the bone tissue and that is designed to supply nutrients to and remove wastes from the osteocytes. This system of canals is known as the **haversian system.**

The large canals through which blood vessels enter and leave a bone are known as **Volkmann's canals.** In general, Volkmann's canals run perpendicular to the bone. The slightly smaller canals buried within the bone matrix are the **haversian canals,** which contain the capillaries through which substances can enter or leave the blood. Each haversian canal is connected by very tiny canals, called **canaliculi,** to the spaces around the osteocytes. The small spaces around the osteocytes are called **lacunae.** Thus nutrients are carried into the interior of a bone by the blood, which flows through the blood vessels located in Volkmann's canals and the haversian canals. The nutrients can then diffuse out of the capillaries and through the canaliculi to the lacunae which surround the osteocytes. From the lacunae, the nutrients are transported across the cell membrane and into the cytoplasm of the osteocytes. Wastes formed in the osteocytes move into the lacunae and through the canaliculi to the haversian canals at which point they enter the blood and are carried away from the bone.

Two types of bone tissue, compact bone and spongy (cancellous) bone, are distinguished on the basis of their density and by the way in which the lamellae of bone matrix are arranged. In **compact bone** the lamellae are arranged in dense concentric rings around the haversian canals. Compact bone has no open spaces other than the small canals and is therefore very strong. In **cancellous bone** the matrix is arranged into interconnecting columns that are separated by marrow-filled spaces. Thus, the cancellous bone is more like a board scaffolding, whereas the compact bone is more like solid concrete. As a result, cancellous bone is considerably lighter but is not quite as strong as compact bone.

## CARTILAGE TISSUE

Cartilage is similar to bone in that it is a strong and firm tissue. However, the intercellular substance of cartilage is not calcified like that of bone. Rather, it is composed of a fairly firm gel which allows the cartilage to be more flexible than bone although not quite as strong. The strong and flexible nature of cartilage, as well as its capability for rapid growth, make it ideally suited for its role as the major substance composing the fetal skeleton. After birth, cartilage still plays a role in the growth of bone, as well as in cushioning bones at the sites at which they join with each other.

There are no blood vessels and canals in cartilage; therefore the metabolic needs of the cartilage cells, called **chondrocytes,** must be met by nutrients that have diffused through the ground substance of cartilage from adjacent tissues.

There are three types of cartilage: hyaline cartilage, elastic cartilage, and fibrocartilage, each of which differs as to the amount and type of intercellular fibers they contain. **Hyaline cartilage,** the most common type, contains predominantly collagen fibers, although other types of fibers and a considerable amount of ground substance are also present. In the fetus, hyaline cartilage forms the bulk of the early skeleton. In an adult, hyaline cartilage remains at the ends of some of the ribs, at the sites where bones join, and in the trachea and larynx. **Elastic cartilage,** as the name implies, is characterized by its elasticity. This is caused by the presence of elastin fiber in the intercellular matrix. It is found in the outer ear and in the larynx. **Fibrocartilage** is characterized by the presence of large amounts of collagen with only a small amount of ground substance. The discs that separate the vertebral bones are composed of fibrocartilage.

## BONE STRUCTURE

Bones are classified on the basis of their shape and structure into four major categories: long bones, short bones, flat bones, and irregular bones.

Each **long bone,** as shown in Figure 5-2, is composed of a central shaft, the **diaphysis,** and two knoblike ends, the **epiphysis.** The entire surface of a long bone, except where it joins (articulates) with another bone, is covered by a fibrous sheath, the **periosteum.** At the end of a long bone, where it joins with other bones, the surface is covered by a thin sheet of cartilage, called the **articular cartilage.** The periosteum is composed of two layers: an outer fibrous layer and an inner osteogenic layer. The inner, osteogenic layer of the periosteum is composed of specialized cells, the **osteoblasts,** which are capable of producing new bone tissue. These cells play an essential role in the widening of bone, which takes place as a child grows, and in the repair of fractured bones.

Dense, compact bone tissue forms the bony matrix, which lies beneath the periosteum in the diaphysis of a long bone. In the wider epiphyseal ends, the bone matrix is composed primarily of spongy bone with only a thin outer layer of compact bone. The use of spongy bone tissue in the epiphysis makes the bone lighter than it would be if it were constructed entirely of compact bone.

**90** The Skeletal System

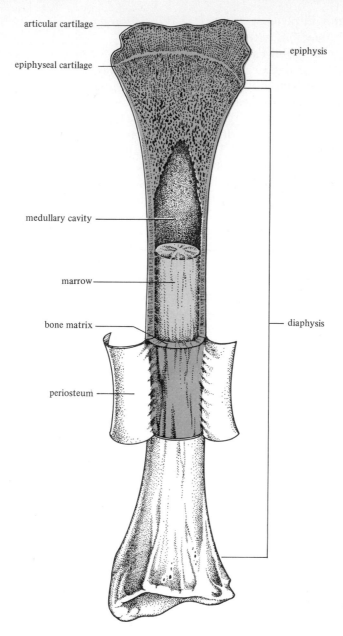

**Figure 5-2** Structure of a long bone.

A long cavity, the **medullary cavity,** is located in the center of the diaphysis of a long bone. This cavity is filled with the bone marrow as are the spaces distributed throughtout the spongy bone of the epiphysis.

There are two types of bone marrow: **yellow marrow,** which is composed of fat-storing adipose tissue, and **red marrow,** which is composed of the hemopoietic tissue that produces the blood cells.

In a child, most of the bones are filled with red marrow; as the child grows, much of this red marrow is converted to yellow marrow. In an adult, red marrow is found only in a few of the skull bones, the vertebrae, the ribs, and in the proximal (closest to the body) epiphyses of the humerus and femur.

**Short bones** are more or less cube-shaped. They are composed of central spongy bone surrounded by a thin layer of compact bone. The bones of the wrists and ankles are short bones.

**Flat bones** are thin and flat. They are composed of a central layer of spongy bone sandwiched between two outer layers of compact bone. Some of the skull bones, the ribs, and the shoulder blade (scapula) are flat bones.

**Irregular bones** are the bones that have more complex shapes than those of the other three types. In general they are similar in composition to short bones and flat bones in that an inner core of spongy bone is surrounded by a thin layer of compact bone.

## BONE GROWTH AND REPAIR

After its first two months of development in the uterus, the fetus has a complete skeleton formed of hyaline cartilage and fibrous membranes. During the next five or six months, this primitive skeleton is converted almost entirely into a bony skeleton. Ossification is brought about by the metabolic activities of the bone-forming osteoblast cells. These cells produce and secrete the intercellular substances, primarily large protein-carbohydrate molecules (mucopolysaccharides) and collagen, which are required for the deposition of calcium salts and the formation of a rigid bony matrix.

In the dense fibrous membranes that form the fetal skull bone, formation occurs by a process known as **intramembraneous bone formation.** This process involves the growth of bone in concentric columns from the center of the membrane to the outside. At birth, this process is not yet complete and small areas, known as **fontanels,** represent membrane that has not yet been converted to bone. Nevertheless these membranes are exceedingly tough and do not break except under the most unusual circumstances.

Hyaline cartilage is converted to bone by a process known as **endochondral bone formation.** In essence, this process involves the enlargement and subsequent death of the cartilage cells. Blood vessels and connective tissue grow into the vacated spaces left by these dead cells. Certain of the connective tissue cells become osteoblasts and produce bone in these spaces. Other connective tissue cells turn into the cells of hemopoietic tissue and produce red bone marrow. The blood vessels become enclosed in haversian canals formed by the bone matrix.

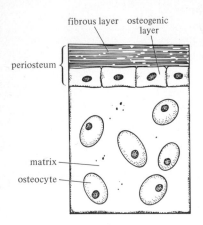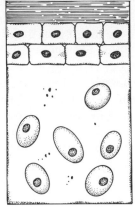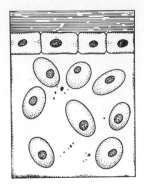

**Figure 5-3** Growth of bone in width.

In a long bone, ossification starts in the center of the diaphysis and proceeds toward each epiphysis. Ossification of the epiphysis usually does not take place until after birth and even then a narrow band of cartilage, the **epiphyseal cartilage,** remains. This cartilage is responsible for the lengthening of bone that takes place throughout childhood and adolescence. When growth stops, this cartilage also is replaced by bone and the only cartilage that remains in the adult is the articular cartilage that coats the bone at the joint.

Bones grow in width as a result of the formation of new bone matrix by the osteoblast cells of the periosteum, as illustrated in Figure 5-3. The formation of new bone at the outer surface of the matrix is accompanied by the destruction of some of the bone at the inner surface of the matrix. This destruction is carried out by specialized cells called **osteoclasts** and is necessary for the enlargement of the medullary cavity in the center of the bone.

Bone formation takes place not only during growth, but after a bone is injured. After a bone is broken or fractured, scar tissue, which is primarily connective tissue, forms at the site of injury. This tissue is then converted to fibrocartilage which fills in the space between the separated bone ends. The cartilage is then gradually replaced by bone formed by osteoblasts from the adjacent periosteum. Casts and traction are applied to healing bones for two basic reasons: to hold the ends that are growing together in place so that a smooth union can be formed, and to prevent movement that might pull the ends apart again before they are solidly joined.

## DEPOSITION AND RELEASE OF BONE CALCIUM

The formation of a calcified bony matrix depends on two factors: the composition of the intercellular material in which calcium is deposited and the concentration of calcium in the body fluids. Because the collagen fibers of bone are similar to those found in other types of connective tissue, some other component of the intercellular substance must be responsible for the fact that calcium will deposit in bone, but not in other types of connective tissue.

The concentration of calcium in the blood is very carefully controlled by the body. If there is not enough calcium in the blood, certain hormones, primarily the hormone **parathormone,** stimulate the osteoclasts to destroy bone tissue, leading to the release of calcium into the blood. This helps restore the blood level of calcium back to normal. Because blood calcium is maintained at the expense of bone calcium, any condition that serves to lower blood calcium will affect the structure of bones. **Rickets** is a disease characterized by weak bones that are very poorly calcified. Rickets is caused by a lack of vitamin D, a substance necessary for the absorption of calcium from the digestive tract into the blood. Pregnant or lactating women on a calcium-poor diet can end up with weak bones, as calcium is removed from the bones to replace the large amounts of calcium which are lost to the child.

## THE SKELETAL BONES

The 206 bones that comprise the adult human skeleton are usually divided for the purposes of study into two parts: the central **axial skeleton,** consisting of the bones of the skull, rib cage, and vertebral column, and the **appendicular skeleton,** consisting of the bones of the shoulder, arms, hips, and legs. These two divisions of the skeleton are illustrated in Figure 5-4.

### Axial Skeleton

The axial skeleton is designed primarily for the support and protection of the vital organs. As befits its protective role the axial skeleton is quite rigid, although the inflexibility of the vertebral column allows it a certain degree of movement.

### The Skull

The skull functions to protect the brain as well as the sensory organs of vision, hearing, smell, and taste. All of the bones of the skull, with the exception of the lower jawbone, are held rigidly in place and cannot be moved. The skull can be divided into two portions: the **cranium,** consisting of the bones that surround the brain, and the **face,** consisting of the bones that protect the orbital cavities, nasal cavity, and oral cavity.

**94** The Skeletal System

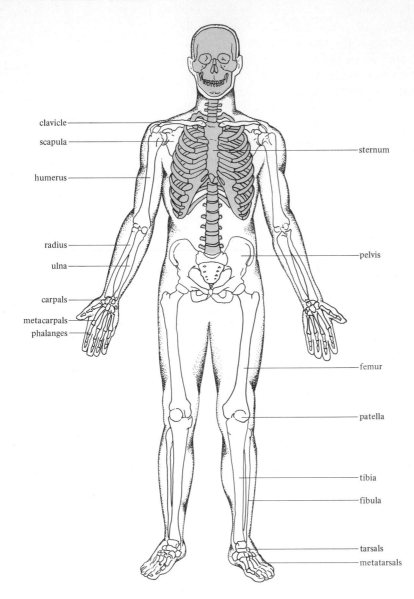

**Figure 5-4** Axial and appendicular bones.

**The cranium.** Figures 5-5 and 5-6 show the bones of the cranium as they are seen from the front and side. The **frontal bone** forms the anterior roof of the cranium, the forehead, and extends downward to form the roof of the orbital and nasal cavities. Two **parietal bones** form the bulk of the roof as well as a portion of the sides of the cranium. The remainder of each side of the cranium is formed by one of the two **temporal bones.** These bones also extend downward and inward to form part of the floor of the cranium. A projection of the temporal bone, the **zygomatic process,** arches

## 95 The skeletal bones

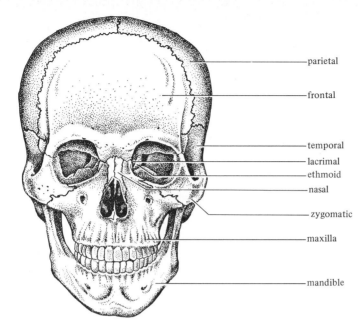

**Figure 5-5** Bones of the skull (frontal view).

forward to join the temporal bone with the cheek bone (zygomatic bone). The arch formed by the zygomatic process and the temporal process of the zygomatic bone is called the **zygomatic arch.** It can be felt just below the eye. A second projection of the temporal bone, the **mastoid process,** extends downward from the lateral,

**Figure 5-6** Bones of the skull (side view).

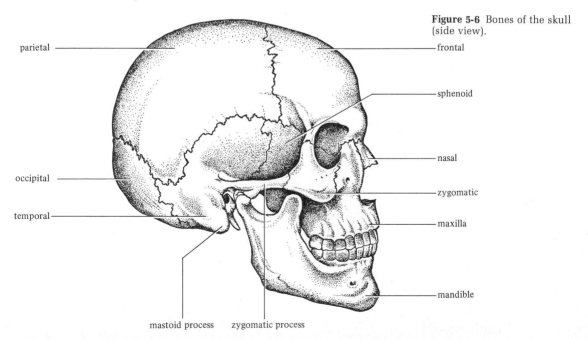

inferior surface. The mastoid process contains air spaces which communicate with the middle ear; these spaces can serve as a route for the spread of infection from the ear to the brain. Inflammation of these spaces is referred to as **mastoiditis.**

The **external auditory meatus,** the external ear canal, can be seen as a depression in the temporal bone, just anterior to the mastoid process. A thin membrane, the tympanic membrane, separates the external ear from the middle and inner ears which are embedded deep within the temporal bone. The **occipital bone** forms the back and much of the base of the cranium.

Two irregular bones, the ethmoid and the sphenoid, complete the cranium. As can be seen in the frontal view of Figure 5-5, the **ethmoid bone** is located in the center of the face between the two orbital cavities. It forms the medial wall of each orbital cavity, the posterior and lateral walls of the upper portion of the nasal cavity, and the top portion of the nasal septum which divides the nasal cavity.

The **sphenoid bone** is shaped somewhat like a bat with two large wings. Located just posterior to the ethmoid, the body of the sphenoid forms part of the floor of the cranium. The two wings curve upward and backward to form the posterior wall of the orbital cavities and part of the side of the head.

Figure 5-7 shows the floor of the cranium. The anterior part of the floor is formed by the frontal bone and a portion of the ethmoid bone; the lateral part of the floor is formed by the two temporal

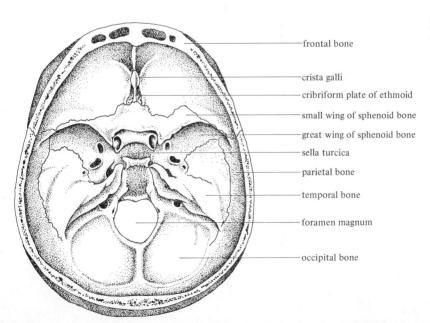

**Figure 5-7** Floor of the cranium as seen from above.

bones; the center of the floor is formed by the sphenoid bone; and the posterior part of the floor is formed by the occipital bone. A depression in the sphenoid bone, the **sella turcica,** forms a bony cavity for the pituitary gland, which projects from the bottom of the brain. The large opening in the occipital bone, the **foramen magnum,** is the site at which the spinal cord enters the cranium to join with the brain.

To summarize, the cranium is a bony case surrounding the brain formed by the frontal, parietal, temporal, occipital, ethmoid, and sphenoid bones. The frontal bone forms the front and the anterior parts of the roof and floor of this bony case; the two parietal bones form the roof and part of the sides; the two temporal bones form the sides and the lateral parts of the floor; the occipital bone forms the back and the posterior part of the floor; the ethmoid forms part of the front and floor; and the sphenoid forms the center of the floor and a part of the front and sides.

**The face.** The upper jaw is formed by the two **maxilla bones.** In addition to forming the upper jaw, these bones curve upward to form the floors of the orbital cavities and part of the lateral walls of the nasal cavities. They also curve inward to form the anterior three-fourths of the bony roof of the mouth, the **hard palate.**

Some babies are born with a genetic defect in which the two maxillae are not completely joined at the roof of the mouth, a condition known as **cleft palate.** The opening between the maxillae communicates with the nasal cavities, presenting serious problems in nursing for the baby because air, as well as milk, is pulled in when the baby swallows.

The two **zygomatic bones,** the cheek bones, form the front of the face and parts of the lateral walls and floors of the orbital cavities. They unite with the zygomatic process of the temporal bone to form the zygomatic arch.

The two **nasal bones** form the upper anterior portions of the nasal cavity, the bridge of the nose. Their shape more or less determines the shape of the nose because the remainder of the anterior wall of the nose is formed from the skin and cartilage.

The **mandible,** which forms the lower jaw, is the largest and heaviest bone of the face and its movement plays an essential role in chewing food and in talking.

Several smaller bones complete the structure of the face. The two **lacrimal bones** are very thin bones, about the size of a fingernail, which are located in the medial, posterior portion of each orbit, adjacent to the nasal bones. Tears are drained from the orbital cavities to the nasal cavity by a thin canal, the **lacrimal canal,** which passes through these bones.

A triangular-shaped bone, the **vomer,** forms the bottom portion of the nasal septum.

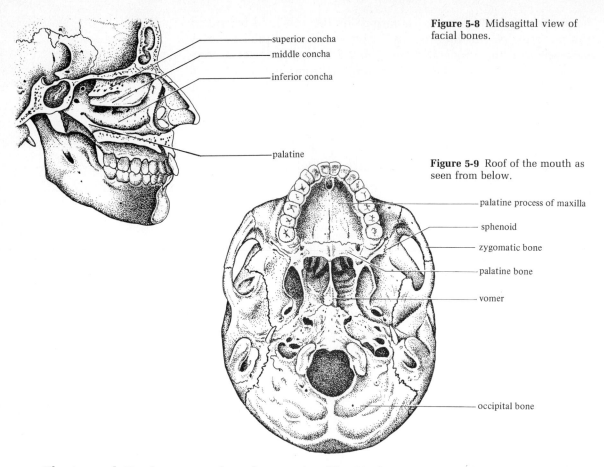

**Figure 5-8** Midsagittal view of facial bones.

**Figure 5-9** Roof of the mouth as seen from below.

The two **palatine bones** are shaped somewhat like L's facing each other (L⅃). The vertical portion forms the lateral wall of the nasal cavity and the horizontal portion forms the inferior wall of the nasal cavity as well as the posterior portion of the hard palate, as can be seen in Figures 5-8 and 5-9.

The **inferior concha** are scroll-shaped bones which project into the nasal cavity from the lateral wall. Along with two projections of the ethmoid bone, the **superior** and **middle conchae**, they divide the nasal cavity into irregular channels.

In summary, the anterior projecting portion of the face is formed by the zygomatic, nasal, maxilla, and mandible bones. A number of smaller bones, the lacrimal, vomer, palatine, and inferior concha, enter into the formation of the orbital, nasal, and oral cavities of the face.

### Fetal Skull

When a child is born, the bones of the skull are not completely fused and the spaces between the bones are covered by a

## 99 The skeletal bones

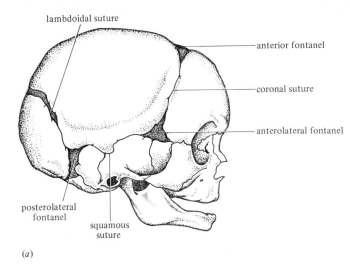
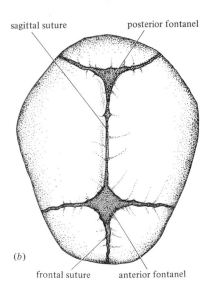

**Figure 5-10** Fetal skull: (a) side view and (b) top view.

tough fibrous membrane, as shown in Figure 5-10. This membrane is located in the narrow **sutures** between the large bones of the cranium and in expanded areas known as **fontanels,** located at the top, back, and sides of the head. The major sutures of the skull are the **frontal suture,** which separates the two frontal bones; the **coronal suture,** which separates the frontal and parietal bones; the **sagittal suture,** which separates the two parietal bones; the **lambdoidal suture,** between the parietal bones and the occipital bone; and the **squamous suture,** between the parietal and temporal bones.

The fontanels are expanded areas where the sutures are joined: the **anterior fontanel** is at the top of the head where the frontal, sagittal, and coronal sutures join; the **posterior fontanel** is at the back of the head where the sagittal and lambdoidal sutures are joined; the anterolateral fontanel is at the side of the head where the coronal and squamous sutures are joined; and the posterolateral fontanels are at the side of the head where the lambdoidal and squamous sutures are joined.

The unossified sutures and fontanels serve two important functions: they allow the cranial bones to override somewhat as the baby's head moves through the narrow birth canal and they permit growth of the skull to accompany the growing brain. Normally the lateral and posterior fontanels close a few months after birth and the anterior fontanel closes after about a year and a half. The sutures do not ossify until growth of the skull is complete. If they ossify too early, the skull will not grow to full size, a condition known as microencephaly. People with microencephaly are usually severely retarded because brain growth is inhibited at the point that the skull ceases to grow.

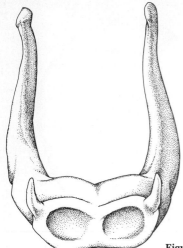

**Figure 5-11** Hyoid bone.

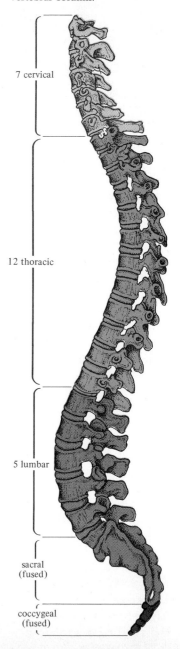

**Figure 5-12** Regions of the vertebral column.

## Hyoid Bone

The **hyoid bone,** as illustrated in Figure 5-11, is located in the neck just below the chin. It is suspended by ligaments from the temporal bone and functions as a point of attachment for some of the muscles that move the tongue in speaking and swallowing.

## Vertebral Column

The **vertebral column,** or backbone, serves as the major longitudinal supporting structure of the axial skeleton and as a protective bony covering for the spinal cord. Its flexibility results from the fact that it is composed of a number of separate bones, the **vertebral bones,** which can move with respect to each other. The vertebral column supports the head from below and serves as a point of attachment for the ribs and hips.

The 26 bones that make up the adult vertebral column can be divided into five regions, as shown in Figure 5-12. The **seven cervical vertebrae** are located in the neck, the **12 thoracic vertebrae** at the back of the chest, the **five lumbar vertebrae** in the small of the back, and the **sacral** and **coccygeal** vertebrae below the top of the hips. In a child there are five sacral and four coccygeal vertebrae; however, these vertebrae fuse into one large triangular-shaped sacral bone and a small, triangular coccygeal bone. The coccygeal bone is the equivalent of the tail bone found in dogs, cats, horses, etc.

Although not all vertebrae are identical, most have certain structural features in common. All but the first two have a large,

**101** The skeletal bones

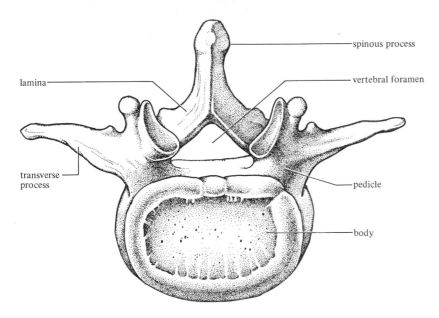

**Figure 5-13** Typical vertebra.

round **body** in the anterior portion, as illustrated in Figure 5-13. An opening in the center, the **vertebral foramen,** forms the protective passageway for the spinal cord. The side and back walls of this passageway are formed by the **pedicles** in front and the **laminae** in the rear. If the laminae are not completely united, the spinal cord or its coverings, or both, can protrude posteriorly, a condition known as *spina bifida*. A projection, the **spinous process,** extends downward from the union of the two laminae.

The first two cervical vertebrae, the **atlas** and the **axis,** are specially designed to support the skull and allow it to rotate. The atlas, named for the Greek god who supported the world, has no body or spinous process as illustrated in Figure 5-14. It is simply a

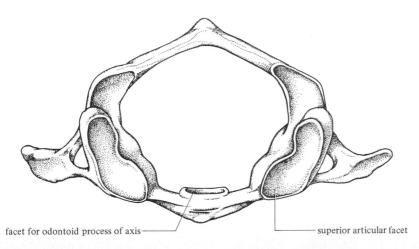

**Figure 5-14** Atlas (top view).

**102** The Skeletal System

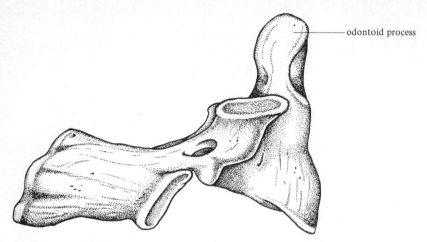

Figure 5-15 Axis (side view).

bony ring in which the occipital bone of the skull rests. The axis contains a special process, the **odontoid process,** which extends upward into the atlas, as illustrated in Figure 5-15. When you turn your head from side to side, the atlas rotates with the skull, using the odontoid process as a pivot.

Figure 5-16 illustrates the manner in which the various vertebrae are held together. Ligaments, which extend almost the entire length of the cord, tie the body, laminae, and spinous process of each vertebra to the corresponding parts of the vertebrae above and below it. A cartilaginous disc, the **intervertebral disc,** is located

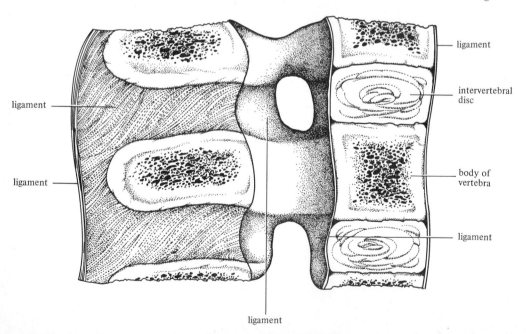

Figure 5-16 Ligaments holding vertebrae together.

between the bodies of each two vertebrae. It functions to absorb any shock applied to the vertebral column. If this disc slips out of place, it can push on the nerves of the spinal cord causing extreme pain. Likewise any injury to the vertebral column can cause extensive damage to the underlying spinal cord. This damage can cause paralysis or even death.

The vertebral column is curved to give it strength and to balance the weight of the various parts of the body. In a newborn baby, the vertebral column is bowed backward. As a child begins to stand, secondary curves develop and the column eventually develops four curves. A hunchback has an exaggerated thoracic curve known as kyphosis, whereas a swayback has a lumbar curve known as lordosis. An exaggerated lateral (sideways) curve is referred to as scoliosis.

### The Thorax

The thorax, as illustrated in Figure 5-17, is a helmet-shaped, bony cage formed by the sternum, the 12 pairs of ribs, and the 12 thoracic vertebrae. A dome-shaped muscle, the diaphragm, curves upward into this cage, separating the thoracic cavity above from the abdominal cavity below. The bones of the thorax protect the heart and the lungs, which are located in the thoracic cavity, as well as the stomach and liver, which are located just below the diaphragm at the top of the abdominal cavity. Expansion and contraction of

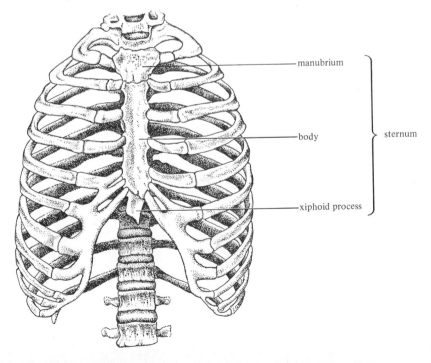

**Figure 5-17** Thorax.

the thoracic cage plays a role in the respiratory movements which move air in and out of the lungs.

The **sternum,** or breastbone, at the front of the thorax is a flat bone which is shaped somewhat like a dagger with the point facing downward. It is divided into three portions: the upper manubrium, the middle body, and the lower, pointed xiphoid process. The xiphoid process is somewhat fragile and can break off relatively easily. This is a complication which occasionally occurs during external massage of the chest after cardiac arrest.

Because it is so close to the skin, the sternum is used as the site from which the bone marrow is removed for biopsy, a procedure that is necessary for the diagnosis of certain blood disorders such as anemia.

The 24 **ribs,** 12 on each side of the thorax, are attached posteriorly to the thoracic vertebrae, a pair of ribs to each of the 12 thoracic vertebrae. Anteriorly the ribs are not attached directly to the sternum; they are linked to the sternum by bands of hyaline cartilage, the **costal cartilages.** The first seven pairs of ribs are called **true ribs** because their cartilages are attached directly to the sternum. The next three pairs of ribs are called **false ribs** because their cartilage attaches to the cartilage of the rib above rather than to the sternum directly. The last two ribs are called **floating ribs** in recognition of the fact that their anterior end is not attached to any other structure.

## Appendicular Skeleton

The appendicular skeleton consists of the bones of the extremities, the arms and legs, as well as the bones of the shoulders and hips which attach the bones of the extremities to the axial skeleton. This portion of the skeleton is designed primarily for movement and for supporting the body in an upright position. In addition the pelvic portion of the appendicular skeleton is designed to protect the internal organs located in the lower portion of the abdominal cavity.

### Shoulder Bones

The bones of the shoulder function to attach the arms to the axial skeleton in such a way as to permit the arms a large degree of movement. As shown in Figure 5-18, the shoulder is composed of two bones, the clavicle and the scapula. The **clavicle,** or **collarbone,** is a thin bone which forms the front of the shoulder. It is attached medially to the sternum and laterally to the scapula. This bone is not very strong and often fractures after a blow to the shoulder, such as occurs when a person falls sideways and lands on his shoulder.

The **scapula,** or shoulder blade, is a large, flat bone located in

## 105 The skeletal bones

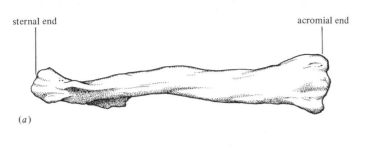

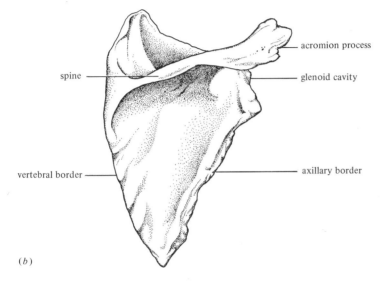

**Figure 5-18** (a) Clavicle (superior view) and (b) scapula (posterior view).

the upper part of the back. A thin process which projects from the superior, lateral surface of the scapula, the **acromion process,** forms the site of attachment between the scapula and clavicle. The acromion process, like the clavicle, is somewhat fragile and likely to break after a blow to the shoulder. Medially the scapula is not joined directly to the ribs; it is somewhat loosely held in place by muscles and tendons. A depression on the lateral surface of the scapula, the **glenoid cavity,** serves as a socket for the attachment of the humerus, the bone of the upper arm.

### Arm Bones

The upper portion of the arm contains one large bone, the **humerus,** illustrated in Figure 5-19. A projection of the proximal (closer to the axial skeleton) end of the humerus, the rounded **head,** fits into the glenoid cavity of the scapula. This forms a ball-and-socket joint, which permits the arm to rotate. The distal portion of the humerus has two smooth depressions, the capitulum and trochlea, which form joints with the two bones of the lower arm, the radius and the ulna. A nerve which passes just below a medial

**106** The Skeletal System

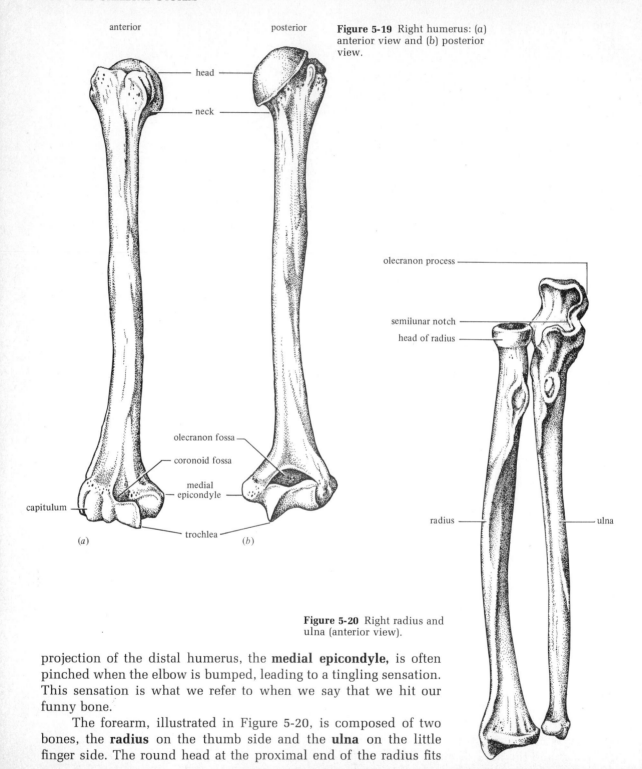

Figure 5-19 Right humerus: (a) anterior view and (b) posterior view.

Figure 5-20 Right radius and ulna (anterior view).

projection of the distal humerus, the **medial epicondyle,** is often pinched when the elbow is bumped, leading to a tingling sensation. This sensation is what we refer to when we say that we hit our funny bone.

The forearm, illustrated in Figure 5-20, is composed of two bones, the **radius** on the thumb side and the **ulna** on the little finger side. The round head at the proximal end of the radius fits into the capitulum of the humerus and the radial notch of the ulna. Distally the radius articulates with two of the carpal bones of the wrist and the knoblike head of the ulna.

The ulna is somewhat longer than the radius with a large hook-shaped process extending from its proximal end. The top, back part of this hook is the **olecranon process,** which forms the protuding posterior surface of the elbow; the smooth inner surface is the **semilunar notch,** which articulates with the trochlea of the humerus. Distally the ulna articulates with the radius, but not with the bones of the wrist. It is separated from these bones by a cartilaginous disc.

When the hand is rotated from the palm-forward-to-a-palm-backward position, the ulna remains fixed and the radius crosses over the ulna, as illustrated in Figure 5-21. During this rotation, the rounded head of the proximal end of the radius turns, more or less like a wheel, in the radial notch of the ulna.

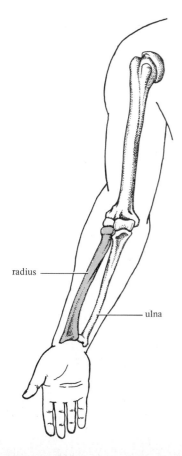
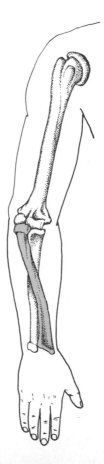

**Figure 5-21** Rotation of the hand.

**108** The Skeletal System

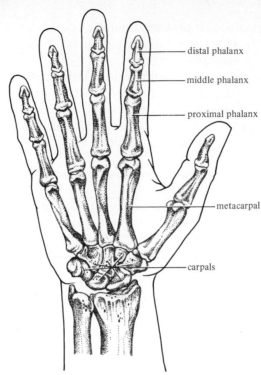

**Figure 5-22** Wrist and hand (anterior view).

### Wrist and Hand

The wrist is composed of eight **carpal** bones arranged in two rows of four as illustrated in Figure 5-22. The two lateral, posterior carpal bones articulate with the radius and the four anterior carpal bones articulate with the **metacarpal bones** which form the hand. Rounded extensions at the distal ends of the slender metacarpals form the knuckles and serve as the points of articulation between the metacarpals and the first row of phalanges, or finger bones.

There are 14 **phalanges:** three in each finger and two in the thumb. The numerous bones and joints of the hand act to make it a very mobile structure which can assume a wide variety of shapes, a design obviously well suited to its function, which is to grasp and hold objects. The fact that the thumb can be bent in such a way as to trap objects between the thumb and the four fingers greatly enhances the grasping ability of the hand.

### The Hip Bones

The pelvic girdle is a thick-boned structure which serves a number of functions: it attaches the bones of the legs to the axial skeleton, it supports the weight of the body when one is in a sitting position, and it protects the internal organs located in the inferior portion of the abdominal cavity such as the bladder, the end of the large intestine, and some of the reproductive organs.

**109** The skeletal bones

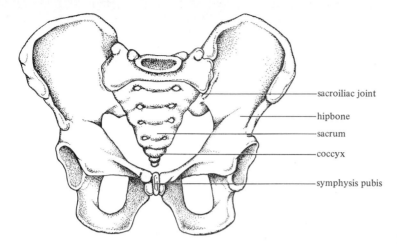

**Figure 5-23** Pelvic girdle.

As illustrated in Figure 5-23, the pelvic girdle is composed of four bones: the two large, rounded **hip bones** that form the sides and front and the sacral and coccygeal vertebrae that form the back. The **symphysis pubis** is the site where the two hip bones join at the front of the pelvis. Posteriorly each hip bone joins with the sacrum to form the **sacroiliac joint.**

In a young child each hipbone is composed of three separate bones: a superior **ilium,** a lateral and inferior **ischium,** and an anterior **pubis.** As the child grows, these three bones fuse into one hipbone, but the names are still used to describe the various parts of the hipbone, as illustrated in Figure 5-24. Thus, the superior

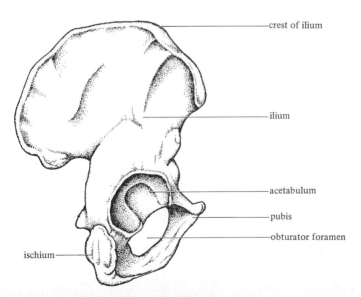

**Figure 5-24** Right hipbone (side view).

**110**  The Skeletal System

scooped portion of each hipbone is the ileum, the inferior, heavy portion, the ischium, the anterior thin portion, the pubis.

Other important parts of the hip include: the **iliac crest,** which is the upper rim of the ileum; the **acetabulum,** a socket at the junction of the ileum, ischium, and pubis, which serves as the point of attachment for the femur bone of the thigh; the **ischial tuberosity,** which extends downward from the ischium to help support the body when one is sitting down; and the **pubic arch,** which is the space between the pubic bones just below the symphysis pubis.

The structure of the pelvis is of particular importance

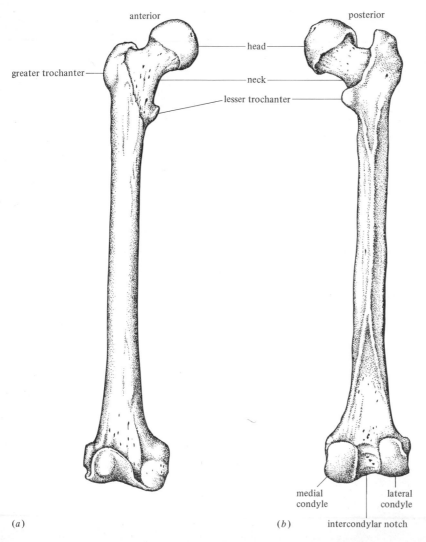

**Figure 5-25** Right femur: (a) anterior view and (b) posterior view.

during the birth of a baby, since the baby must pass through the pelvic opening. In obstetrics the pelvis is divided by the **pelvic brim** into a superior wide **false pelvis** and an inferior narrow **true pelvis.** The pelvic brim consists of the curved space extending from the top of the sacrum along the inner ridges of the ileum to the top of the symphysis pubis. Thus the false pelvis is the space between the flared portions of the ileum and the anterior muscular wall of the abdomen.

The true pelvis is bounded by the symphysis pubis and pubic arch in front, the pubis and ischium on the sides, and the sacral and coccygeal vertebrae in back. The dimensions of the true pelvis must be large enough to accommodate the passage of the baby through the birth canal.

**The Legs**

Because the bones of the legs must be able to support the weight of the body when one is standing, they are considerably larger than the bones of the arms. The **femur,** or thigh bone, is the longest and heaviest bone of the body. As can be seen in Figure 5-25, the rounded head at the proximal end of the femur fits into the acetabulum of the hip to form a ball-and-socket joint. The long **neck** of the femur attaches the head to the shaft. Because it joins the shaft at a sharp angle, the neck is sometimes broken after a fall in which a person lands on his hip. At its distal end, the femur has two smooth processes on its posterior surface, the **medial and lateral condyles,** which serve as the point of attachment between the femur and the tibia, one of the bones of the lower leg.

The **patella,** or knee cap, illustrated in Figure 5-26, is embedded in a tendon which extends from the quadriceps femoris, a muscle located in the thigh. This bone serves to protect the anterior surface of the knee joint. Injury to the knee is common, particularly in sports such as football where the knee is subject to forces applied in a direction opposite to that in which the knee bends.

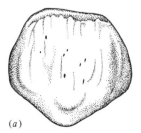

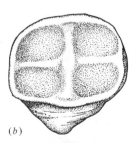

**Figure 5-26** Patella: (a) anterior view and (b) posterior view.

(a)  (b)

The lower leg is composed of two bones, the **tibia** and **fibula,** illustrated in Figure 5-27. The tibia, or shinbone, is the larger of the two and located anteriorly and medially to the smaller fibula. Two large bulges at the proximal end of the tibia, the **medial and lateral condyles,** articulate with the condyles of the femur. A hook-shaped extension from the distal end of the tibia, the **medial malleolus,** forms the articulation between the tibia and the ankle. The medial malleolus can be felt as the bony projection on the inner surface of the ankle.

The fibula articulates proximally with the tibia, but not with the femur, and thus does not help support the weight of the body. An extension of the distal portion of the fibula, the **lateral malleolus,** forms the bony projection on the lateral surface of the ankle.

The lateral malleolus of the fibula and the medial malleolus of the tibia form a socket which fits around the talus, one of the ankle bones.

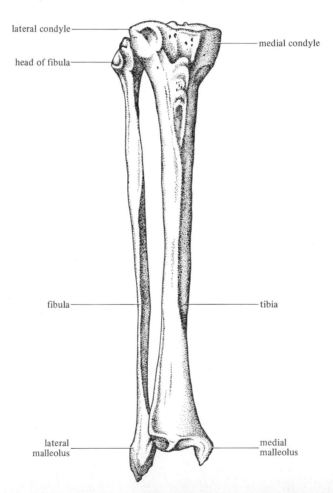

**Figure 5-27** Tibia and fibula (anterior view).

The seven **tarsal,** or ankle, bones are shown in Figure 5-28. The **talus** is a somewhat rounded bone which transfers the weight of the body from the lower leg to the foot. Located below and behind the talus is the large **calcaneus,** or heel bone, on which the talus rests. The calcaneus also serves as the point of attachment for the Achilles tendon which extends downward from the gastrocnemius muscle of the calf. Five smaller tarsal bones and the five metatarsals form the remainder of the foot.

In order to better support the weight of the body, the tarsal and metatarsal bones of the foot are shaped into two arches: a longitudinal arch passing from heel to toe and a transverse arch passing from side to side, as illustrated in Figure 5-29. These arches are

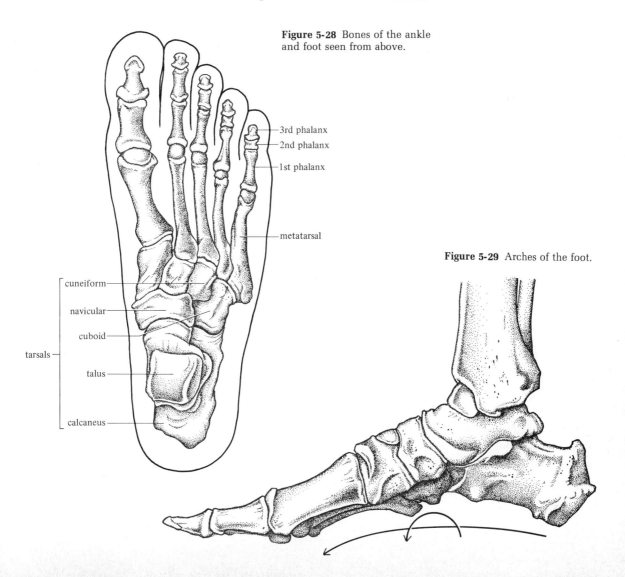

**Figure 5-28** Bones of the ankle and foot seen from above.

**Figure 5-29** Arches of the foot.

maintained by muscles, tendons, and ligaments. If these supporting structures weaken, the arch falls and the foot becomes flat.

The toes, like the fingers, are composed of 14 bones referred to as **phalanges.** There are two phalanges in the large toe and three in each of the other toes.

### Articulations Between Bones (Joints)

The articulations, or joints, function to attach the bones to one another, the structure of each joint determining to a large extent the amount and type of movement which can take place between the bones which are attached at that joint. Although a variety of complex systems have been developed for the classification of joints, basically, the various joints can be divided into two main types: the synarthroses and the diarthroses.

The **synarthroses** are joints in which the space between the bones is filled with a band of fibrous connective tissue or cartilage which functions to hold the bones together. These joints permit little or no motion between the attached bones. The joints between the bones of the cranium are synarthroses in which a thin layer of fibrous connective tissue attaches the bones.

Another type of synarthrotic joint is the one between adjacent vertebrae where a cartilagenous disc ties the vertebrae together in such a way that only a small amount of movement is possible. Large movements of the vertebral column are brought about by small movements between each pair of adjacent vertebrae.

The **diarthroses,** or **synovial joints,** are joints in which the space between the bones is filled with fluid. Figure 5-30 illustrates the structure of a synovial joint. The bones are held together by a capsule of fibrous connective tissue, which is lined on its inner surface by a serosal membrane, the **synovial membrane. Synovial fluid** is secreted into the cavity between the bones by the epithelial cells on the interior side of the synovial membrane.

The synovial fluid acts as a lubricant which permits the articulating bones to move smoothly past each other. A layer of articular cartilage covers each bone at the joint; this cartilage functions to absorb any shock that occurs if the bones happen to press against each other.

The synovial joints permit varying amounts of motion depending on the particular structure of the joint. For example, ball-and-socket joints, like the ones formed between the humerus and scapula and between the femur and hipbone, permit a large amount of motion in almost any direction. On the other hand, hinge joints, like the elbow and knee, allow a more limited degree of motion.

Unfortunately, diseased joints are both quite common and quite painful. The general term **arthritis** refers to the inflammation of a joint, a condition which can be brought about by any of a

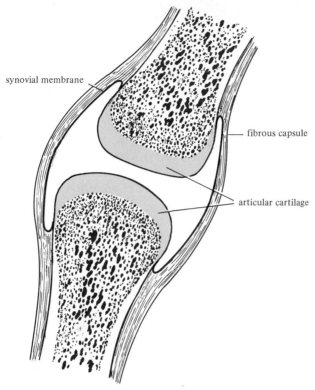

**Figure 5-30** Structure of a synovial joint.

number of causes. **Osteoarthritis** is an inflammation of the joints that occurs as the result of the degeneration of the cartilage which covers bones at their articulating surface. In **gout,** uric acid, a nitrogen waste product, builds up in the blood and begins to deposit in the cartilage of joints, causing the joints to become inflamed. **Rheumatoid arthritis,** a disease that affects one out of every 50 people in this country, is probably the most common form of joint disease. It manifests itself as a very painful swelling of joints, particularly those of the hands, wrists, knees, and feet. Although the cause of rheumatoid arthritis has not yet been determined, it is currently speculated that it may be brought about either by a chronic infection or by a deficiency in the person's system of immunity in which they form antibodies against their own joints. Diseases in which the body forms antibodies against itself are called autoimmune diseases; these diseases will be discussed in Chapter 10.

## TYPES OF MOVEMENT

A special set of terms is often used to describe the type of movement taking place at a particular joint. These terms can best be understood if one starts with the body in the anatomical position, standing upright with the arms resting at the sides and the palms facing forward.

**Flexion** is any anterior movement away from the anatomical position, such as bending the head forward, bending the arm at the elbow, or curling the fingers. The only exception to this is flexion at the knee or toes which is a backward movement. In the anatomical position, the feet are already flexed; further flexion of the feet is called **dorsiflexion.**

**Extension** is the reverse of flexion; it is the return to the anatomical position after flexion has taken place. Example of extension would be straightening the head after it has bent forward, straightening the arm at the elbow or uncurling the fingers. Continuation of extension past the anatomical position is called hyperextension, as exemplified by bending the head backward or lifting the arms backward. Extension of the foot past the anatomical position, as occurs when we stand on our tiptoes, is called **plantar flexion.**

**Abduction** is lateral movement away from the midline of the body, such as lifting the arms sideways.

**Adduction** is the opposite of abduction; it is movement in a medial direction toward the midline of the body.

**Circumduction** is a circular motion of a body part. Examples of this are drawing a circle with the arm straight or rolling the head forward, sideways, backward, and then to the other side.

**Rotation** is a pivoting around the longitudinal axis of the body as exemplified by turning the head from side to side to say "no" or moving the trunk from side to side when dancing the twist.

**Supination** is the turning of the palm forward or upward.
**Pronation** is turning the palm backward or downward.
**Inversion** is turning the sole of the foot inward.
**Eversion** is turning the sole of the foot outward.
**Protraction** is lowering the jaw or sticking out the tongue.
**Retraction** is raising the jaw or pulling in the tongue.

## OBJECTIVES FOR THE STUDY OF THE SKELETAL SYSTEM

At the end of this unit you should be able to:

1. State at least four functions of the skeletal system.
2. Describe the structure of the bone matrix.
3. State the function of the haversian system.
4. Distinguish between compact and spongy bone.
5. Describe how cartilage differs from bone.
6. Name the three basic types of cartilage.
7. State the location within a long bone of the following parts: diaphysis, epiphysis, periosteum, articular cartilage, matrix, and medullary cavity.
8. Distinguish between yellow marrow and red marrow.

9. Name the band of cartilage that is responsible for the growth of long bones in length.
10. Describe how a long bone grows in width.
11. Name the hormone that stimulates the release of calcium from the bones to the blood.
12. State the function of vitamin D.
13. Distinguish between the axial skeleton and the appendicular skeleton.
14. State the location of the following bones and bone processes: frontal, parietal, temporal, mastoid process, occipital, ethmoid, sphenoid, maxilla, zygomatic, zygomatic arch, nasal, mandible, lacrimal, palatine, conchae, hyoid, vertebral, sternum, ribs, clavicle, scapula, acromion process, glenoid cavity, humerus, radius, ulna, olecranon process, carpals, metacarpals, phalanges, ileum, ischium, pubis, iliac crest, acetabulum, femur, neck of the femur, tibia, fibula, medial malleolus, lateral malleolus, tarsals, talus, calcaneous, and metatarsals.
15. State the location of the anterior and posterior fontanels.
16. State the location of the following cerebral sutures: coronal, sagittal, lambdoidal, and squamous.
17. Name the five regions of the adult vertebral column and state the number of vertebrae found in each.
18. Name the first two cervical vertebrae.
19. Distinguish between synarthrosis and diarthrosis.
20. State the location of the following parts of a synovial joint: fibrous capsule, synovial membrane, articular cartilage, and synovial fluid.
21. Describe the type of motion that takes place during flexion, extension, hyperextension, abduction, adduction, circumduction, rotation, supination, pronation, inversion, eversion, protraction, and retraction.

# 6
# The Muscular System

INTRODUCTION
STRUCTURE OF A SKELETAL
   MUSCLE CELL
MECHANISM OF CONTRACTION
ENERGY FOR MUSCULAR CONTRACTION
FORCE OF CONTRACTION
THE SKELETAL MUSCLES

# INTRODUCTION

The skeletal system provides a firm supporting framework for the body. Movement of this framework from one position to another is brought about by the various skeletal muscles that make up the **muscular system.**

Each of the skeletal muscles is attached by tendons to two or more bones separated by a joint, as illustrated in Figure 6-1. When a skeletal muscle contracts, the bones on one side of the joint are pulled toward those on the other side. This pulling force can act either to move the bones or to counter a force pulling the bones in the opposite direction.

Contractions that actually move bones, such as those used in walking, lifting an object, or chewing food, are called **isotonic contractions.** Contractions that counter opposing forces, but do not cause movement, are called **isometric contractions.** An example of isometric contraction would be the contraction of muscles in the back and legs necessary for holding the body in a standing position. The contraction of these muscles acts to counter the force of gravity, which would otherwise pull the body down.

**Figure 6-1** Movement at a joint.

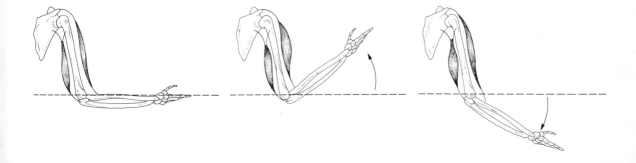

## Parts of a Muscle

A skeletal muscle consists of a bundle of parallel muscle cells, called **muscle fibers,** wrapped in a connective tissue sheath, as illustrated in Figure 6-2. The connective tissue sheath, which contains the nerves and blood vessels of the muscle, is composed of three interconnected portions: an outer **epimysium** that surrounds the entire muscle, a **perimysium** that extends inward from the epimysium to surround small groups of muscle cells, and an **endomysium** that extends from the perimysium to form a thin layer of connective tissue around each of the individual muscle cells.

The connective tissue of a muscle serves as a connecting link between the individual muscle cells, which are attached to the inner endomysium, and the tendons, which are attached to the

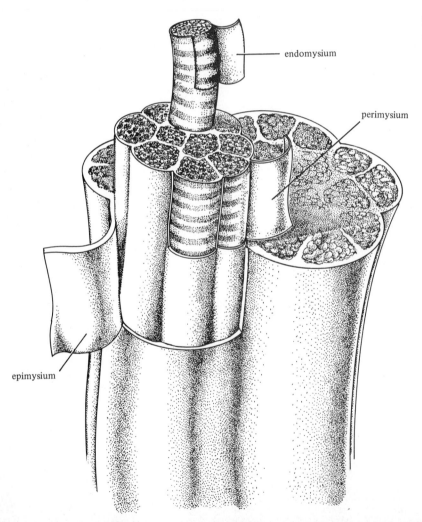

**Figure 6-2** Connective tissue sheath of a muscle.

## STRUCTURE OF A SKELETAL MUSCLE CELL

outer epimysium. When the muscle cells contract, they pull on the connective tissue which, in turn, pulls on the tendons. The tendons, of course, then pull on the bones.

## STRUCTURE OF A SKELETAL MUSCLE CELL

In order to understand the mechanism by which muscle cells contract we must first look at the structure of these cells. Muscle cells are usually long, cylindrical cells that contain a number of nuclei. Although many of the organelles in a muscle cell are similar to those found in other cells, they are often given special names in a muscle cell. Thus, the surrounding cell membrane of the muscle cell is called the **sarcolemma** and the cytoplasm of a muscle cell is called the **sarcoplasm.**

Embedded in the sarcoplasm of a muscle cell are tiny crossbanded fibers known as myofibrils. Because the crossbands of each myofibril are in line with those of adjacent myofibrils, the entire muscle cell has a banded or striated appearance, as illustrated in Figure 6-3. These striations are the reason that skeletal muscle is also referred to as **striated muscle.**

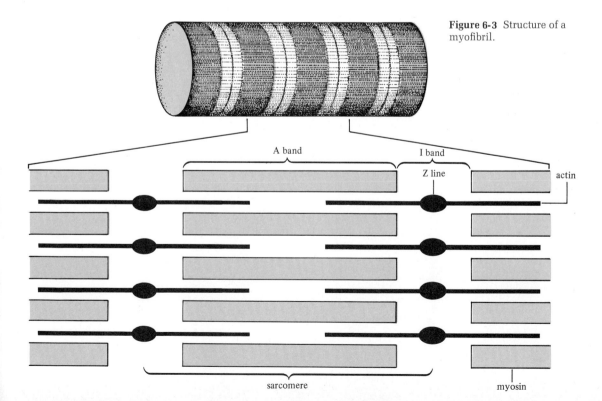

**Figure 6-3** Structure of a myofibril.

The myofibrils are composed of the proteins **actin** and **myosin,** which are responsible for the contraction of muscle cells. The crossbanded appearance of a myofibril is a result of the arrangement of the actin and myosin proteins within it. The wide **A band** is the length of the thicker myosin molecules.

The darker portion of the A band corresponds to the region in which myosin and actin molecules overlap, and the lighter central **H region** of the A band corresponds to the portion of the myosin molecules which do not overlap with the actin molecules. Alternating with the A bands are the lighter **I bands** which correspond to the unoverlapped portions of the thin actin molecules. A thin dark line, the **Z line,** runs through the middle of each I band. The space between successive Z lines of a myofibril is called a **sarcomere;** this is the basic contractile unit of the muscle cell. When a muscle cell contracts, the actin filaments slide past the myosin filaments and the sarcomere becomes shorter.

In summary, muscles are composed of individual muscle cells, the muscle fibers, wrapped in connective tissue. The myofibrils are tiny fibers embedded in the sarcoplasm of each muscle cell. These myofibrils are composed of the contractile proteins actin and myosin, which are arranged in an overlapping pattern into sarcomeres. Contraction of the individual sarcomeres leads to contraction of the muscle cells.

Before discussing the sequence of events that leads to the shortening of the sarcomeres, we must consider one more feature of muscle cell structure. As can be seen in Figure 6-4, the myofibrils of a muscle cell are surrounded by membrane-lined tubules. These tubules are part of two different systems: the **sarcoplasmic reticulum,** which forms an internal network of intercommunicating canals that extend the length of a sarcomere; and the **t system** composed of tubules that extend into the cell along the Z lines.

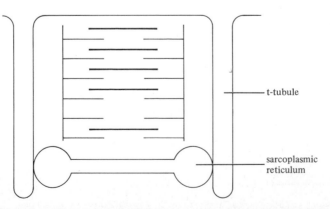

**Figure 6-4** Sarcoplasmic reticulum and t-tubules.

The canals of the sarcoplasmic reticulum do not communicate with the exterior of the cell; they end in blind sacs located at the Z lines at each end of the sarcomere, immediately adjacent to the t-tubules. However, the t-tubules which are formed by an inward growth (invagination) of the cell membrane do communicate with the exterior of the cell and thus are filled with extracellular fluid.

## MECHANISM OF CONTRACTION

Shortening of the sarcomere, and therefore contraction of the muscle, depends on the actin filaments sliding past the myosin filaments. But how is this brought about? What forces move the actin filaments relative to the myosin filaments and what sets these forces in motion?

In order for a skeletal muscle cell to contract, it must first be stimulated by a special type of nerve cell, known as a **motor neuron.** As illustrated in Figure 6-5, the motor neuron is separated by a small space from the sarcolemma of the muscle cell. The area where the motor neuron and the muscle cell are in close proximity to each other is called the **neuromuscular junction.**

The mechanism by which a motor neuron stimulates a muscle cell involves the release of a chemical transmitter, **acetylcholine,** from the end of the motor neuron. Once released, acetylcholine diffuses to the muscle cell and combines with the sarcolemma of the muscle in such a way as to initiate the contractile process. After contraction has been initiated, the acetylcholine is destroyed by an enzyme known as **cholinesterase.**

In the next chapter we will discuss activation of the motor neuron and the process by which it releases acetylcholine, for now we are going to look at the mechanism by which acetylcholine stimulates the muscle to contract.

The response of a muscle cell to stimulation by acetylcholine

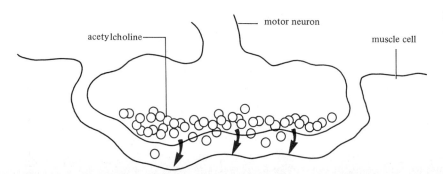

**Figure 6-5** Neuromuscular junction.

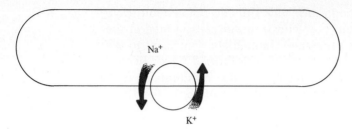

**Figure 6-6** Active transport of $Na^+$ and $K^+$.

depends on certain properties of the muscle cell membrane. When the muscle cell is resting, that is, not contracting, the muscle membrane, like all cell membranes, actively transports $Na^+$ from the inside of the cell to the extracellular fluid and $K^+$ from the extracellular fluid to the inside of the cell, as illustrated in Figure 6-6.

As a result of this active transport, the concentration of $Na^+$ in the extracellular fluid is higher than the concentration of $K^+$ in the extracellular fluid. Although the resting muscle cell membrane is not very permeable to either $Na^+$ or $K^+$, it is more permeable to $K^+$ than to $Na^+$. Consequently, more $K^+$ ions diffuse out of the cell, leading to a net movement of positive charge from the inside of the cell to the outside of the cell.

As positive charge moves from the inside of the cell to the outside of the cell, the muscle cell membrane becomes **polarized**, that is, the outside of the cell becomes positive with respect to the inside.

When acetylcholine combines with the muscle cell membrane at the neuromuscular junction, it causes a change in the structure of the membrane at that point, such that it becomes quite permeable to both $Na^+$ and $K^+$. Once this happens, $Na^+$ moves into the cell at a rate equal to that at which $K^+$ moves out and the muscle cell membrane becomes depolarized at the neuromuscular junction.

Depolarization at the neuromuscular junction initiates a wave of depolarization which spreads along the entire muscle cell membrane, as illustrated in Figure 6-7. Figure 6-7a shows the situation in which the membrane is depolarized at the neuromuscular junction, but is polarized over the remainder of the cell. The inside of the segment of membrane adjacent to the neuromuscular junction is negative with respect to the inside of the membrane at the neuromuscular junction. Likewise the outside of the segment of membrane adjacent to the neuromuscular junction is positive with respect to the outside of the membrane at the neuromuscular junction.

Because positive charge is attracted to negative charge, there will be a movement of positive charge from the outside of the adjacent segment to the outside of the neuromuscular junction and from the inside of the neuromuscular junction to the inside of the

## 125 Mechanism of contraction

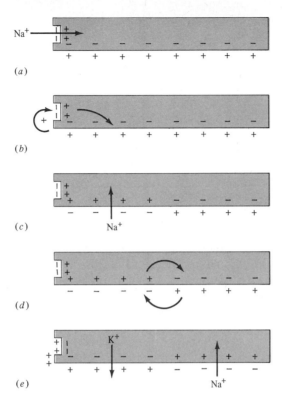

Figure 6-7 Spread of depolarization along the muscle cell membrane.

adjacent segment, as illustrated in Figure 6-7b. As a result of this movement, the outside of the adjacent segment becomes less positive and the inside becomes less negative, and the adjacent segment is somewhat depolarized.

Once a segment of muscle cell membrane becomes somewhat depolarized, the membrane undergoes structural changes such that it suddenly becomes very permeable to $Na^+$. This allows $Na^+$ to rush into the cell at this segment, as illustrated in Figure 6-7c. Movement of $Na^+$ across the membrane at this segment depolarizes it further to the point where it becomes positive inside and negative outside.

As soon as this happens, the entire sequence is repeated at the next segment of membrane. It becomes slightly depolarized by the spread of positive charge toward the inside and away from the outside, $Na^+$ rushes in, and it becomes positive on the inside and negative on the outside. The depolarization of each segment of membrane stimulates the depolarization of the next segment of membrane in such a way that depolarization spreads over the entire muscle cell membrane, as illustrated in Figure 6-7d,e.

The spread of depolarization is analogous to what happens if one stands a row of bricks on end arranged very close together and then pushes over the first brick. This brick will knock over the next brick which will, in turn, knock over the next brick, etc., until all of the bricks are knocked down.

Because muscle cells must be used over and over again, depolarization of the membrane is followed by repolarization: the return of the membrane to its original polarity. Immediately after $Na^+$ rushes through a segment of membrane, the membrane once again becomes impermeable to $Na^+$, but is permeable to $K^+$. As $K^+$ moves out of the cell, the outside again becomes positive with respect to the inside. The $Na^+$ that enters during depolarization and the $K^+$ that leaves during repolarization are returned to their original sites by active transport.

A wave of polarization followed by a wave of repolarization is referred to as an **impulse** or an **action potential.** As you will see in the next chapter, impulses are important not only in the contraction of muscle cells, but also in the communication of information by nerve cells. At this point, it must seem that we have digressed from what we were originally talking about. We started out discussing muscle contraction and ended up talking about impulses. What is the link between an impulse in a muscle cell and contraction of the cell?

The impulses which are initiated by acetylcholine spread from the sarcolemma to the t-tubules which extend into the interior of the muscle cells. Depolarization of the t-tubules stimulates the adjacent sarcoplasmic reticulum to release calcium which is stored inside the sarcoplasmic reticular canals, as illustrated in Figure 6-8. The released $Ca^{2+}$ combines with and inactivates a protein called troponin which in its active form inhibits the interaction of actin and myosin.

At the present time, the exact mechanism by which the actin

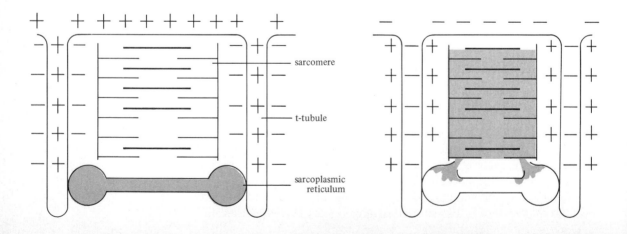

**Figure 6-8** Release of calcium from the sarcoplasmic reticulum.

**127** Mechanism of contraction

filaments are moved relative to the myosin molecules is not completely understood. The current theory is that there are cross bridges extending from the myosin molecules to the actin molecules and in the presence of ATP and $Ca^{2+}$, these cross bridges bend in such a way as to pull the actin filaments relative to the myosin filaments. This theory is illustrated in Figure 6-9.

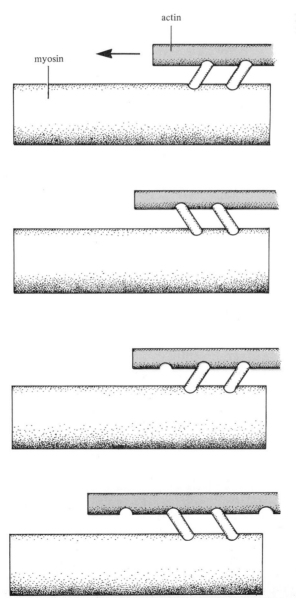

**Figure 6-9** Movement of actin relative to myosin.

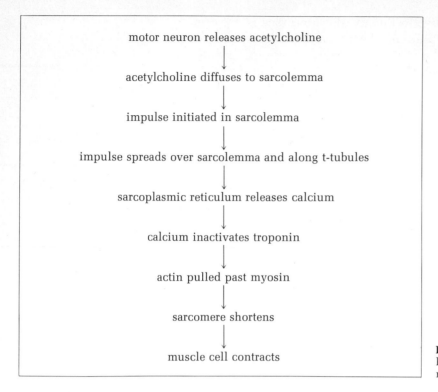

**Figure 6-10** Summary of events leading to the contraction of a muscle.

The entire sequence of events that leads to the contraction of a muscle is summarized in a flow chart shown in Figure 6-10. A motor neuron stimulates a muscle cell to contract by releasing acetylcholine which then diffuses to the sarcolemma. This initiates a wave of depolarization which spreads over the entire sarcolemma and via the t-tubules to the interior of the muscle cell. Depolarization of the t-tubules stimulates the sarcoplasmic reticulum to release $Ca^{2+}$ which in turn inactivates the inhibitory protein troponin.

Once troponin is inactivated, actin and myosin are free to combine, and in the presence of ATP, the actin molecules are pulled past the myosin molecules with the consequent shortening of the sarcomere. Contraction of the sarcomeres leads to contraction of the muscle cells. These pull on the surrounding connective tissue in such a way as to contract the entire muscle.

Relaxation of a muscle is brought about by the active transport of $Ca^{2+}$ back into the canals of the sarcoplasmic reticulum. As calcium is removed from the sarcoplasm, troponin once again inhibits the combination of actin and myosin.

### Paralysis

A muscle will become paralyzed if any of the steps in the contractile process are interfered with. The most common causes of paralysis are dysfunctions of the motor neuron or neuromuscular junction. Motor neuron dysfunction can result from infection, such as **poliomyelitis;** it can also result from damage to the brain or spinal cord caused by cardiovascular disease or physical injury. Once a motor neuron ceases to function, the muscle cells can no longer be voluntarily contracted.

Under circumstances in which some of the motor neurons remain functional, physical therapy can be used to prevent atrophy of the muscles and to train the patient to use the remaining motor neurons in controlling the muscle.

**Myasthenia gravis** and **muscular dystrophy** are both diseases in which the neuromuscular junction does not function properly. In these diseases, nerve impulses that reach the ends of the motor neurons do not initiate an impulse in the muscle cells. Current research seems to indicate that myasthenia gravis involves abnormal nerve endings and that muscular dystrophy involves some form of inhibition of the combination of acetylcholine with the sarcolemma.

Drugs can also interfere with muscular contraction. For example, **curare** combines with the sarcolemma in such a way as to block the action of acetylcholine. Curare was originally used as a poison, but is now used in lower doses to relax the skeletal muscles during surgery.

## ENERGY FOR MUSCULAR CONTRACTION

Muscular contraction, like all other types of cellular work, requires energy. When a muscle cell is stimulated by a motor neuron an enzyme converts the high-energy compound adenosine triphosphate (ATP) into adenosine diphosphate (ADP) and the released energy is used to slide the actin filaments past the myosin filaments. Because the muscle cells often require large amounts of ATP during short periods of time, they must store energy in a form that can be rapidly converted into ATP. Energy is stored in muscle cells in the form of **glycogen** and **creatine phosphate** molecules.

At a time when the muscle cell is resting, some of the glucose molecules that enter the cell are converted to glycogen, as shown in Figure 6-11. Other glucose molecules are broken down and used to form ATP. This ATP then reacts with a compound called **creatine** to form **creatine phosphate** (CP) by the reaction

$$C + ATP \longrightarrow CP + ADP$$

The whole point of this reaction is to transfer energy from ATP which cannot be stored in muscle cells to CP which can be stored.

The glycogen and creatine phosphate that are formed when

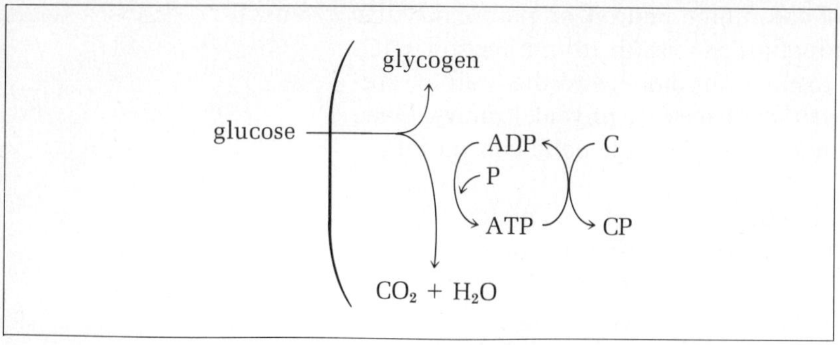

## 130 THE MUSCULAR SYSTEM

**Figure 6-11** Storage of energy in a muscle cell.

the muscle is relaxed are used to form ATP when the muscle contracts. At the onset of muscular contraction, the creatine phosphate converts ADP to ATP by the reaction

$$CP + ADP \longrightarrow C + ATP$$

The creatine phosphate, however, is depleted shortly after the muscle begins to contract and additional ATP must be formed from the glycogen or some other energy source. The manner in which this ATP is formed depends on the type of muscular activity that is taking place. During intense muscular activity, such as running up a flight of stairs, most of the ATP is formed by glycolysis, as illustrated in Figure 6-12. Glycogen is rapidly broken down to pyruvic acid with the resultant formation of a small amount of ATP from each glycogen molecule that is used.

However, as oxygen cannot be supplied to the cell fast enough during intense muscular activity to allow oxidative phosphorylation to take place, the pyruvic acid is converted to a compound called **lactic acid.** Although this pattern of metabolism can supply ATP rapidly it is quite limited. Each glycogen molecule leads to the formation of only a few ATP molecules; therefore large amounts of glycogen must be used as a consequence of which large quantities of lactic acid are formed. As lactic acid accumulates in the muscle cell, it interferes with the ability of the cell to contract, a process

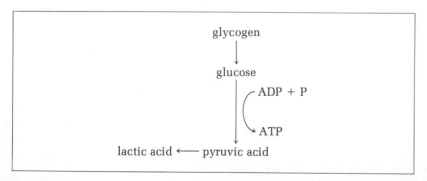

**Figure 6-12** ATP formation during intense muscular activity.

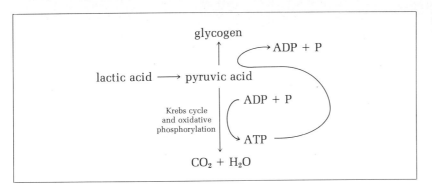

**Figure 6-13** Reconversion of lactic acid to glucose.

known as fatigue. The pain associated with fatigued muscles is probably caused by stimulation of nerve endings by the lactic acid.

Once a period of intense muscular activity is over, the accumulated lactic acid is either broken down or converted back to glycogen, as illustrated in Figure 6-13. All of the lactic acid is first converted back to pyruvic acid. Some of this pyruvic acid is used to form ATP via the reactions of the Krebs cycle and oxidative phosphorylation. The ATP that is formed is then used to convert the bulk of the pyruvic acid back to glycogen.

Removal of the accumulated lactic acid requires oxygen; therefore a period of rapid breathing always follows a period of strenuous activity. The amount of oxygen required to remove all of the accumulated lactic acid is referred to as the **oxygen debt.**

During more moderate muscular activity such as standing or walking, oxygen can be supplied to the muscle cells fast enough to allow the formation of ATP by oxidative phosphorylation. Thus in moderate muscular activity, lactic acid does not usually accumulate in the muscle cells. Present research seems to indicate that some muscle cells, the red fibers, are specialized for moderate muscular activity whereas others, the white fibers, are specialized for intense activity.

## FORCE OF CONTRACTION

The force with which a muscle contracts is adjusted to the task the muscle must perform. For example, in lifting a heavy suitcase, the muscles of the arm are contracted much more forcefully than in lifting a book of matches. Control over the force of muscular contraction is exerted by the motor neurons which stimulate the muscle.

The individual motor neurons are wrapped together into a **motor nerve** in much the same way that the individual muscle fibers are wrapped together into a muscle. Each motor neuron stim-

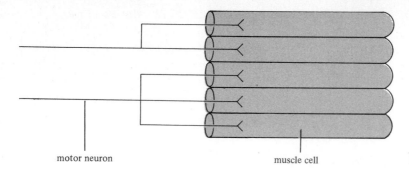

motor neuron    muscle cell

**Figure 6-14** Motor units.

ulates one or more muscle fibers, as illustrated schematically in Figure 6-14. A motor neuron plus the muscle fibers it stimulates is referred to as a **motor unit.**

The overall strength of muscular contraction depends upon two factors: the number of muscle fibers contracting and the force with which each contracts. Each of these factors can be controlled by the motor nerve in such a way as to adjust the force of contraction to the situation.

The number of muscle fibers that contract is determined by the number of motor neurons that stimulate the muscle. As the number of motor neurons stimulating the muscle is increased, so the number of muscle fibers contracting is also increased, leading to a more forceful contraction.

The frequency at which a motor neuron stimulates a muscle fiber determines both the strength and duration of contraction in that fiber. Figure 6-15 shows the response of a muscle to a single stimulation, called a **twitch.** This response consists of a latent phase between the time the muscle is stimulated and the time the muscle contracts followed by a single contraction. When the muscle is stimulated a certain number of fibers contract and then relax.

If the muscle fibers are stimulated a second time before they relax from the first contraction, a second contraction will be added to the first, as illustrated in Figure 6-16. The total response to the two stimuli which are close together is larger than the response to a single stimulus and thus this type of contraction is referred to as **summation.**

The muscle fibers contract with maximum force when they are stimulated many times in rapid succession. This type of contraction, known as **tetanus,** is shown in Figure 6-17. The maximal contraction of tetanus is sustained until the stimulus is removed or the muscle fatigues.

The term **tetanus** is also used to describe an infectious disease that leads to tetanic contractions in skeletal muscles all over the

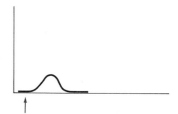

**Figure 6-15** Twitch.

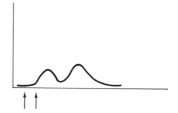

**Figure 6-16** Summation.

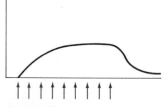

**Figure 6-17** Tetanus.

body. Because the jaw muscles are often the first ones affected, the disease is also referred to as lockjaw.

In summary, the force of muscular contraction is normally controlled by the motor neuron which stimulates the muscle. The overall contractile force can be increased by increasing the number of motor neurons stimulating the muscle and by increasing the frequency at which they stimulate the muscle.

Increasing the number of motor neurons that stimulate the muscle increases the number of muscle fibers that contract whereas increasing the frequency of stimulation increases the force with which the individual fibers contract.

The force with which a muscle contracts depends on conditions that affect the muscle itself as well as on the way it is stimulated. If a muscle is used frequently, the size and strength of the muscle fibers increase. This is referred to as **hypertrophy** of the muscle, a condition that can be seen in its extreme form in the muscles of weightlifters. On the other hand, muscles that are not used, or used infrequently, become small and weak, a condition known as muscular atrophy.

The metabolic conditions inside of the muscle cells also affect the force of contraction. If nutrients and oxygen are not adequately supplied, or if lactic acid and other wastes are not removed fast enough, muscle cells will become fatigued and will no longer be able to contract properly. For example, as you climb stairs, lactic acid begins to build up in the muscles of your legs interfering with the ability of these muscles to lift your body up more stairs. Interestingly, the amount of blood that can flow through a muscle is increased if the muscle is used frequently. This is one of the reasons the muscles of a person who gets a great deal of exercise fatigue far less easily than do those of a person who gets very little exercise.

**Figure 6-18** Relationship between muscle length and force of contraction.

As can be seen in Figure 6-18, the force of contraction is influenced by the length of the muscle when it begins to contract. Up to a certain point, increasing the initial length of a muscle increases the force of contraction. Further increase past this point leads to a diminished ability to contract.

In general, skeletal muscles are at their optimal length when they are in their resting, relaxed position. The relationship between initial length and force of contraction is also of great importance in the heart.

## THE SKELETAL MUSCLES

The various skeletal muscles form a layer of muscular tissue that surrounds the underlying bony skeleton as illustrated in Figures

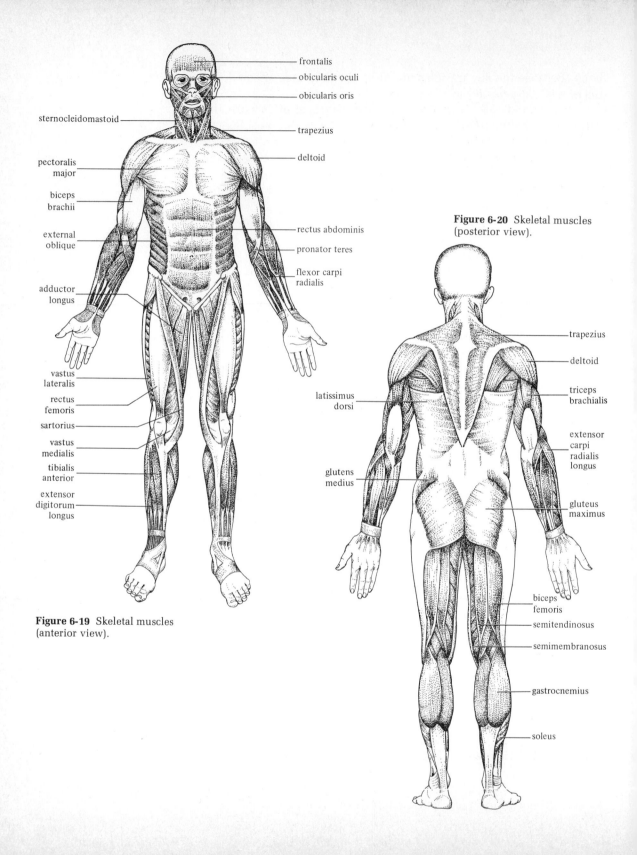

**Figure 6-19** Skeletal muscles (anterior view).

**Figure 6-20** Skeletal muscles (posterior view).

6-19 and 6-20. There are over 600 of these muscles; together they account for about 40% of the body weight.

Movement of any part of the body depends on the coordinated efforts of a number of different muscles. For any particular motion the involved muscles can be classified into three groups. **Prime movers** are the muscle(s) that exert the major pulling force. **Synergists** are muscles that contract at the same time as the prime mover, aiding the prime mover either by exerting an additional pull or by stabilizing the part of the body being moved. **Antagonists** are muscles that relax as the prime mover is contracting. These are muscles that exert a pull in a direction opposite to that of the prime mover. In certain situations, such as holding your head up, muscles that pull in opposite directions contract at the same time.

In this section we are going to describe the location and action of some of the more important skeletal muscles. Information about these muscles, as well as about a number of other muscles, is summarized in Table 6-1 at the end of this chapter.

Before beginning our description, however, we should look at the basis on which most of the muscles are named. The following is a list of features that are used in naming muscles; each feature is followed by the name of a muscle in which it is used.

1. Location: quadriceps **femoris** (in the upper leg)
2. Shape: **trapezius** (like a trapezoid)
3. Number of divisions: **biceps** brachii (two divisions)
4. Points of attachment: **sternocleidomastoid** (the sternum and the mastoid process)
5. Action: **pronator** teres (turns hand backward)
6. Direction of fibers: **rectus** abdominus (straight fibers)

As you can see, often more than one feature is used in naming a muscle. The quadriceps femoris has four divisions and is located in the femoral region. The rectus abdominus has straight fibers and is located in the abdomen.

In learning the location of the various muscles, you should realize that the muscle is not usually located within the part it moves. Hence the biceps brachii is located in the upper arm, but acts to flex the lower arm. You can usually get a good idea of the action of a particular muscle by observing its points of attachment. The attachment to the bone which is moved is called the **insertion** of the muscle whereas the attachment to the bone which remains fixed is called the origin. However, in many cases both bones can be moved and these terms lose their meaning.

### Muscles of Facial Expression

Figure 6-21 illustrates some of the more important muscles of facial expression. The **epicranius,** which stretches over the top of

**136** The Muscular System

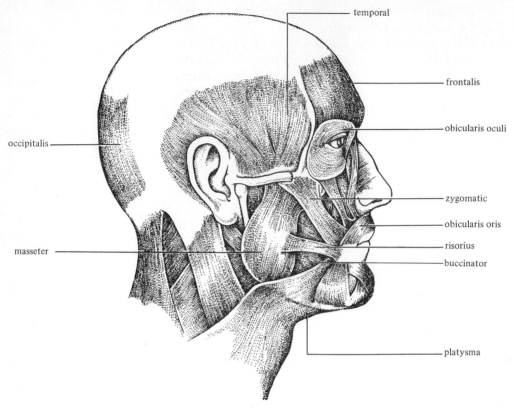

Figure 6-21 Muscles of the face.

the head, consists of two portions: the posterior **occipitalis** and the anterior **frontalis.** Contraction of the occipitalis pulls the skin of the skull backward, whereas contraction of the frontalis raises the eyebrows and wrinkles the skin of the forehead.

The **orbicularis oculi** forms a circle of muscle around the orbital cavity. It is responsible for constricting the opening of the orbital cavity, that is, squinting. This muscle is used to protect the eye from excess light and foreign particles.

The **levator palpebrae superioris** lifts the upper eyelid to "open" the eyes. It is located beneath the eyelid.

The **orbicularis oris** forms a circle of muscle around the mouth. Constriction of this muscle pulls the lips together and pushes them forward; it is one of the muscles used in kissing.

The **buccinator** is located in the lower part of the cheeks. It pulls the corners of the mouth sideways and flattens the cheeks. This muscle is used to hold food in position for chewing. It can also be used for the expulsion of air from the oral cavity as in whistling or playing a wind instrument.

The **zygomatic** muscles extend from the zygomatic bones to the corners of the mouth. They are used to pull the corners of the mouth sideways and upward as in laughing or smiling.

**137** The skeletal muscles

The **platysma** lies in the front of the neck and pulls the corners of the mouth downward as in expressions of sadness or fright.

### Movement of the Eye

There are six muscles that attach the eye to the orbital cavity, as illustrated in Figure 6-22. These muscles make it possible to look in different directions without moving the head. The four rectus

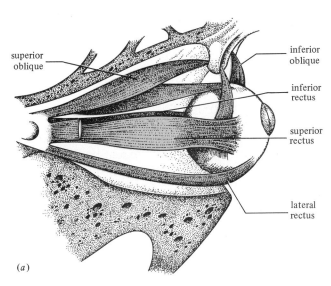

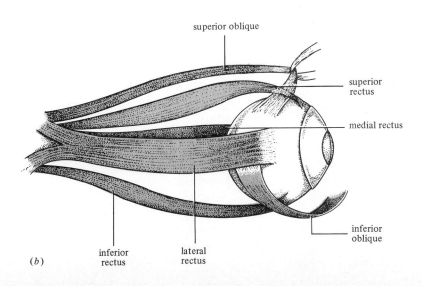

**Figure 6-22** Muscles of eye movement: (a) top view and (b) side view.

muscles—**superior rectus, inferior rectus, medial rectus,** and **lateral rectus**—pull the eye up and down and from side to side. Two other muscles, the **superior and inferior obliques,** extend from the medial surface of the orbital cavity to attach sideways to the eye. These muscles aid in rotating the eyes.

### Muscles of Mastication

Figure 6-23 shows the muscles that move the mandible. The **temporal** muscle covers the **temporal** bone and extends under the zygomatic arch to attach to the mandible. This muscle can be felt by clenching your teeth and touching the side of your head. The **masseter** extends from the zygomatic arch to the mandible. It can be felt by clenching your teeth and touching your cheeks. The temporal and masseter muscles are used to pull the mandible upward, that is, to close the mouth and clench the teeth. If these muscles are relaxed, the jaws hang open. The **pterygoid** muscles located at the corners of the jaw are used to move the jaw from side to side.

### Muscles that Move the Head

Movement of the entire head is controlled to a large extent by two sets of muscles: the two sternocleidomastoid at the front of the neck and the two semispinalis capitis muscles at the back of the neck.

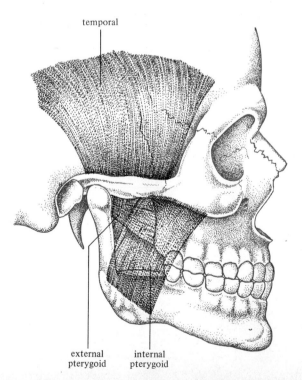

**Figure 6-23** Muscles that move the mandible.

**139** The skeletal muscles

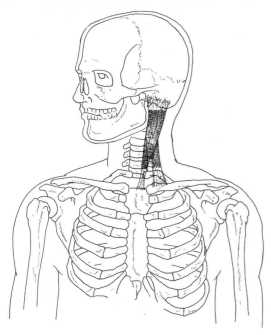

**Figure 6-24** Sternocleidomastoid.

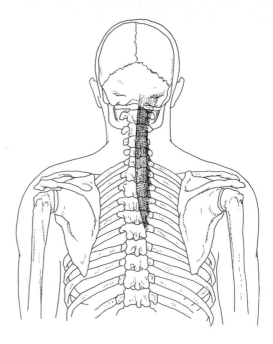

**Figure 6-25** Semispinalis capitis.

The **sternocleidomastoid** muscles, illustrated in Figure 6-24, pass from the sternum and the medial portion of the clavicle to the mastoid process of the temporal bone. If both muscles are contracted at the same time, the head is pulled toward the chest—it is flexed. If only one sternocleidomastoid muscle is contracted, the head is pulled toward the shoulder on that side, and the face is turned toward the opposite side. Contraction of the right sternocleidomastoid would pull the head toward the right shoulder and turn the face toward the left side. Injury or disease of the sternocleidomastoid can lead to torticollis (wry neck), a condition in which the head is permanently bent toward one shoulder with the face pointing in the opposite direction.

The **semispinalis capitis** muscles, illustrated in Figure 6-25, connect the lower thoracic and upper cervical vertebrae with the occipital bone. Contraction of both semispinalis capitis muscles pulls the head backward. Contraction of one semispinalis capitis muscle has basically the same action as contraction of the sternocleidomastoid on the same side; that is, the head is pulled toward that shoulder and the face is turned toward the opposite side.

### Movement of the Arm and Shoulders

The **trapezius** muscles connect the clavicle and scapula with the occipital bone, cervical vertebrae, and thoracic vertebrae, as

## 140 THE MUSCULAR SYSTEM

illustrated in Figure 6-26. Contractions of these muscles can move the shoulders into various positions; for instance, the trapezius muscles are involved in shrugging the shoulders or pulling the shoulders back. The trapezius muscles can be felt as prominant ridges above the scapula.

The **latissimus dorsi,** shown in Figure 6-27, connects the humerus to the lower vertebrae and the pelvis. As can be seen, the latissimus dorsi covers the lower thoracic, as well as the lumbar region of the back. Contraction of this muscle pulls the arm downward and backward; it adducts and extends the arm. This muscle is used in activities that involve pulling the arm down forcefully, such as swimming or chopping wood.

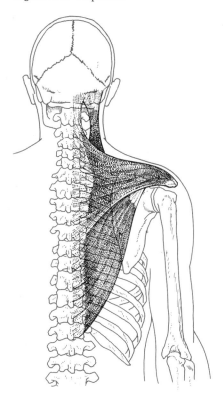

**Figure 6-26** Trapezius.

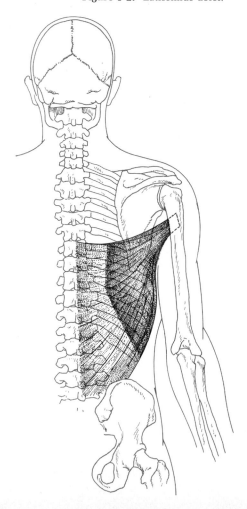

**Figure 6-27** Latissimus dorsi.

**141** The skeletal muscles

The **deltoid** muscle, illustrated in Figure 6-28, extends over the shoulder joint connecting the clavicle and scapula with the humerus. It can be felt by raising your arm sideways and touching your shoulders. The action of the deltoid depends on the part of the muscle that contracts. Contraction of the entire muscle raises the arm laterally (abduction); contraction of the anterior portion pulls the arm forward whereas contraction of the posterior portion pulls the arm backward.

The **pectoralis major** covers the anterior portion of the chest connecting the clavicle and sternum to the humerus, as illustrated in Figure 6-29. This muscle is used to pull the arm forward from the anatomical position. It is also used to adduct the arm.

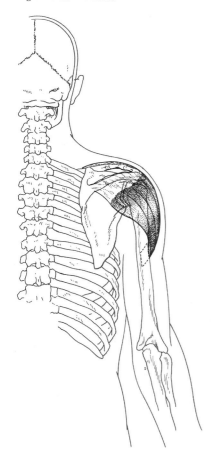

**Figure 6-28** Deltoid.

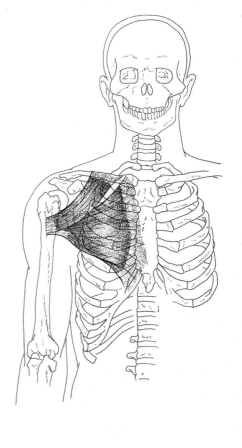

**Figure 6-29** Pectoralis major.

**142** The Muscular System

### Movement of the Forearm and Hand

The **biceps brachii** is a large muscle that covers the anterior surface of the humerus, as illustrated in Figure 6-30. This is the muscle boys like to flex when they "make a muscle." The biceps connects the scapula with the proximal end of the radius and upon contraction flexes the lower arm. Contraction of the biceps is also used to supinate the lower arm (turn the palm forward).

Flexion of the lower arm is aided by the **brachialis** muscle, which extends from the anterior surface of the humerus to the proximal end of the ulna, as illustrated in Figure 6-31.

The **triceps brachii** is the muscle that extends the lower arm. This muscle covers the posterior surface of the humerus and connects the scapula and proximal end of the humerus with the ulna, as illustrated in Figure 6-32.

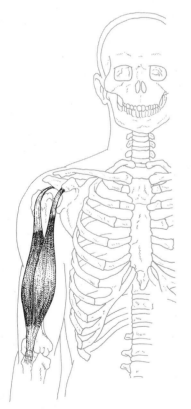

**Figure 6-30** Biceps brachii.

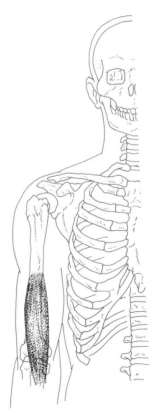

**Figure 6-31** Brachialis.

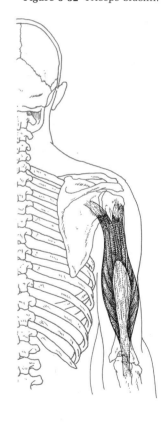

**Figure 6-32** Triceps brachii.

**143** The skeletal muscles

The majority of the muscles that move the hand and fingers are located in the forearm with tendons extending to the bones which they move. Some of these muscles are illustrated in Figure 6-33 and are described in Table 6-1.

**Figure 6-33** Muscles of the forearm: (a) anterior view and (b) posterior view.

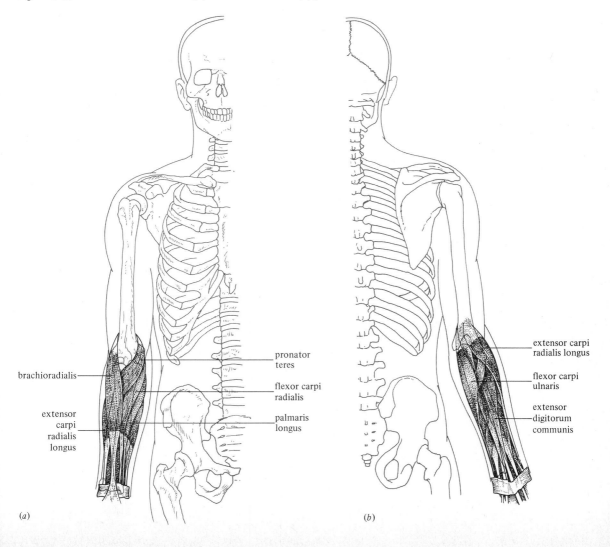

## Muscles of Respiration

The major respiratory muscles are the diaphragm, external intercostal, and internal intercostals shown in Figure 6-34.

The **diaphragm** separates the thoracic cavity from the abdominal cavity. When relaxed, the diaphragm curves upward into

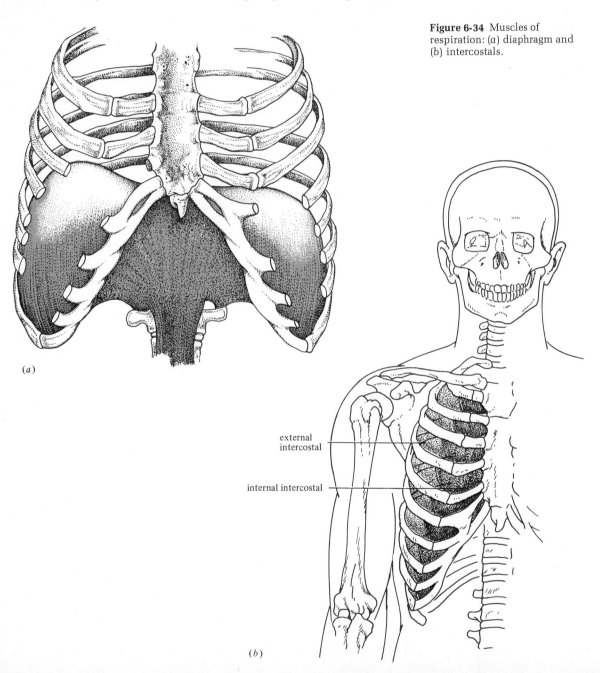

**Figure 6-34** Muscles of respiration: (*a*) diaphragm and (*b*) intercostals.

the thorax in the shape of a bell. Contraction of the diaphragm flattens it downward with the consequent enlargement of the thoracic cavity. This enlargement causes air to move into the lungs as we will describe in Chapter 14.

The **external and internal intercostals** connect each of the first 11 ribs to the rib below it. As the names indicate, the external intercostals lie outside the internal intercostals. The external intercostals act to enlarge the diaphragm during inspiration whereas the internal intercostals contract the diaphragm during forced expiration.

**The Abdominal Wall**

Part of the abdominal cavity is protected by the lower vertebrae and by the bones of the pelvic girdle. The remainder of the abdominal cavity is protected by a thick muscular wall. On the posterior surface this wall is formed by the latissimus dorsi, a muscle we have already described, and a number of smaller muscles.

The lateral walls of the abdominal cavity are formed by three sets of muscles arranged in layers. As illustrated in Figure 6-35, the

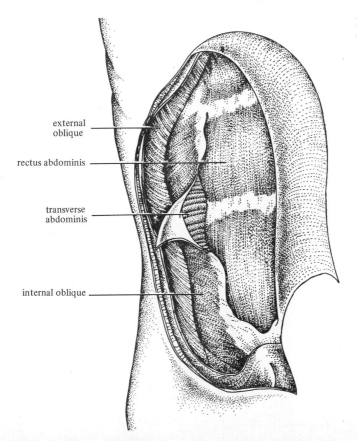

**Figure 6-35** Muscles of the abdominal wall.

**146** The Muscular System

outer layer is formed by the **external obliques,** the middle layer by the internal obliques, and the inner layer by the **transverse abdominis** muscles. Extensions of these three sets of muscles form a fourth set of muscles, the **rectus abdominis** muscles, on the anterior surface of the abdomen. The rectus abdominis muscles connect the pubic portion of the pelvis with the sternum and ribs.

The four sets of muscles—the external oblique, internal oblique, transverse abdominis, and rectus abdominis muscle—which form the lateral and anterior walls of the abdominal cavity, carry out two other functions in addition to the protection of the organs in the abdominal cavity. Constriction of these muscles increases abdominal pressure and thus aids in expelling substances from the abdominal cavity as in urination, defecation, vomiting, and childbirth. These muscles also pull the trunk forward by flexing the vertebral column.

### Muscles that Move the Leg

The **gluteus maximus** is a large muscle that forms a major part of the buttock connecting the pelvis to the femur, as illustrated in Figure 6-36. Contraction of the gluteus maximus extends the thigh as in walking up stairs or getting up after stooping. The gluteus maximus is also used to support the weight of the body when standing. Because of its large size, the gluteus maximus is a convenient site for intramuscular injections.

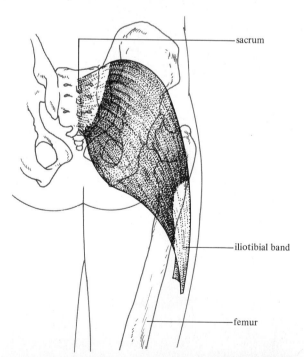

**Figure 6-36** Gluteus maximus.

**147** The skeletal muscles

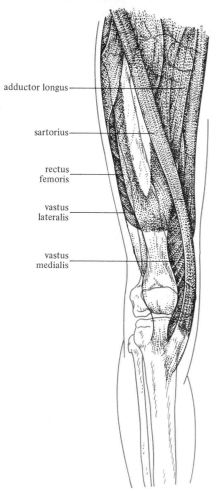

**Figure 6-37** Muscles of the upper leg (anterior view).

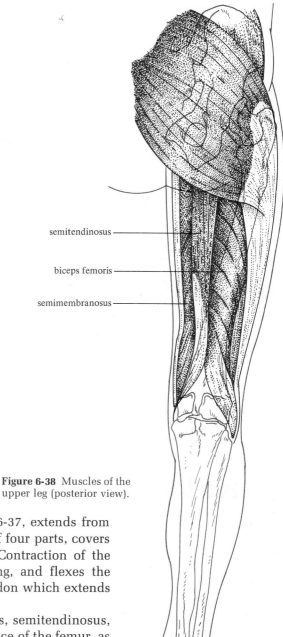

**Figure 6-38** Muscles of the upper leg (posterior view).

The **quadriceps femoris,** shown in Figure 6-37, extends from the pelvis to the tibia. This muscle, consisting of four parts, covers the anterior and lateral surface of the femur. Contraction of the quadriceps extends the lower leg, as in kicking, and flexes the thigh. The patellar bone is embedded in the tendon which extends from the quadriceps to the tibia.

The **hamstring muscles**—the biceps femoris, semitendinosus, and semimembranosus—cover the posterior surface of the femur, as illustrated in Figure 6-38. These muscles are antagonists to the quadriceps in that they are used to flex the lower leg. You can feel the tendons of the hamstring muscles just behind the knee.

The foot is pulled upward toward the shin by the **tibialis anterior muscle,** which is located along the lateral surface of the tibia, as illustrated in Figure 6-39. This muscle is used in walking in order to allow the heel to touch the ground before the toes.

The posterior calf of the lower leg is made up of two muscles, the **soleus** and **gastrocnemius,** which are attached by a common tendon, the Achilles tendon, to the calcaneous bone of the heel. These muscles are illustrated in Figure 6-40. Contraction of the soleus and gastrocnemius muscles raises the heel during walking, jumping, or standing on the toes.

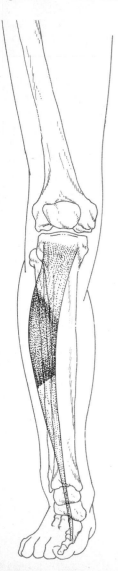

**Figure 6-39** Tibialis anterior.

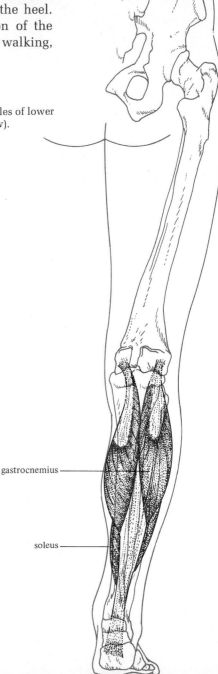

**Figure 6-40** Muscles of lower leg (posterior view).

# 149 The skeletal muscles

**Table 6-1** Skeletal Muscles

| Muscle | Origin | Insertion | Action |
|---|---|---|---|
| *Muscles of Facial Expression* | | | |
| Epicranius | | | |
|   Frontalis | Connective tissue at top of skull | Skin above orbit | Raises eyebrow and draws scalp forward |
|   Occipitalis | Mastoid process and occipital bone | Connective tissue at top of skull | Pulls scalp backward |
| Orbicularis oculi | Medial portion of orbit | Eyelids | Closes eye |
| Levator palpebrae superiorus | Optic foramen | Upper eyelid | Raises lid |
| Obicularis oris | Skin and muscle of lips | Corners of mouth | Closes and pushes out lips |
| Zygomatic | Zygomatic bone | Orbicularis oris muscle | Raises the corner of mouth and pulls it sideways |
| Buccinator | Lateral portion of mandible and maxilla | Orbicularis oris muscle | Draws cheek and lips against teeth |
| Platysma | Fascia covering deltoid and pectoralis major muscles | Corner of mouth and mandible | Pulls corner of mouth down |
| Risorius | Fascia around parotid gland | Corners of mouth | Pulls corner of mouth laterally |
| Levator labii superioris | Lower edge of orbit | Upper lip | Raises upper lip |
| Depressor labii inferioris | Mandible | Lower lip | Lowers lower lip |
| *Muscles that Move Eyeball* | | | |
| Superior rectus | Center of orbital cavity | Superior surface of eyeball | Pulls eyeball upward |
| Lateral rectus | Center of orbital cavity | Lateral surface of eyeball | Pulls eyeball laterally |
| Inferior rectus | Center of orbital cavity | Inferior surface of eyeball | Pulls eyeball downward |
| Medial rectus | Center of orbital cavity | Medial surface of eyeball | Pulls eyeball medially |
| Superior oblique | Center of orbital cavity | Lateral surface of eyeball | Pulls eyeball downward and laterally; rotates eyeball |
| Inferior oblique | Floor of orbital cavity | Lateral surface of eyeball | Pulls eyeball upward and laterally; rotates eyeball |
| *Muscles that Move the Mandible and Tongue* | | | |
| Masseter | Zygomatic arch | Lateral portion of mandible | Raises mandible and pulls it forward |
| Temporal | Temporal bone | Corner of mandible | Raises mandible and pulls it inward |
| External pterygoid | Great wing of sphenoid bone | Corner of mandible | Pulls mandible forward and laterally |
| Internal pterygoid | Maxilla, sphenoid, and palatine bones | Angle of mandible | Raises mandible and pulls it laterally |
| Styloglossus | Styloid process of temporal bone | Lateral surface of tongue | Pulls tongue upward and backward |
| Genioglossus | Mandible | Inferior surface of tongue and hyoid bone | Pulls tongue downward and forward |
| *Muscles that Move the Head* | | | |
| Semispinalis capitis | Lower cervical and upper thoracic vertebrae | Occipital bone | Extends head and pulls it to side |
| Sternocleidomastoid | Sternum and clavicle | Mastoid process | Flexes head and pulls it to opposite side |
| Longus capitus | Lower cervical and upper thoracic vertebrae | Base of occipital bone | Flexes head |

**Table 6-1** (Continued)

| Muscle | Origin | Insertion | Action |
|---|---|---|---|
| \multicolumn{4}{c}{Muscles that Move the Shoulder} | | | |
| Trapezius | Occipital bone, seventh cervical, and thoracic vertebrae | Lateral portion of clavicle and scapula | Raises, lowers, and adducts scapula; raises clavicle |
| Levator scapulae | First four cervical vertebrae | Medial portion of scapula | Raises scapula |
| Rhomboideus major | Second to fifth thoracic vertebrae | Medial portion of scapula | Pulls scapula upward and medially |
| Rhomboideus minor | Last cervical and first thoracic vertebrae | Upper, medial portion of scapula | Pulls scapula medially |
| Serratus anterior | Upper nine ribs | Lower, medial portion of scapula | Rotates scapula upward |
| Pectoralis minor | Second to fifth ribs next to cartilages | Coracoid process of scapula | Pulls scapula forward and downward |
| \multicolumn{4}{c}{Muscles that Move the Upper Arm} | | | |
| Pectoralis major | Sternum, clavicle, and cartilages of second to sixth ribs | Humerus near greater tubercle | Adducts, flexes, and rotates arm medially |
| Latissimus dorsi | Lower thoracic, lumbar, and sacral vertebrae; crest of ilium | Proximal, anterior surface of humerus | Adducts, extends, and rotates arms medially; lowers shoulder |
| Deltoid | Clavicle and scapula | Lateral surface of humerus | Abducts arm; aids in flexion and extension of arm |
| Teres major | Inferior angle of scapula | Anterior, proximal portion of humerus | Adducts and extends arm |
| Supraspinatus | Above spine of scapula | Greater tubercle of humerus | Abducts arm |
| Infraspinatus | Below spine of scapula | Greater tubercle of humerus | Rotates arm laterally |
| \multicolumn{4}{c}{Muscles that Move the Lower Arm} | | | |
| Biceps brachii | Caracoid process and upper border of glenoid cavity of scapula | Proximal end of radius | Flexes forearm; supinates forearm and hand |
| Brachialis | Anterior, distal portion of humerus | Proximal end of ulna | Flexes forearm |
| Brachioradialis | Distal portion of humerus | Proximal position of radius | Flexes forearm |
| Triceps brachii | Scapula and posterior surface of humerus | Olecranon of ulna | Extends forearm |
| Pronator teres | Medial epicondyle of humerus and proximal portion of ulna | Lateral surface of radius | Pronates forearm |
| Supinator | Lateral epicondyle of humerus and proximal portion of ulna | Proximal portion of radius | Supinates forearm |
| \multicolumn{4}{c}{Muscles that Move the Hand and Fingers} | | | |
| Flexor carpi radialis | Medial epicondyle of humerus | Second metacarpal | Flexes hand |
| Palmaris longus | Medial epicondyle of humerus | Connective tissue of palm | Flexes hand |
| Flexor carpi ulnaris | Medial epicondyle of humerus and proximal portion of ulna | Carpals and fifth metacarpal | Flexes and adducts hand |
| Extensor carpi radialis longus | Distal, lateral portion of humerus | Second metacarpal | Extends and abducts hand |
| Extensor carpi ulnaris | Lateral epicondyle of humerus and posterior portion of ulna | Fifth metacarpal | Extends and adducts hand |
| Flexor digitorum superficialis | Distal portion of humerus and proximal portions of radius and ulna | Middle phalanges | Flexes fingers between 1st and 2nd phalanges |
| Flexor digitorum profundus | Anterior surface of ulna | Distal phalanges | Flexes fingers between 2nd and 3rd phalanges |
| Extensor digitorum | Lateral epicondyle of humerus | Phalanges | Extends phalanges |

**151** The skeletal muscles

Table 6-1 (Continued)

| Muscle | Origin | Insertion | Action |
|---|---|---|---|
| *Muscles of Respiration* | | | |
| Diaphragm | Lumbar vertebrae, cartilages of lower six ribs, end of sternum | Central tendon of diaphragm | Enlarges thorax in vertical direction |
| External intercostals | Inferior surface of each rib | Superior surface of rib below | Elevates ribs; enlarges thorax |
| Internal intercostals | Inner surface of each rib | Superior surface of rib below | Pulls ribs together; contracts thorax |
| *Muscles of Anterior Abdominal Wall* | | | |
| External oblique | Lower eight ribs | Linea alba and iliac crest | Compresses abdomen; flexes and rotates vertebral column |
| Internal oblique | Iliac crest and inguinal ligament | Lower three ribs and pubic bone | Compresses abdomen; flexes and rotates vertebral column |
| Rectus abdominis | Pubic bone | Sternum, cartilages of fifth to seventh ribs | Compresses abdomen and flexes vertebral column |
| Transversus abdominis | Lower six ribs and iliac crest | Linea alba and pubis | Compresses abdomen |
| *Muscles of the Pelvic Floor* | | | |
| Levator ani | Pubis and ischium | Coccyx | Raises anus; supports pelvic organs |
| Coccygeus | Ischium | Sacrum and coccyx | Supports pelvic organs |
| External anal sphincter | Attached to the skin surrounding anus and to the coccyx | | Holds anus closed |
| Bulbocavernosus | Ventral portion of root of penis | Connective tissue surrounding root of penis | Compresses urethra during ejaculation |
| *Muscles of Back that Move Vertebral Column* | | | |
| Sacrospinalis | Posterior surface of ribs; cervical, thoracic, lumbar and sacral vertebrae, ilium | Mastoid process of temboral bones, ribs, cervical, thoracic, and sacral vertebrae | Extends and bends laterally the head and vertebral column |
| Quadratus lumborum | Iliac crest and last three lumbar vertebrae | First four lumbar vertebrae and twelfth rib | Extends and bends laterally the vertebral column |
| *Muscles that Move the Thigh* | | | |
| Gluteus maximus | Ilium, sacrum, and coccyx | Proximal end of femur and iliotibial band | Extends thigh and rotates it laterally |
| Gluteus medius | Ilium | Greater trochanter of femur | Abducts thigh |
| Gluteus minimus | Ilium | Greater trochanter of femur | Abducts thigh and rotates it laterally |
| Iliopsoas | Ilium; twelfth thoracic and first five lumbar vertebrae | Lesser trochanter of femur | Flexes thigh |
| Adductor longus | Pubis | Anterior surface of femur | Adducts and flexes thigh |
| Adductor brevis | Pubis | Anterior surface of femur | Adducts and flexes thigh |
| Adductor magnus | Pubis and ischium | Anterior surface of femur | Adducts thigh |

**Table 6-1** (Continued)

| Muscle | Origin | Insertion | Action |
|---|---|---|---|
| *Muscles that Move Leg at Knee* | | | |
| Quadriceps femoris | | | |
|   Rectus femoris | Ilium | Tibia via common tendon of quadriceps | Flexes thigh and extends lower leg |
|   Vastus medialis | Anterior surface of femur | Tibia via common tendon of quadriceps | Extends lower leg |
|   Vastus lateralis | Anterior surface of femur | Tibia via common tendon of quadriceps | Extends lower leg |
|   Vastus intermedius | Anterior, lateral surface of femur | Tibia via common tendon of quadriceps | Extends lower leg |
| Hamstrings | | | |
|   Biceps femoris | Ischium and femur | Lateral condyle of tibia and head of fibula | Flexes lower leg; extends thigh |
|   Semitendinosus | Ischium | Proximal, medial surface of tibia | Flexes lower leg; extends thigh |
|   Semimembranosus | Ischium | Medial condyle of tibia | Flexes lower leg; extends thigh |
| Gracilis | Pubis and ischium | Medial portion of tibia | Flexes lower leg and adducts thigh |
| Sartorius | Ilium | Proximal, medial surface of tibia | Flexes lower leg and thigh |
| *Muscles that Move the Foot and Toes* | | | |
| Gastrocnemius | Condyles of the femur | Calcaneus | Flexes lower leg and plantar flexes foot |
| Soleus | Proximal ends of fibula and tibia | Calcaneus | Plantar flexes foot |
| Tibialis anterior | Lateral condyle of tibia | First metatarsal and cuneiform | Dorsally flexes and inverts foot |
| Tibialis posterior | Proximal, posterior surface of tibia and fibula | Tarsals and metatarsals | Plantar flexes and inverts foot |
| Peroneus longus | Proximal ends of tibia and fibula | First metatarsal and cuneiform | Plantar flexes and everts foot |
| Peroneus brevis | Lateral surface of fibula | Fifth metatarsal | Dorsally flexes and everts foot |
| Peroneus tertius | Distal portion of fibula | Fourth and fifth metatarsals | Dorsally flexes and everts foot |
| Flexor digitorum longus | Tibia | Distal phalanges | Flexes toes |
| Exterior digitorum longus | Proximal ends of tibia and fibula | Middle and distal phalanges | Extends toes |

## OBJECTIVES FOR THE STUDY OF THE MUSCULAR SYSTEM

At the end of this unit you should be able to:

1. State the function of the skeletal muscular systems.
2. Distinguish between isotonic and isometric contractions.
3. Define the term muscle fiber.
4. Name the three portions of the connective tissue sheath of a muscle and state the relative location of each.
5. State the function of connective tissue in muscle.

## Objectives

6. Describe the following parts of a muscle cell: sarcolemma, sarcoplasm, myofibrils, sarcomere, sarcoplasmic reticulum, and t-tubules.
7. Name the two proteins that compose the myofibrils.
8. Explain the role of the motor neuron, acetylcholine, and cholinesterase in muscular contraction.
9. State the distribution of $Na^+$, $K^+$, and electrical charge across the resting muscle cell membrane.
10. Describe the process by which an impulse or action potential moves along the muscle cell membrane.
11. Describe the response of the sarcoplasmic reticulum to an impulse.
12. Explain the role of calcium in muscular contraction.
13. Explain the role of actin and myosin in muscular contraction.
14. Name the two forms in which muscle cells store energy.
15. Explain under what circumstances muscles use the glycolytic conversion of glycogen to lactic acid as a means of forming ATP.
16. Define the term oxygen debt.
17. Describe two different ways in which a motor nerve can stimulate a more forceful muscular contraction.
18. Distinguish on the basis of the strength and duration of contraction between the following types of muscular contraction: twitch, summation, and tetanus.
19. Explain the relationship between the initial length of a muscle and the force with which it contracts.
20. Distinguish the roles of a prime mover, synergists, and antagonists.
21. State the location and one action of each of the following muscles: epicranius, obicularis oculi, levator palpebrae superiorus, obicularis oris, buccinator, zygomatic, platysma, superior rectus, inferior rectus, medial rectus, superior oblique, inferior oblique, temporal, masseter, pterygoid, sternocleidomastoid, semispinalis capitis, trapezius, latissimus dorsi, deltoid, pectoralis major, biceps brachii, brachialis, triceps brachii, diaphragm, external intercostals, internal intercostals, external obliques, internal obliques, transverse abdominis, rectus abdominis, gluteus maximus, quadriceps femoris, hamstrings, tibialis anterior, soleus, gastrocnemius.

# 7

# The Nervous System

INTRODUCTION
DIVISIONS OF THE NERVOUS SYSTEM
CELLS OF THE NERVOUS SYSTEM
MECHANISM OF NEURAL ACTIVITY
SPINAL REFLEX
STRUCTURE OF THE SPINAL CORD
SPINAL NERVES
THE BRAIN
CRANIAL NERVES
PROTECTION OF THE
   CENTRAL NERVOUS SYSTEM

# INTRODUCTION

The basic function of the nervous system is to detect changes in the internal and external environments and to direct the appropriate response to these changes. If the skin comes in contact with something sharp or hot, that part of the body must be moved so that it is not damaged. If you are walking along and there is an object in your path, your direction must be changed so that you don't bump into it and hurt yourself. If your blood pressure begins to fall, steps must be taken to raise it.

In order to carry out its function, the nervous system must perform four basic types of activities.

1. It must detect changes that occur inside and outside the body.
2. It must relay information about these changes to a central site.
3. It must fit this information together with information from other sources.
4. It must direct the appropriate response to these changes.

Different components of the nervous system are specialized for carrying out these different types of activities. Changes that occur inside and outside the body are responded to by special cells called **receptor cells.** For example, there are receptor cells that respond to light, sound, and temperature in the external environment; and there are receptor cells that respond to the amount of oxygen and the blood pressure in the internal environment. Each type of receptor cell is usually sensitive to only one kind of stimulus. Thus, the receptor cells of the eye are sensitive to light, but not to sound.

The central site at which information from various receptors is brought together is known as the **central nervous system** (CNS). After receiving information from various sources, the central nervous system puts the information together and then directs the appropriate response to the information. Thus, if one picks up a hot frying pan that contains a meal that took three hours to prepare, the central nervous system must fit together information about how much damage the heat can do to the hand and how important the meal is and then decide whether or not the frying pan should be dropped. The central nervous system is composed of the **spinal cord,** contained within the vertebral column, and the **brain,** contained within the bones of the cranium.

Information is transmitted from the receptors to the central nervous system via **afferent (sensory) pathways.** For example, an afferent pathway will transmit information about light from the receptor cells in the eye to the central nervous system. Likewise another afferent pathway will transmit information about blood pressure from the blood pressure receptors to the central nervous system.

We have stated that once the central nervous system fits together (integrates) information from the various receptors, it must direct the body's response to this information. The central nervous system is connected with the structures that produce a response by **efferent (motor) pathways.** Structures that actually produce a response are called **effectors.** For example, the skeletal muscles are effectors since they move the bones; the heart is an effector since it moves the blood; and the pancreas is an effector since it releases the hormone insulin, which can allow sugar to enter cells.

## DIVISIONS OF THE NERVOUS SYSTEM

The nervous system can be divided into two basic divisions: the central nervous system and the peripheral nervous system. We have already stated that the central nervous system consists anatomically of the brain and spinal cord. The function of these structures is to integrate information from different sources and to direct the appropriate response to this information. In addition, the central nervous system is responsible for conscious sensation and for the storage of information for future use.

The **peripheral nervous system** consists of the afferent pathways which bring information into the central nervous system and the efferent pathways which carry information out of the central nervous system. The peripheral nervous system is subdivided into the voluntary nervous system and the autonomic nervous system. The **voluntary nervous system** connects the central nervous system with the receptors that respond to external stimuli and with the skeletal muscles which move the body. The **autonomic nervous system** connects the central nervous system with

# 157 Cells of the nervous system

the receptors that respond to internal stimuli and with effectors that can change the internal environment.

## CELLS OF THE NERVOUS SYSTEM

The nervous system is composed of two basic types of cells: neurons and neuroglial cells. **Neurons** are the cells responsible for carrying out the essential functions of the nervous system. These cells are capable of responding to stimuli and of transmitting information from one part of the body to another. The neuroglial cells function to support, insulate, and possibly nourish the neurons.

A neuron, as illustrated in Figure 7-1, consists of a cell body

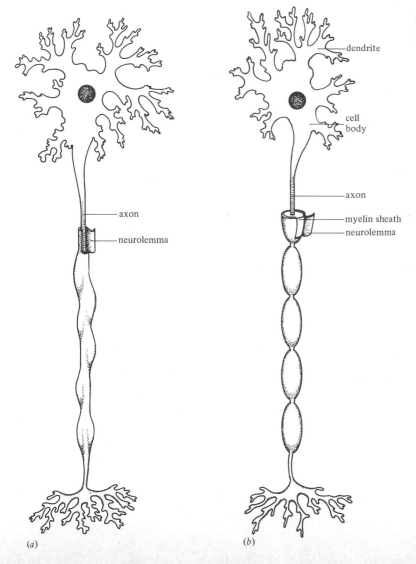

**Figure 7-1** Structure of a neuron: (*a*) unmyelinated and (*b*) myelinated.

and processes which extend outward from the cell body. The cell body contains the various organelles necessary for protein synthesis and ATP production. Thus, the cell body contains the nucleus, ribosomes, endoplasmic reticulum, and mitochondria. In neurons, the ribosomes and endoplasmic reticulum are referred to as nissl bodies. Running throughout the cell body and into the various processes are numerous fine tubes called neurofibrils. The primary function of the cell body is to produce and maintain the extending processes.

The processes, called fibers, which extend outward from the cell body are divided into two types: dendrites and axons. **Dendrites** respond to stimuli and conduct impulses toward the cell body whereas axons transmit impulses from the cell body to the opposite end of the neuron. Neurons may contain large numbers of

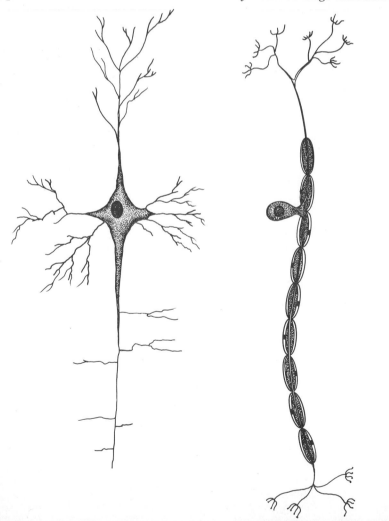
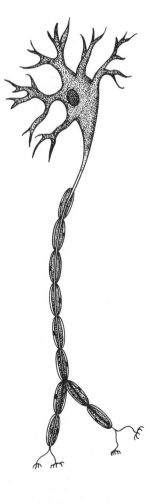

**Figure 7-2** Different-shaped neurons.

dendrites; however, each neuron contains only one axon. The cell shown in Figure 7-1 has short dendrites and a long axon. This is not true for all neurons. As can be seen in Figure 7-2, neurons can have many different shapes. In some neurons, the axon is longer than the dendrites; in others, the dendrites are longer than the axon; and in still others, the dendrites and axon are approximately the same length. The difference between a dendrite and an axon is that a dendrite conducts impulses toward the cell body and an axon conducts impulses away from the cell body.

In the peripheral nervous system, most of the longer processes are protected by a sheath formed by special cells known as **Schwann cells.** This sheath, called the **neurolemma,** functions to protect the neural processes and to provide a pathway through which damaged fibers can be replaced with new fibers growing from the cell body. Figure 7-3 shows a Schwann cell forming a sheath around the axon of a neuron.

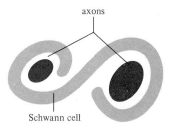

**Figure 7-3** Schwann cell forming neurolemma around two neurons.

Some neural processes contain a second sheath, called a **myelin sheath,** located below the neurolemma. The myelin sheath is formed by a Schwann cell which wraps many times around the neural process, as illustrated in Figure 7-4. As the Schwann cell wraps around the neuron, its cytoplasm is squeezed to the outer layer. The thicker outer layer of the Schwann cell forms the neurolemma of myelinated fibers.

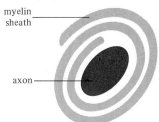

**Figure 7-4** Schwann cell forming myelin sheath.

The myelin sheath of a nerve process is not continuous over the entire length of the process. It is interrupted at intervals by gaps known as **nodes of Ranvier.** These gaps represent the spaces between successive Schwann cells.

The **neuroglial cells** of the nervous system far outnumber the neurons. These cells are found interspersed among the neurons; they serve to support the neurons and attach them to blood vessels and adjoining connective tissue. In addition to supporting the neurons, the neuroglial cells insulate the neurons from each other so that the electrical activity of one neuron does not affect the electrical activity of another neuron. Some of the neuroglial cells are capable of phagocytosis and thus serve to protect the neurons against infectious agents. Other functions, such as memory and the metabolic nourishment of neurons, have been attributed to neuroglial cells, but these functions are still subject to debate.

## MECHANISM OF NEURAL ACTIVITY

The function of neurons is to respond to stimuli and to transmit information from one part of the body to another. They carry out these functions through the formation and conduction of action potentials. An action potential, as described in Chapter 6, is a wave of depolarization and repolarization which starts at one end of the cell and passes to the opposite end. In neurons, the action potential travels from dendrite to cell body to axon.

Given that information is transmitted from one end of a neuron to the other, two questions naturally arise: What starts the action potential in the first place? What happens once the action potential reaches the other end of the cell? Let's look at these questions in reverse order. The terminal end of each neuronal axon is separated from another neuron, a muscle cell, or a gland cell by a small space known as a synapse, as illustrated in Figure 7-5.

The terminal end of the axon is slightly swollen at the synapse into a structure known as a **synaptic knob.** Located within the synaptic knob are tiny vesicles which contain **chemical transmitter.** When an action potential reaches the synaptic knob, it causes the chemical transmitter to be released into the synaptic space. At the present time, it is not known how depolarization of the synaptic knob causes the release of chemical transmitter.

Once released, the transmitter diffuses across the synapse and combines with the membrane of the postsynaptic cell, as illustrated in Figure 7-6. The effect of the transmitter on the postsynaptic cell depends on the particular transmitter and cell involved. For example, at the synapse between a motor neuron and a muscle cell, the transmitter acetylcholine depolarizes the muscle cell setting in motion the sequence of events that lead to muscular contraction.

At a synapse between two neurons, the transmitter released by the presynaptic neuron may either increase or decrease the chances of an action potential in the postsynaptic neuron. In order for an action potential to be initiated in the postsynaptic neuron it must be

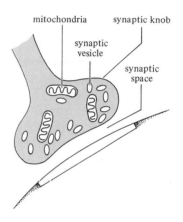

**Figure 7-5** Synapse.

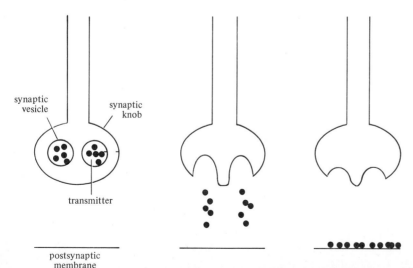

**Figure 7-6** Diffusion of chemical transmitter across a synapse.

depolarized from its resting potential to a lower threshold potential. Once any part of the postsynaptic cell membrane is depolarized to the threshold level, an action potential will be initiated and will pass over the entire membrane to the opposite end of the cell. Transmitters that depolarize the postsynaptic membrane closer to threshold, as illustrated in Figure 7-7, are called **excitatory transmitters.** The depolarization caused by an excitatory transmitter is called an **excitatory postsynaptic potential** (EPSP). On the other hand, transmitters that increase the degree of polarization across the postsynaptic cell membrane away from threshold, as illustrated in Figure 7-8, are called **inhibitory transmitters.** The increased polarization caused by an inhibitory transmitter is referred to as an **inhibitory postsynaptic potential** (IPSP). Each presynaptic neuron can release only one particular type of chemical transmitter. Whether this particular chemical transmitter acts as an excitatory transmitter or an inhibitory transmitter depends on both the nature of the transmitter and the particular postsynaptic cell. Thus, a particular chemical transmitter may act as an excitatory transmitter at one synapse and as an inhibitory transmitter at another synapse.

Whether or not an action potential is initiated in a postsynaptic neuron depends on the combined effects of excitatory and inhibitory transmitters. As illustrated in Figure 7-9, each postsynaptic neuron has synapses with many presynaptic neurons.

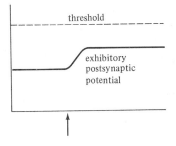

**Figure 7-7** Excitatory postsynaptic potential (EPSP).

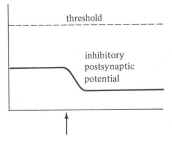

**Figure 7-8** Inhibitory postsynaptic potential (IPSP).

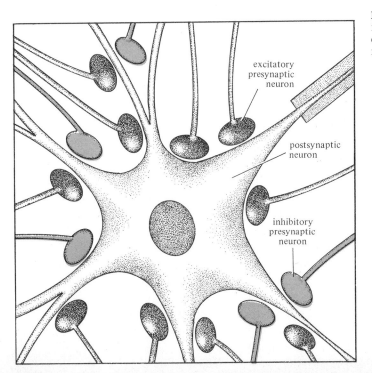

**Figure 7-9** Excitatory and inhibitory presynaptic neurons converging on a postsynaptic neuron.

Some of these presynaptic neurons are excitatory whereas others are inhibitory. Depending on the situation the excitatory neurons may be active, the inhibitory neurons may be active, both may be active, or neither may be active. When the amount of excitatory transmitter sufficiently exceeds the amount of inhibitory transmitter, the potential of the postsynaptic membrane will be lowered to threshold and an action potential will be initiated.

This answers in part the first of our original two questions, that is, what initiates a neuronal action potential in the first place. An action potential can be initiated in a neuron by excitatory chemical transmitters released by other neurons. However, in some neurons, action potentials are initiated by a different mechanism. In neurons that serve a receptor function, action potentials are initiated by the particular type of stimulus to which the neurons are sensitive. For example, certain neurons that have endings in the skin are sensitive to pressure. When the skin is touched, an action potential is initiated in these neurons. To sum up, we could say that functionally the nervous system is composed of neurons and synapses. Information is carried from one end of a neuron to the other by action potentials. Information is carried from one neuron to the next by chemical transmitters which diffuse across the synapse between the neurons. The basic function of the neuron is to transmit information from one part of the body to another, whereas the basic function of the synapse is to bring together and integrate information from a number of different sources.

## SPINAL REFLEX

The way in which the nervous system functions can be seen in its simplest form in a basic spinal reflex. Examples of spinal reflexes would be the knee jerk elicited when a physician taps the knee during a physical examination, or pulling the hand away from a hot object. Figure 7-10 illustrates the neurons involved in pulling the hand away from a hot object. This reflex involves three types of neurons: **afferent neurons,** which connect the skin of the hand with the spinal cord, **interneurons,** located entirely within the spinal cord, and **efferent neurons,** which connect the spinal cord with the biceps, the muscle that can pull the hand away from the object.

The heat initiates action potentials that travel along the afferent neurons to the spinal cord. At the spinal cord, excitatory transmitters released by the afferent neurons initiate action potentials in the interneurons. The interneurons in turn release excitatory transmitters which initiate action potentials that travel along the efferent neurons to the biceps. When action potentials reach the end of the efferent neurons, acetylcholine is released. The ace-

**163** Spinal reflex

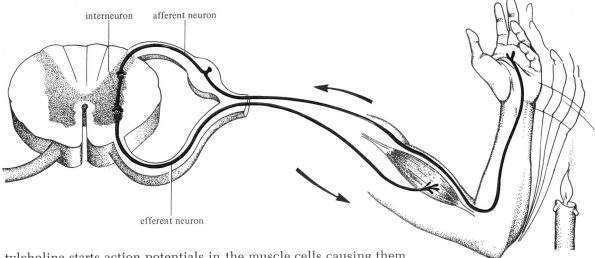

**Figure 7-10** Neurons involved in pulling hand away from hot object.

tylcholine starts action potentials in the muscle cells causing them to contract. Contraction of the biceps pulls the hand away from the hot object.

The entire response is considered a reflex since stimulation of the afferent neuron automatically leads to contraction of the muscle. No voluntary effort is required; that is, the reflex is independent of the brain. If we touch a hot object, we of course feel pain; but careful measurement has shown that the pain occurs after the hand has been moved. The reflex depends only on the afferent neurons, interneurons, and efferent neurons.

The knee-jerk reflex elicited during a physical examination involves only two types of neurons, as illustrated in Figure 7-11. This reflex involves afferent neurons which connect stretch receptors in the muscles of the thigh with the spinal cord and efferent neurons which connect the spinal cord with the quadriceps muscle. There

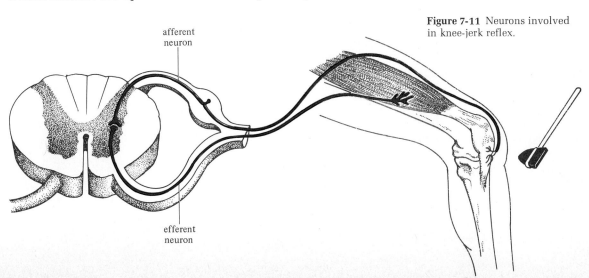

**Figure 7-11** Neurons involved in knee-jerk reflex.

are no interneurons involved. When the knee is tapped, the tendons which pass over the knee pull on the muscles of the thigh. Action potentials are initiated in the stretch receptors of the thigh muscles and pass over the afferent neurons to the spinal cord. The afferent neurons release excitatory transmitters which initiate action potentials in the efferent neurons. Acetylcholine released by the efferent neurons causes the quadriceps to contract. This type of reflex plays an important role in adjusting the strength of muscle contraction to suit the task being performed.

## STRUCTURE OF THE SPINAL CORD

As illustrated in Figure 7-12, the spinal cord is a long narrow structure which extends from the brain to the bottom edge of the 1st lumbar vertebra. The cord is enclosed within the vertebral foramen of the vertebrae and is thus protected by bone. A pair of spinal nerves emerges from the cord through the intervertebral foramen between each two successive vertebrae. These nerves contain the afferent and efferent neurons which connect the spinal cord with the skin, skeletal muscles, and internal organs.

Seen in cross section, as illustrated in Figure 7-13, the spinal cord is an oval structure about the diameter of the little finger. Dorsal and ventral grooves, the **dorsal median fissure,** and the **ventral median fissure** partially divide the cord into right and left halves. Spinal nerves are connected to the cord by the dorsal and ventral roots. The **dorsal roots** contain the afferent neurons which carry information into the cord, whereas the **ventral roots** contain the efferent neurons which carry information out of the cord. A swelling of the dorsal roots, the **dorsal root ganglion,** contains the cell bodies of the afferent neurons.

The spinal cord itself is divided into two regions: an inner H-shaped region of **gray matter** and an outer surrounding region of **white matter.** Composed primarily of synapses, cell bodies, and short interneurons, the core of gray matter is subdivided into **dorsal, lateral, and ventral horns.**

The white matter is composed of myelinated neuronal fibers which transmit information up and down the cord. These fibers are organized into **tracts,** which are groups of fibers carrying a specific type of information. For example, one tract consists of fibers carrying temperature information to the brain and another consists of fibers carrying touch information to the brain. Specific tracts and the types of information they transmit will be discussed in Chapter 8. The various tracts of white matter are combined into three major columns (or funiculi) of white matter: the **anterior, lateral,** and **posterior columns.**

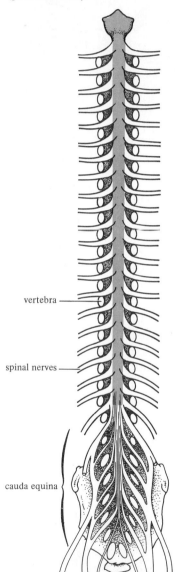

**Figure 7-12** Spinal cord (posterior view).

vertebra

spinal nerves

cauda equina

**165** Structure of the spinal cord

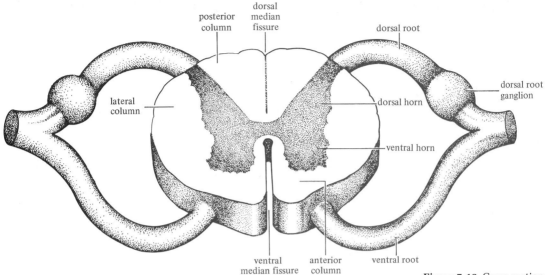

**Figure 7-13** Cross section of the spinal cord.

The functions of various portions of the spinal cord are illustrated in Figure 7-14. In this figure, we have shown some of the neurons that are involved in withdrawing the hand from a hot object and transmitting information about the heat to the brain. The

**Figure 7-14** Location of different types of neurons within the spinal cord.

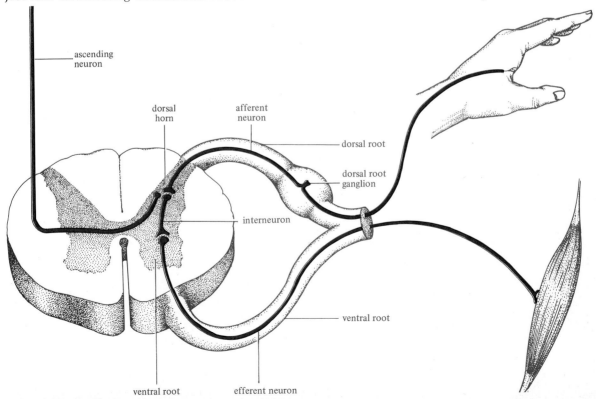

afferent neurons which are stimulated by the heat enter the cord through the dorsal root and synapse with interneurons in the dorsal horn. The cell bodies of the afferent neurons are located in the dorsal root ganglion. The interneurons, located entirely within the gray matter of the cord, extend from the end of the afferent neuron to the cell body of the efferent neuron in the ventral horn. The efferent neuron leaves the cord through the ventral root and extends to the biceps which can move the hand away from the hot object. However, in addition to synapsing with the short interneuron, the afferent neuron also synapses with an ascending neuron that crosses the cord in the gray matter and then passes into the white matter to travel up the cord to the brain. This neuron informs the brain about the heat.

## SPINAL NERVES

Just as individual muscle cells are combined together into muscles, the individual neurons of the peripheral nervous system are combined together into nerves. As can be seen in Figure 7-15, a nerve consists of a bundle of neurons wrapped in a connective tissue sheath. The outer layer of connective tissue, the **epineureum,** extends inward to form the **perineureum** which surrounds groups of nerve fibers. Extending from the perineureum into spaces between the individual nerve cells is a very thin layer of connective tissue, the **endoneureum.** Blood vessels and fat cells are also found within the nerves.

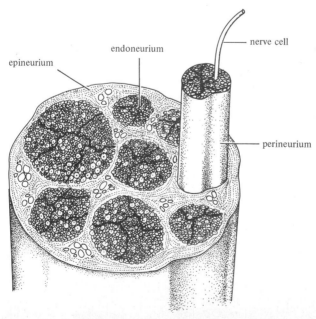

**Figure 7-15** Structure of a nerve.

# Spinal nerves

The peripheral nerves which carry information to and from the central nervous system can be divided into two major groups — the **cranial nerves,** which connect with the brain, and the **spinal nerves,** which connect with the spinal cord.

There are 31 pairs of spinal nerves, each named according to the region from which it leaves the spinal cord, as illustrated in Figure 7-16. However, since the spinal cord only extends to the 1st lumbar vertebra, the lumbar, sacral, and coccygeal spinal nerves must pass downward through the spinal canal before exiting at their proper level. The lower end of the spinal cord and its attached, descending spinal nerves looks something like a horse's tail. For this reason, it is referred to as the **cauda equina** (which is "horse's tail" in Latin).

As can be seen in Figure 7-17, each spinal nerve is attached to the spinal cord by the dorsal and ventral roots. These roots join to form the spinal nerve which emerges from within the vertebral column through the space between adjacent vertebrae, known as the intervertebral foramen. After emerging from the vertebral column, the spinal nerve divides two major branches, the **posterior ramus** and the **anterior ramus.** The posterior ramus contains sensory neurons coming from the skin of the back and motor neurons which supply the skin and muscles of the arms, legs, and anterior and lateral surfaces of the trunk. Thoracic and lumbar spinal nerves give rise to a third branch, the **white ramus,** which contains neurons that are part of the autonomic nervous system.

The anterior rami of most spinal nerves do not go directly to the skin and muscles. With the exception of 11 of the thoracic spinal nerves, the anterior rami of the spinal nerves enter into the formation of complicated nerve networks known as **plexuses.**

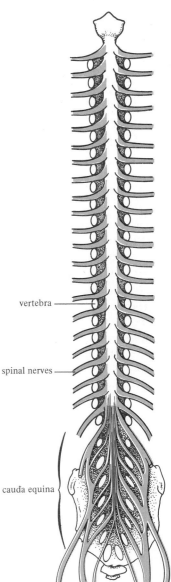

Figure 7-16 Spinal nerves leaving cord.

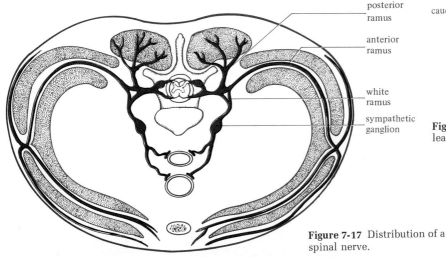

Figure 7-17 Distribution of a spinal nerve.

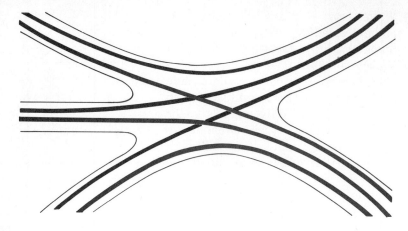

**Figure 7-18** Schematic illustration of a plexus.

Basically a plexus, as illustrated in Figure 7-18, is a site at which neurons are redistributed within nerves. A number of spinal nerves enter the plexus and a number of nerves going to more distal portions of the body emerge from the plexus. Neurons which are grouped together in a particular spinal nerve are not necessarily grouped together in one of the nerves emerging from the plexus. The nerves which emerge from the plexus divide into smaller and smaller branches which supply the skin and muscles of the various parts of the body. These smaller nerves are often named in accordance with the area of the body they supply. For example, the radial nerve runs along the radius.

Figure 7-19 illustrates the four plexuses of the peripheral nervous system, the spinal nerves which enter into their formation, and the distal nerves which emerge from them. The **cervical plexus** is formed by the first four cervical nerves. Nerves which emerge from this plexus supply the lower part of the head, the neck, and the shoulders. An exceedingly important nerve, the **phrenic nerve,** also emerges from this region. Formed from nerve fibers of the 3rd, 4th, and 5th cervical nerves, the phrenic nerve extends downward and is the motor nerve for the diaphragm. Any injury to the spinal cord above the level at which the fibers which form the phrenic nerve emerge from the cord will lead to death. This is because there will no longer be any way of stimulating the diaphragm to contract.

The **brachial plexus** is formed by the 5th, 6th, 7th, and 8th cervical nerves and the 1st thoracic nerve. This plexus gives rise to the nerves which supply the arms, hands, and parts of the chest and upper back.

The lumbosacral plexus is formed by lumbar and sacral spinal nerves. Emerging from this plexus are two prominent nerves which supply the skin and muscles of the leg. The **femoral nerve** emerges from the lumbar portion of the plexus and supplies the anterior

**169** Spinal nerves

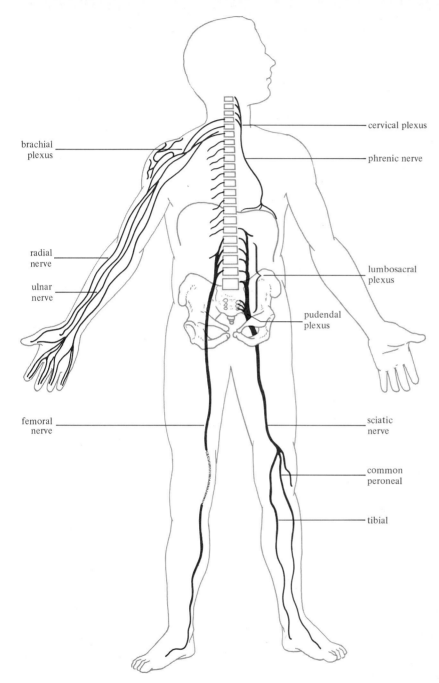

**Figure 7-19** Major plexuses and nerves.

muscles of the thigh such as the quadriceps femoris. The **sciatic nerve,** the largest nerve in the body, emerges from the sacral region of the plexus and passes along the back of the thigh, where it supplies the hamstring muscles. Just above the thigh the sciatic nerve divides into a posterior tibial nerve and an anterior common **peroneal nerve.** These nerves supply the lower leg and foot.

The **pudendal plexus** is formed from the 2nd, 3rd, and 4th sacral spinal nerves. The largest nerve emerging from this plexus is the **pudendal nerve** which supplies the skin, muscles, and other structures within the perineum, the lower portion of the pelvic cavity.

## THE BRAIN

The two parts of the central nervous system—the brain and spinal cord—are continuous with each other. That is, many neuronal processes ascend from the spinal cord into the brain or descend from the brain into the spinal cord. The brain is simply the part of the central nervous system that is located within the bones of the skull.

As can be seen in Figure 7-20, the brain is composed of a central core of tissue, the **brainstem,** and two outgrowths, the **cerebellum** and the **cerebrum.** When the intact brain is viewed from the side, much of the central brainstem is obscured by the large overhanging cerebrum. However, in a midsagittal section the entire brainstem can be seen clearly.

The portion of the brainstem closest to the spinal cord is called the **medulla oblongata.** Located within the medulla are areas of gray matter, called **nuclei,** which act as control centers for the vital functions of circulation and respiration. As will be discussed in Chapter 13, the cardiovascular center receives information concerning the blood pressure through afferent fibers and in turn adjusts the blood pressure to its proper level via efferent fibers to the heart and blood vessels. Thus, the cardiovascular center acts to make sure that the blood pressure is always appropriate to the situation. Likewise, the respiratory center, as will be discussed in Chapter 14, receives information concerning the level of oxygen, carbon dioxide, and $H^+$ in the body fluids and in turn adjusts respiration in such a way that these substances can be maintained at their normal concentration. Because the control of respiration and circulation is essential to the maintenance of life, injury to the medulla, such as can be caused by a blow to the base of the skull, will usually prove fatal. In addition to the centers which control circulation and respiration, the medulla also contains centers which control swallowing, vomiting, coughing, and sneezing.

It should be recalled that gray matter is composed of synapses and cell bodies and that synapses are the sites of integration within the nervous system. Thus the various control centers of the medulla are composed of gray matter. The medulla, however, also contains

# The brain

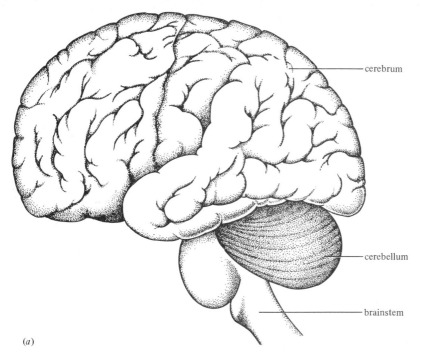

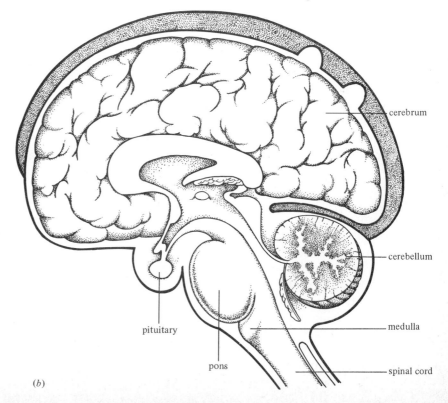

**Figure 7-20** The brain: (a) intact sagittal and (b) midsagittal sections.

white matter; that is, tracts of myelinated neuronal processes carrying information up and down the cord. Many of these tracts cross, or **decussate,** at the level of the medulla. As we shall see in the next chapter, all of the sensory information from the left side of the body ultimately goes to the right hemisphere of the brain and all of the sensory information from the right side of the body ultimately goes to the left hemisphere of the brain. Likewise the muscles of the right side of the body are controlled primarily by the left side of the brain and the muscles of the left side of the body are controlled primarily by the right side of the brain. The medulla is one of the major sites where both sensory and motor neurons cross from one side of the nervous system to the other.

Located just above the medulla in the brainstem is the **pons.** Tracts of white matter pass through the pons linking the medulla to the cerebrum and cerebellum. The gray matter of the pons contains the nuclei of the 5th, 6th, 7th, and 8th cranial nerves. All of the cranial nerves will be discussed in a subsequent section of this chapter. The pneumotaxic center in the pons aids in the regulation of respiration.

The **midbrain** is the continuation of the brainstem above the pons. On its ventral surface it contains two tracts of white matter, the cerebral peduncles. These tracts carry information to and from the higher brain centers. Located on the dorsal surface of the midbrain are four areas of gray matter: the two superior and the two inferior colliculi. The superior colliculi form part of the visual pathway and the inferior colliculi form part of the auditory pathway.

Running through the brainstem, from the medulla to the midbrain, is a central core of gray and white matter known as the **reticular formation.** The reticular formation includes the respiratory and circulatory centers in the medulla as well as a number of areas of gray matter which act to help regulate muscular activity. A portion of the reticular formation, known as the **reticular activating system,** acts to control sleep, wakefulness, and alertness. It does this through its neuronal connections with the cerebral cortex, the portion of the brain responsible for consciousness.

Superior to the midbrain are two important areas of gray matter: the **hypothalamus** and **thalamus.** As can be seen in Figure 7-21, the hypothalamus forms the floor and part of the lateral walls of the 3rd ventricle, one of the cavities of the brain which will be discussed in a later section of this chapter. The right and left thalami form the bulk of the lateral walls of the 3rd ventricle. The hypothalamus, the two thalami, and the 3rd ventricle form the **diencephalon.**

The hypothalamus plays a role in a wide variety of activities related to the maintenance of a stable internal environment. It has control centers that regulate hunger, water balance, and body temperature. The hunger center in the hypothalamus receives information about the body's need for food and gives rise to a feeling of hunger when food is needed. Information concerning the concentration and volume of the body fluids is sent to the water-balance center. This center in turn can control the concentration and volume of urine as well as give rise to a sensation of thirst. The temperature center in the hypothalamus receives information about the body temperature and controls responses that can raise body temperature if it begins to fall and lower body temperature if it begins to rise.

Through its control over the autonomic nervous system and the secretory activity of the pituitary gland, the hypothalamus plays a role in the regulation of most of the internal organs. For example, the hypothalamus controls the secretion of pituitary hormones which in turn control the secretion of reproductive hormones. Thus, the hypothalamus plays an important role in regulating the menstrual cycle.

Because the hypothalamus receives input from the cerebral

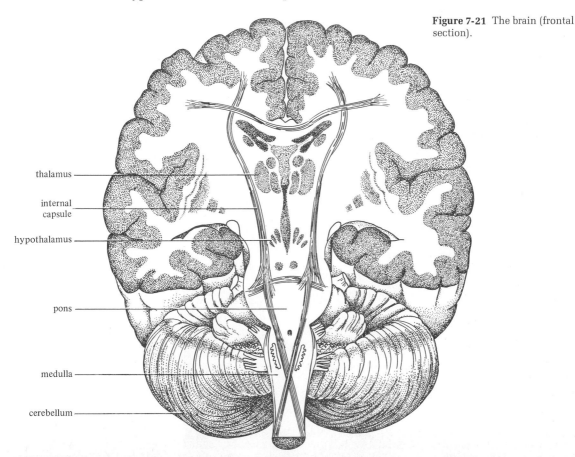

**Figure 7-21** The brain (frontal section).

**174** The Nervous System

cortex, it serves as a link between "psychological" activity and physical activity. For example, anxiety can cause increased activity in certain areas of the hypothalamus which influence acid secretion by the stomach. In this situation, the hypothalamus may act as the link between anxiety and an ulcer.

Finally, the hypothalamus is thought to play an important role in basic emotions such as anger, fear, sexual drive, and pleasure. The ways in which the hypothalamus and higher levels of the cerebrum interact to produce emotions is very poorly understood at the present time.

The major function of the thalamus is to act as a relay center which sorts out sensory information before it is sent on to the cerebral cortex. Present research also seems to indicate that the thalamus is responsible to some degree for the awareness of sensation, particularly as to whether something is pleasant or unpleasant.

**The Cerebellum**

As indicated in Figure 7-22, the cerebellum is situated dorsal to the brainstem at the level of the pons. It is somewhat oval with two large lateral masses, the **cerebellar hemispheres,** and a smaller central portion, the **vermis.** The outer shell of the cerebellum, the **cerebellar cortex,** is composed of gray matter. Information is brought to and from this gray matter by an underlying series of branching tracts of white matter, the **arbor vitae** (tree of life). Three large, paired tracts of white matter, the inferior, middle, and supe-

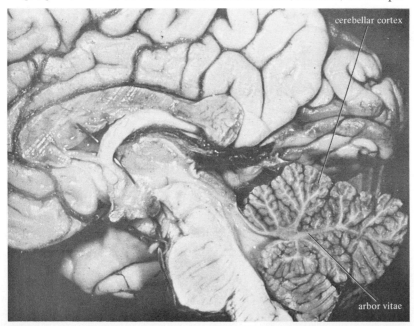

**Figure 7-22** Cerebellum. (Photo by Lester Bergman & Assoc.)

rior peduncles, link the cerebellum with the brainstem. Through these tracts the cerebellum receives information from and sends information to the brain and spinal cord.

The function of the cerebellum is to help coordinate muscular activity. To carry out this function, it receives information from the muscles, tendons, joints, and vestibular (balance) organ of the inner ear concerning the position and degree of contraction of the various parts of the body. At the same time, it receives information from the cerebral cortex concerning the type of movement desired. The cerebellum acts on this information to adjust the commands of the cerebral cortex in such a way that voluntary movements are carried out smoothly and balance is maintained. Injury to the cerebellum leads to a loss of coordination and equilibrium as well as to jerky movements. For example, a person with cerebellar injury would walk in a reeling fashion as if intoxicated.

**The Cerebrum**

The cerebrum, which covers the top, back, and sides of the brainstem, is by far the largest part of the brain. The cerebrum is divided by a longitudinal groove, the **longitudinal fissure,** into right and left **cerebral hemispheres.** The two cerebral hemispheres are joined by tracts of white matter, the **corpus callosum.**

As can be seen in Figure 7-23, each cerebral hemisphere is it-

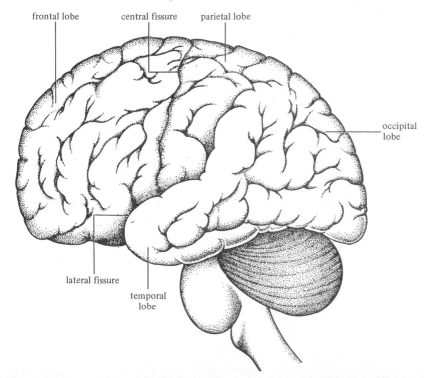

**Figure 7-23** Lobes of the cerebrum.

self divided into four lobes. The lobes—**frontal, temporal, occipital,** and **parietal**—are named for the bone that lies over them. Thus the frontal lobe is at the front, the temporal lobe at the side, the parietal lobe at the top, and the occipital lobe at the back. The lobes are partially divided from each other by grooves, or fissures, which extend downward from the surface of the brain. A central groove, the **central fissure (of Rolando),** separates the frontal and parietal lobes. The **lateral fissure (of Sylvius)** divides the frontal and parietal lobes from the temporal lobe below. A third groove, the **parieto-occipital fissure,** separates the temporal and parietal lobes in front from the occipital lobe behind.

When the brain is looked at through a frontal section, Figure 7-24, it can be seen that each cerebral hemisphere is divided into three layers. The outer layer, the **cerebral cortex,** is composed of gray matter. Below the cerebral cortex is an area of white matter which consists of neuronal tracts that carry information to and from the cerebral cortex. Deep within each cerebral hemisphere are areas of gray matter, known as the **basal ganglia.**

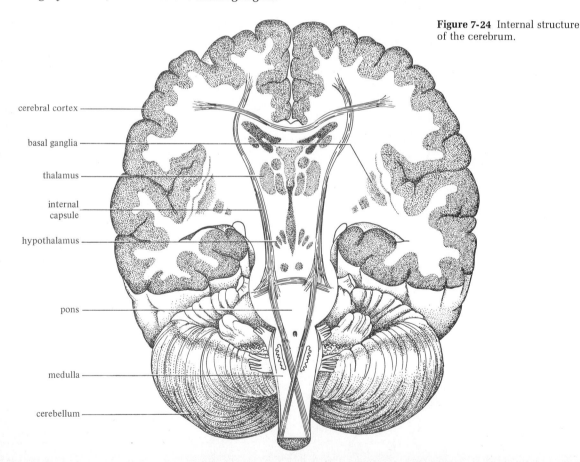

**Figure 7-24** Internal structure of the cerebrum.

The major basal ganglia are the substantia nigra, the caudate nucleus, and the lentiform nucleus, which is subdivided into the putamen and pallidum. Basal ganglia play an important role in the control of voluntary movement and damage to them can result in an uncontrolled, jerky motion. Huntington's chorea (St. Vitus dance) and Parkinson's disease are both examples of diseases in which the basal ganglia are damaged.

The middle layer of white matter in the cerebrum consists of myelinated nerve tracts carrying information to and from the cerebral cortex as well as from one part of the cortex to another. One particularly large tract, the **internal capsule,** passes between the caudate and lentiform nuclei of the basal ganglia. This tract contains both sensory and motor fibers linking the cerebral cortex with lower portions of the central nervous system.

The outer layer of the cerebrum, the **cerebral cortex,** is responsible for conscious sensation and for the control of voluntary movement. In order to increase the surface area of the cerebral cortex, it is folded into a series of ridges and grooves. Each ridge is referred to as a **gyrus** and each groove is referred to as a **sulcus.** This folding pattern allows a larger amount of cerebral cortex than could be afforded if the surface of the brain were smooth.

Conscious sensation comes about as a result of the activity of neurons in the cerebral cortex. Various portions of the cortex are responsible for different types of sensation. As illustrated in Figure 7-25, the parietal lobe is responsible for sensation arising from the skin. As we shall see in the next chapter afferent pathways from receptors in the skin ultimately bring sensory information to the parietal cortex. Information from different areas of the skin is sent to different parts of the parietal cortex. The temporal cortex is responsible for auditory (sound) and olfactory (smell) sensations,

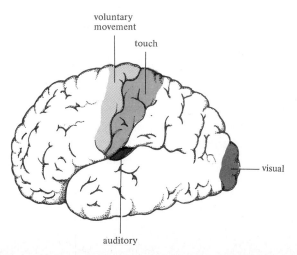

**Figure 7-25** Primary sensory and motor areas of the cortex.

whereas the occipital cortex is responsible for visual sensations.

Voluntary movement is initiated and controlled by neurons in the portion of the frontal cortex located just in front of the central fissure. Some neurons descend from this area directly to the efferent neurons which leave the spinal cord and stimulate the skeletal muscles. Other neurons pass to the basal ganglia and cerebellum and help to coordinate muscular movements into a smooth pattern. Motor pathways will be discussed in more detail in the next chapter. A specialized area of the motor cortex, **Broca's area,** is responsible for the ability to form words in both writing and speaking.

The areas of the cortex not directly involved in primary sensations are known as **association areas.** These areas link together information from the various senses so that the brain can make decisions on the basis of all information. They are also responsible for memory, that is, the storage of information for future use. The exact way in which information is stored is not understood at the present time.

It seems that the cerebral cortex, particularly the frontal lobe, interacts with the lower areas of the brain to give rise to the various emotions. For example, injury to the frontal cortex can cause a person to become very disassociated. The person is conscious of various physical sensations and thoughts, but is unable to attach emotional significance to them. These facts were used as the basis of performing frontal lobotomies, a severing of frontal neurons, on mental patients who suffered severe anxiety. Although no longer performed on mental patients, the operation is still sometimes performed on cancer patients suffering from intractable pain. After the operation the patients seem to still feel the pain, but it no longer bothers them!

## CRANIAL NERVES

As mentioned earlier in this chapter, the peripheral nervous system consists of nerves that bring information to and from the central nervous system. The spinal nerves are the portion of the peripheral nervous system that bring information to and from the spinal cord. The **cranial nerves** are a portion of the peripheral nervous system that bring information directly to or from the brain. There are 12 cranial nerves, all of which arise from the undersurface of the brainstem. They are numbered in the order in which they are attached to the brainstem from top to bottom. Individual cranial nerves will be described in the sections of the book in which their function is relevant. The names and functions of all the cranial nerves are summarized in Table 7-1.

**Table 7-1** Cranial Nerves

| Number | Name | Origin in Brain | Function |
|---|---|---|---|
| I | Olfactory | Olfactory bulb | Sensory for smell |
| II | Optic | Thalamus and midbrain | Sensory for vision |
| III | Oculomotor | Midbrain | Motor for movement of eyelid and eyeball |
| IV | Trochlear | Midbrain | Motor for movement of eyeball |
| V | Trigeminal | Pons | Sensory for touch and temperature from skin of face, teeth, and mucous membranes of the nose and mouth<br>Motor for chewing |
| VI | Abducens | Medulla | Motor for movement of eyeball |
| VII | Facial | Medulla | Sensory for taste<br>Motor for facial expression and salivation |
| VIII | Acoustic | Medulla | Sensory for hearing and balance |
| IX | Glossopharyngeal | Medulla | Sensory for taste<br>Motor for swallowing and salivation |
| X | Vagus | Medulla | Sensory from heart, blood vessels, digestive tract, and respiratory tract<br>Motor to heart, digestive tract, and respiratory tract |
| XI | Accessory | Medulla and spinal cord | Motor to sternocleidomastoid and trapezius |
| XII | Hypoglossal | Medulla | Sensory from muscles of tongue<br>Motor for movement of tongue |

## PROTECTION OF THE CENTRAL NERVOUS SYSTEM

In addition to being enclosed in bone, the central nervous system is protected by a sheath of connective tissue known as the **meninges.** As can be seen in Figure 7-26, the meninges are divided into three layers: an outer tough layer known as the **dura mater;** a middle, more delicate layer, the **arachnoid layer,** which is shaped somewhat like a cobweb; and an inner layer, the **pia mater,** which

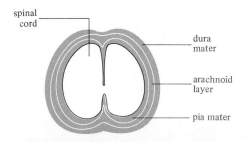

**Figure 7-26** Meninges.

adheres closely to all the folds and grooves of the cerebral cortex. The pia mater contains many of the blood vessels that supply the brain and spinal cord.

Further protection for the central nervous system is provided by the **cerebrospinal fluid.** This fluid provides a cushion which both surrounds and runs throught the middle of the central nervous system. The exterior cushion is provided by the cerebrospinal fluid which fills the subarachnoid space of the meninges. This is the space between the arachnoid layer and pia mater. The interior cushion is formed by the cerebrospinal fluid which fills the ventricles of the brain and the central canal of the spinal cord.

As shown in Figure 7-27, the four ventricles are cavities located deep within the brain. The two **lateral ventricles** extend sideward and backward, somewhat like wings, within the two cerebral hemispheres. They are joined by a narrow opening, the interventricular foramen, with the **3rd ventricle** which is a rather thin cavity that passes between the right and left thalami and above the hypothalamus. The **4th ventricle** is a diamond-shaped structure located anterior to the cerebellum and posterior to the pons and medulla. It is connected to the 3rd ventricle by the **cerebral aqueduct** and opens into the central canal through the **foramen of**

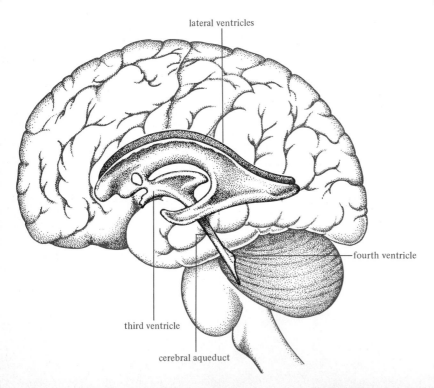

**Figure 7-27** Ventricles of the brain.

**181** Protection of the central nervous system

**Magendie** and into the subarachnoid space by the two **foramina of Luschka.**

Figure 7-28 illustrates the formation, circulation, and drainage of cerebrospinal fluid. The fluid is formed from the blood at special networks of capillaries, the **choroid plexuses,** located in the walls of the lateral and 3rd ventricles. Its composition is quite different from that of blood or ordinary tissue fluid. Fluid formed in the lateral ventricles flows through the interventricular foramen into the 3rd ventricle. From the 3rd ventricle, fluid flows through the cerebral aqueduct to the 4th ventricle. A relatively small amount of fluid flows from the 4th ventricle through the foramen of Magendie into the central canal. The bulk of the fluid flows through the foramina of Luschka into the subarachnoid space surrounding both the brain and spinal cord.

Eventually, the cerebrospinal fluid is drained out of the subarachnoid space of the brain and into the blood. Any obstruction to the flow of cerebrospinal fluid will lead to a buildup behind the obstruction. For example, a tumor may close off the cerebral aqueduct leading to the accumulation of fluid in the lateral and 3rd ventricles. The accumulated fluid presses upon and damages the

**Figure 7-28** Flow of cerebrospinal fluid.

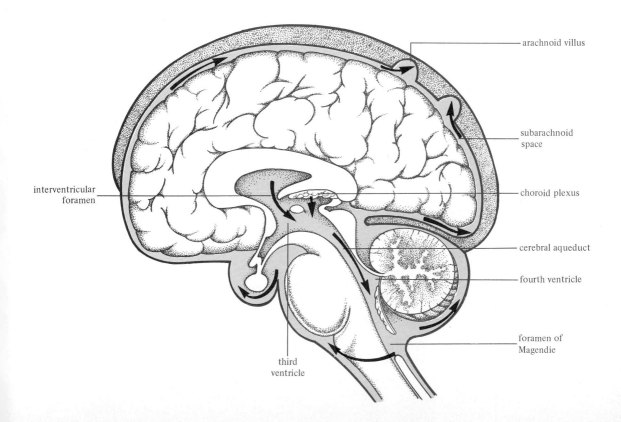

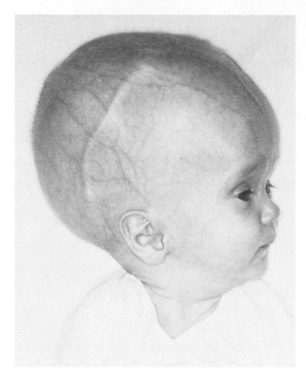

**Figure 7-29** Hydrocephalus. (Photo © Robert Ford, RBP, 1975, Camera M.D. Studios, Inc.)

surrounding brain tissue. As illustrated in Figure 7-29, the accumulation of fluid in the brain of an infant can lead to an enlargement of the entire skull, a condition known as hydrocephalus.

Examination of the cerebrospinal fluid is carried out on samples of fluid withdrawn through a needle inserted into the lumbar region of the vertebral cavity, a process known as **lumbar puncture.** The spinal cord ends at the level of the first lumbar vertebra, whereas the arachnoid and dura mater layers of the meninges extend to the sacral region of the cord. For this reason, the area within the lumbar region of the vertebral cavity is subarachnoid space filled with cerebrospinal fluid. Because this region does not contain the spinal cord but does contain cerebrospinal fluid, fluid can be withdrawn without the chance of injuring the cord.

## OBJECTIVES FOR THE STUDY OF THE NERVOUS SYSTEM

At the end of this unit you should be able to:

1. State four functions of the nervous system.
2. Describe the function of receptor cells, afferent pathways, the central nervous system, and efferent pathways.
3. State the function of the peripheral nervous system.
4. Name the two divisions of the peripheral nervous system.

## Objectives

5. Draw and label a typical neuron including the following parts: nucleus, cell body, nissl bodies, neurofibrils, dendrites, and axons.
6. Describe how a Schwann cell forms a myelin sheath and neurolemma.
7. Name the space between successive Schwann cells.
8. Name at least two functions of neuroglial cells.
9. Explain how information is transmitted along neurons.
10. Describe the structure of a synapse in terms of the synaptic knob, synaptic vesicles, and synaptic cleft.
11. Explain the function of chemical transmitters.
12. Describe the mechanism of a spinal reflex.
13. Identify from a drawing the following parts of the spinal cord: dorsal root, ventral root, dorsal horn, ventral horn, gray matter, and white matter.
14. State the function of the neurons found in the following parts of the spinal cord: dorsal root, ventral root, gray matter, and white matter.
15. Name the three portions of the connective tissue sheath of a nerve and state the relative location of each.
16. Describe the cauda equinae.
17. Identify from a drawing the following portions of a spinal nerve: posterior ramus, anterior ramus, and white ramus.
18. Explain the function of a plexus.
19. State the location of the following plexuses: cervical, brachial, lumbar, sacral, and pudendal.
20. State from which plexuses the following nerves originate: phrenic, radial, ulnar, femoral, sciatic, pudendal.
21. Name the three basic divisions of the brain.
22. State the function and anatomical location of the following parts of the brainstem: medulla, pons, reticular formation, hypothalamus, and thalamus.
23. Describe the location of the following parts of the cerebellum: cerebellar hemispheres, vermis, cerebellar cortex, arbor vitae, peduncles.
24. State the function of the cerebellum.
25. State the location of the frontal, parietal, temporal, and occipital lobes of the cerebral hemispheres.
26. State the location of the following fissures of the cerebrum: longitudinal, central, and lateral.
27. State the location of the cerebral cortex, internal capsule, and basal ganglia.
28. Name the areas of the cerebrum responsible for: tactile sensation, auditory sensation, olfactory sensation, visual sensation, voluntary movement, and speech.
29. Name the three layers of the meninges.
30. State the location of the four cerebral ventricles.
31. Describe the formation, circulation, and drainage of cerebrospinal fluid.

# 8
# Senses and Motor Pathways

TOUCH AND PRESSURE
TEMPERATURE AND PAIN
PROPRIOCEPTION
TASTE
SMELL
THE EYE AND VISION
THE EAR AND HEARING
BALANCE
MOTOR PATHWAYS TO SKELETAL MUSCLE
AUTONOMIC NERVOUS SYSTEM

We stated in the previous chapter that the primary function of the nervous system is to detect changes in the internal and external environments and to direct the appropriate response to these changes. In this chapter we are going to look at some of the sensory receptors that respond to stimuli in the external environment. We are also going to look at the anatomical pathways by which sensory information about the external environment is sent to the cerebral cortex and by which the cerebral cortex sends motor information to the skeletal muscles. The receptors that respond to changes in the internal environment will be discussed in the later chapters as they become relevant. For example, the receptors that respond to blood pressure will be discussed in Chapter 13 in relation to the regulation of blood pressure. However, in this chapter we will take a general look at the autonomic nervous system, which helps regulate the internal environment.

## TOUCH AND PRESSURE

The three basic types of receptors that respond to touch or pressure are illustrated in Figure 8-1. Nerve endings that surround hair follicles are activated when the hair is bent relative to the skin. **Meissner's corpuscles** and **Pacinian corpuscles** are nerve endings wrapped in a sheath of connective tissue. Meissner's corpuscles are

**Figure 8-1** Touch receptors: (a) nerve endings around hair follicles, (b) Meissner's corpuscles, and (c) Pacinian corpuscles.

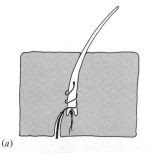
(a)

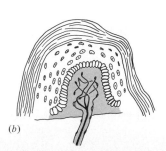
(b)

(c)

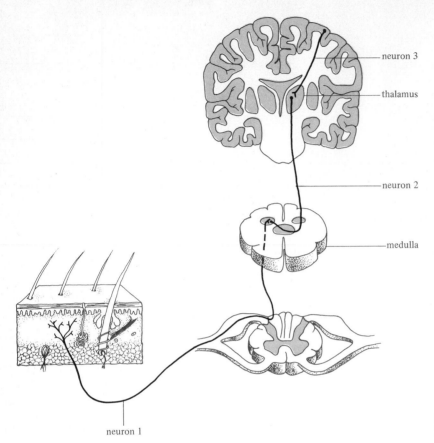

**Figure 8-2** Uncrossed pathway for touch information.

located at the boundary of the epidermis and dermis and are most numerous in hairless portions of the skin such as the palms and lips. Pacinian corpuscles are located deep in the dermis or subcutaneous tissue and thus are only activated by heavy pressure against the skin.

The nerve endings of all three types of touch receptors are parts of afferent neurons which extend all the way to the spinal cord. Thus once an action potential is generated in the nerve endings of one of the receptors it will travel along the afferent neuron to the spinal cord.

There are two major pathways by which touch information moves up (ascends) the spinal cord to the brain. The uncrossed pathway, illustrated in Figure 8-2, consists of three neurons: an afferent neuron that branches and ascends in the posterior column to the medulla, a second neuron that crosses at the medula and ascends through the brainstem to the thalamus, and a third neuron that connects the thalamus with the parietal cortex. In this pathway

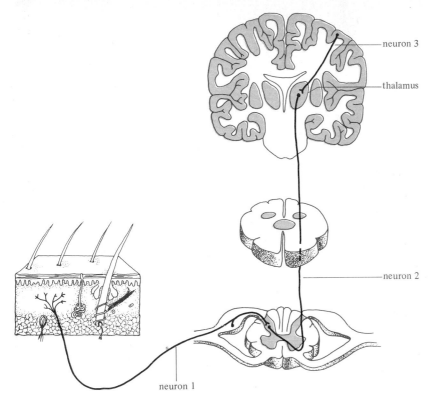

**Figure 8-3** Crossed pathway for touch information.

information ascends on the same side of the cord as the part of the body touched. At the medulla the information crosses to ascend to the cortex on the other side.

The crossed pathway, illustrated in Figure 8-3, consists of the afferent neuron, a second neuron which crosses the cord and ascends in the **ventral spinothalamic tract** to the thalamus, and a third neuron which ascends from the thalamus to the parietal cortex. This pathway differs from the uncrossed pathway in that information ascends on the side of the cord opposite to the side of the body touched. In both pathways information goes to the side of the brain opposite the side of the body touched. However, in one pathway, the uncrossed pathway, information ascends the cord and then crosses, whereas in the other pathway, the crossed pathway, information crosses the cord and then ascends.

Touch information from the head does not enter the spinal cord, but enters the brain directly—at the level of the pons. The afferent neurons carrying sensory information from the skin of the head form the fifth cranial nerve, the **trigeminal nerve.**

## TEMPERATURE AND PAIN

Specialized receptor structures in the skin respond to heat, cold, and pain. These receptors are illustrated in Figure 8-4. The corpuscles of Krause and Ruffini are connective tissue covered nerve endings which respond to cold and warmth, respectively. **Free nerve endings,** the most abundant type of receptor in the skin, seem to be responsive to painful stimuli of any type.

The sensory pathway from temperature and pain receptors to the cerebral cortex is quite similar. This pathway is illustrated in Figure 8-5. It consists of an afferent neuron from the receptor to the spinal cord, a second neuron that crosses the cord and ascends in the **lateral spinothalamic tract** to the thalamus, and a neuron that connects the thalamus with the parietal cortex. One can see that this pathway is similar to the crossed touch pathway, except that pain and temperature information ascends in the lateral spinothalamic tract while touch information ascends in the ventral spinothalamic tract.

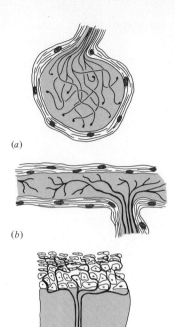

**Figure 8-4** Receptors for temperature and pain: (a) corpuscles of Krause, (b) corpuscles of Ruffini, and (c) free nerve endings.

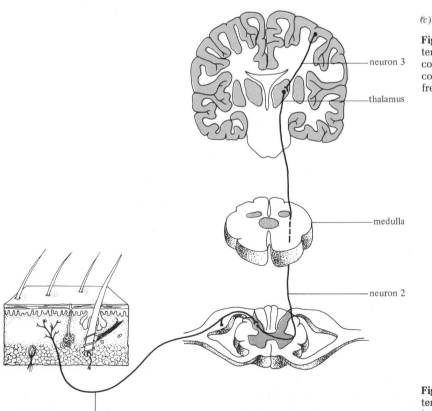

**Figure 8-5** Pathway for temperature and pain information.

# PROPRIOCEPTION

In order to have coordinated movements it is necessary for the brain to receive information about the location and state of contraction of the various muscles. Receptors sensitive to the location and state of contraction of muscles are known as **proprioceptors.** The proprioceptors are located in muscles, tendons, and joints. Information is sent from the proprioceptors to the cerebral cortex leading to a conscious sense of body position and muscular tension and to the cerebellum, which can use the information as an aid in coordinating body movements.

The sensory pathways from proprioceptors to the cerebral cortex and cerebellum are illustrated in Figure 8-6. The pathway to the cerebral cortex is similar to the uncrossed touch pathway. It consists of an afferent neuron that branches and ascends through the posterior column to the medulla, a second neuron that crosses to the other side of the brainstem and ascends to the thalamus, and a third neuron that connects the thalamus with the parietal cortex.

The pathway from proprioceptors to the cerebellum consists

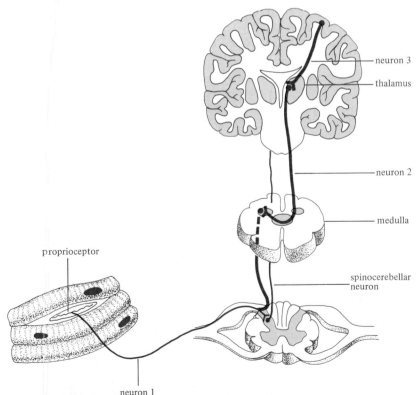

**Figure 8-6** Pathways for proprioceptive information.

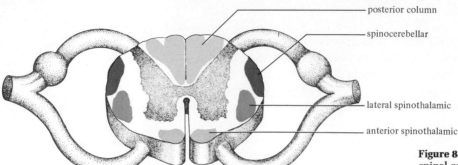

**Figure 8-7** Major tracts of the spinal cord as seen in cross section.

of the afferent neuron into the cord and a neuron which ascends in the **dorsal** or **ventral spinocerebellar** tracts to the cerebellum. This pathway differs from the others we have discussed so far in that it is a two-neuron pathway rather than a three-neuron pathway. The tracts through which the pathways for touch, temperature, pain, and proprioception ascend in the spinal cord are illustrated in Figure 8-7.

## TASTE

The receptors for taste are the **taste buds** located within the papillae, tiny projections on the surface of the tongue. As illustrated in Figure 8-8, each taste bud opens to the outside through a tiny pore bounded by the taste receptor cells. Neuronal endings are intermingled among the receptor cells. For a chemical to activate the taste receptor cells it must first be dissolved and then enter the pore. Within the pore the chemical interacts with tiny hairs that

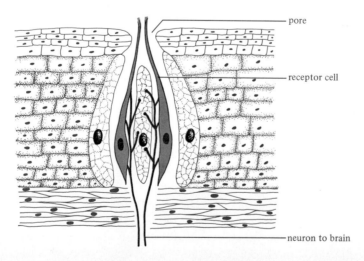

**Figure 8-8** Taste bud.

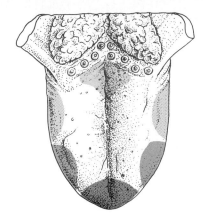

**Figure 8-9** Distribution of different types of taste buds on the tongue.

project from the receptor cells into the pore. The stimulating chemical interacts with the hair in such a way as to change the potential across its membrane. This change in potential in the receptor cells initiates impulses in the adjacent neuronal endings and these impulses travel along the neurons to the brain. The mechanisms by which the stimulating chemicals interact with the hairs and by which the receptor cells stimulate the adjacent neurons are not understood at the present time.

Different taste buds respond to different types of chemicals. The four basic tastes are sweet, sour, salt, and bitter. Taste buds that respond to these different types of stimulation are localized to some degree in different regions of the tongue, as illustrated in Figure 8-9. The back of the tongue is more sensitive to bitter, the sides to sour and salt, and the tip to salt and sweet. It should be kept in mind that the taste of a particular piece of food derives from varying degrees of stimulation of the different types of taste receptors. Information is carried from the taste receptors to the brainstem by the **facial** (seventh) and **glossopharyngeal** (ninth) cranial nerves. The facial nerve brings taste information from the anterior two-thirds of the tongue to the pons, and the glossopharyngeal nerve brings taste information from the posterior third of the tongue to the medulla. From synapses at the pons and medulla the information is sent along a second set of neurons to the thalamus. Finally, a third set of neurons carries taste information from the thalamus to the parietal lobe of the cortex.

## SMELL

The receptor cells for smell are located in the upper portion of the nasal cavity. As illustrated in Figure 8-10, these receptor cells are embedded in the mucous membrane that lines the cavity. Inter-

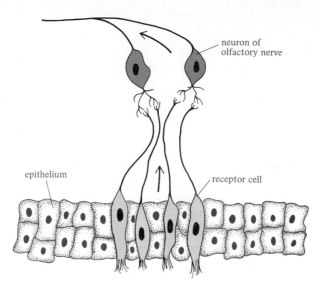

**Figure 8-10** Olfactory receptor.

spersed among the receptor cells are mucus-secreting glands. In order for a gas to stimulate the receptor cells it must first be dissolved in the mucus. As is the case with taste, the dissolved substance then interacts with hairs that extend from the receptor cells in such a way as to change the potential across their membrane. This, in turn, initiates impulses in adjacent neuronal endings and the impulses are sent along the **olfactory** (first) cranial nerve to the anterior portion of the brainstem. From the brainstem impulses are sent to the temporal cortex which contains the basic sensory area for smell.

## THE EYE AND VISION

### Protection of the Eye

The receptor organs for light are the two eyes. Each eye is protected to a large extent by the bones of the orbital cavity in which it is contained. A layer of fatty connective tissue is located in the space between the eye and the bones of the cavity. This helps cushion the eye and prevent damage which might be caused by external pressures. The presence of this fat is graphically demonstrated when a person loses weight rapidly and the eyes sink deep into the orbital cavities.

Protection for the front of the eye is provided by the palpebrae, or eyelids. Obviously the eyelids must be open if light is to enter the eye and stimulate the visual process. However, if a physical object is moving toward the eye the eyelids can be reflexively

**193** The eye and vision

closed to keep the object from damaging the eye. The eyelashes which extend from the eyelids aid in filtering dust that might otherwise enter the eye. Inflammation of the sebaceous glands associated with the hair follicles of the eyelashes leads to a condition known as **sty.** A mucous membrane, the **conjunctiva,** lines the interior surface of the eyelids and portions of the anterior surface of the eye. The mucus secreted by this membrane permits the eyelid to slide smoothly over the eyeball. Inflammation of the conjunctiva is known as **conjunctivitis,** or pink eye. The anterior surface of the eyeball is moistened and cleansed by the tears continuously secreted by the **lacrimal glands.** As illustrated in Figure 8-11, the lacrimal glands are located at the superior, lateral surface of each eyeball. The watery secretions of the lacrimal gland are conducted to the adjacent, anterior surface of the eye by tiny lacrimal ducts. After passing across the anterior surface of the eye the tears drain into the **lacrimal canals** located at the medial edge of the eyeball. From the lacrimal canals the tears pass into the **lacrimal sac** and then through the **nasolacrimal duct** to the interior portion of the nasal cavity. Normally the rate of tear secretion by the lacrimal glands is such that all of the secreted tears are drained into the

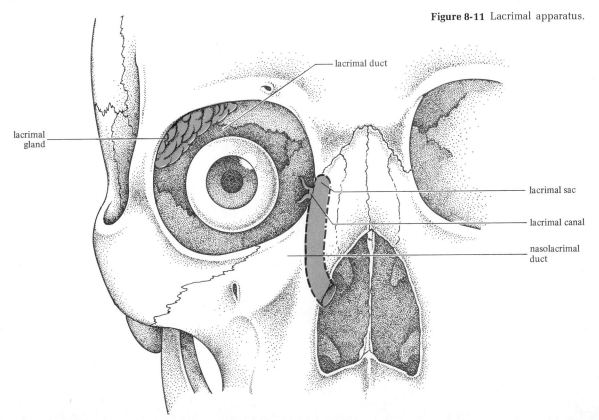

**Figure 8-11** Lacrimal apparatus.

nasal cavity. However, for unknown reasons, the rate of tear secretion is tremendously increased during periods of emotional upset. In this situation all of the fluid cannot be drained into the nasal cavity and some spills over onto the cheeks.

### Structures of the Eye

Figure 8-12 illustrates the structure of the eye. The wall of the eye is divided into three layers: an outer layer composed of fibrous connective tissue and epithelium, a middle layer that contains blood vessels and smooth muscle, and an inner layer that contains the receptor cells sensitive to light. The outer layer is composed of two portions: the transparent **cornea** which covers the central anterior portion of the eye and the white opaque **sclera** which covers the remainder of the eye. Extrinsic muscles which move the eye (see Chapter 6) are attached to the sclera. The sclera also functions to protect the interior structures of the eye. Behind the cornea is the

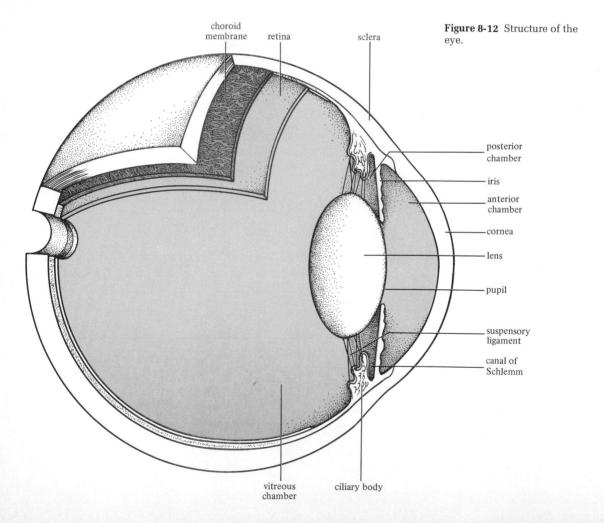

**Figure 8-12** Structure of the eye.

**anterior chamber** which is filled with a watery fluid known as the aqueous humor. The **iris,** a pigmented, muscular structure shaped somewhat like a donut, is situated at the rear of the anterior chamber. Depending on the particular pigment present in the iris a person will have brown eyes, blue eyes, hazel eyes, etc. Light must pass through the **pupil,** the hole in the center of the iris, to eventually reach the light-sensitive cells deep within the eye. Two muscles in the iris determine the size of the pupil, and thus the amount of light which can pass through it. Contraction of the circular **sphincter muscle** of the iris reduces the size of the pupil, whereas contraction of the radial **dilator muscle** of the iris enlarges the size of the pupil.

The **lens** which functions to focus the light onto the receptor cells is located just behind the pupil and iris. It is held in place by the **suspensory ligaments** that extend from the lens to the **ciliary body,** an extension of the middle layer of the eye at the region where the cornea and sclera join. The shape of the lens can be changed by changing the tension of the suspensory ligaments. This is accomplished through the action of the **ciliary muscles** located within the ciliary body. The manner in which the ciliary muscles adjust the shape of the lens and help focus the incoming light will be discussed in a later section of this chapter.

The space bounded by the iris in front and the lens, suspensory ligaments, and ciliary body behind is known as the **posterior chamber.** It is continuous through the pupil with the anterior chamber, and like the anterior chamber it is filled with aqueous humor. The aqueous humor, which is similar in many respects to cerebrospinal fluid, is formed at capillaries in the ciliary body. Because the aqueous humor is being continuously formed it must also be continuously drained. Aqueous humor is drained from the anterior chamber back to the circulatory system through the **canal of Schlemm** located just in front of the ciliary body. If the canal of Schlemm becomes blocked the aqueous humor will accumulate in the anterior and posterior chambers. This will cause an increase in the interocular pressure, the pressure within the eye, a condition referred to as **glaucoma.** The increased pressure damages the receptor cells and the neurons carrying visual information to the brain and thus greatly interferes with vision.

The large space behind the lens is the **vitreous chamber.** It is filled with a jellylike substance, the vitreous humor, that differs from aqueous humor in that once formed it remains in the vitreous chamber. It is not formed and drained continuously. At the back of the vitreous chamber is the **retina,** the inner layer of the eye wall which contains the receptor cells that respond to light. The structure of the retina and the mechanism by which the receptor cells respond to light will be discussed in a subsequent section. Between the retina and the sclera is the **choroid membrane.** This is the por-

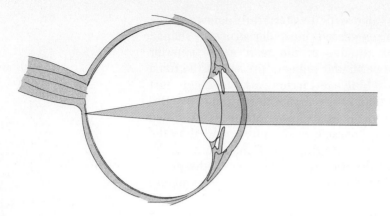

Figure 8-13 Focusing of light on the retina.

tion of the middle layer of the wall which contains many of the blood vessels of the eye.

**Passage of Light to the Retina**

In order for an object to be seen clearly the rays of light coming from the object must pass through the interior of the eye and be brought to focus on the retina. As illustrated in Figure 8-13, focusing depends upon bending all of the light rays which arise from the same point on the object in such a way that they strike the same point on the retina. The bending of light rays is referred to as **refraction.**

The degree to which the incoming light must be bent depends upon the distance of the object. In order to bring a close object into focus it is necessary to bend the incoming light rays to a much larger extent than is necessary when viewing a distant object. In the eye the shape of the lens determines the degree to which the incoming light is bent. When we view a near object the curvature of the lens is increased so that the object can be focused on the retina, a process known as **accommodation.**

Figure 8-14 illustrates how the shape of the lens is changed as one adjusts from looking at a distant object to looking at a close object. In both cases the incoming light is bent as it crosses the cornea. When viewing a distant object the lens is relatively flat and does not bend the light much more than it has already been bent by

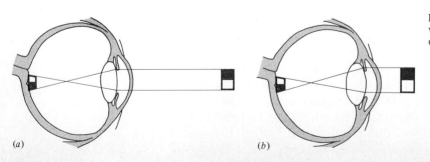

Figure 8-14 Shape of lens when focusing on (a) distant object and (b) close object.

the cornea. The lens is held flat as a result of the large amount of tension in the suspensory ligaments that attach the lens to the ciliary body. As one switches from looking at a distant object to looking at a close object the tension in the suspensory ligaments is reduced. This causes the front surface of the lens to bulge forward as a result of its elasticity. The lens thus becomes much more curved and adds a considerable amount of refraction to the incoming light.

Relaxation of the suspensory ligaments is accomplished by contraction of the ciliary muscles. These muscles are attached in such a way that upon contraction they pull the ciliary body forward and slightly inward. This reduces the distance between the lens and ciliary body and thus the tension in the suspensory ligaments which attach the two. Once this tension is reduced the lens bulges forward because of its internal elasticity. In other words the ciliary muscles are relaxed when viewing distant objects and contracted when viewing close objects. This is one of the reasons that reading can be a strain on the eyes.

Anything that interferes with the passage and focusing of light will obviously interfere with vision. The most common type of interference with the passage of light is **cataract,** in which the lens becomes opaque. This condition can be corrected by the surgical removal of the opaque lens and the use of special glasses.

Focusing problems can arise for a number of different reasons. If the surface of the cornea is not spherical, vertical lines will be bent to a different degree than horizontal lines and vision will be blurred. This condition, known as **astigmatism,** can be corrected by glasses that are curved in such a way as to compensate for the irregularity of the cornea. **Myopia,** or nearsightedness, is caused by either too elongated an eye or too thick a lens. In either case, as illustrated in Figure 8-15, the image falls in front of the retina. Myopia can be corrected by a concave lens, one thicker at the edges

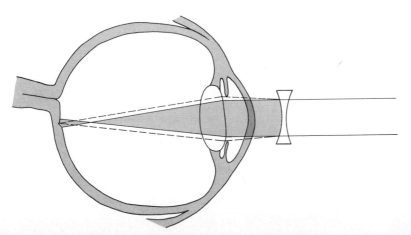

**Figure 8-15** Myopia and corrective lens.

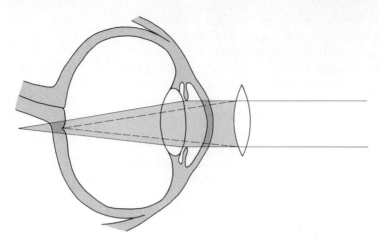

Figure 8-16 Hyperopia and corrective lens.

than at the center. Such a lens will diverge the incoming light so that the image will fall on the retina. **Hyperopia,** or farsightedness, is the opposite of myopia. In this case, as illustrated in Figure 8-16, the lens is too thin or the eye too short. This causes the image to fall behind the retina. Hyperopia can be corrected by a convex lens, one thicker at the center than at the edges. Convex lens bend the incoming light toward the center and thus bring the image forward to the retina. With increasing age the lens tends to lose its elasticity. This decreases the degree to which it will bulge forward when the tension in the suspensory ligaments is reduced. In other words the eye loses its ability to accommodate to viewing close objects. **Presbyopia,** or old-sightedness, is treated by a convex lens like those used for farsightedness.

### Receptor Function of the Retina

There are two types of receptor cells in the eye: the **rods** and the **cones.** As illustrated in Figure 8-17, the cones are more concentrated in the center of the retina, the **macula lutea,** and the rods are

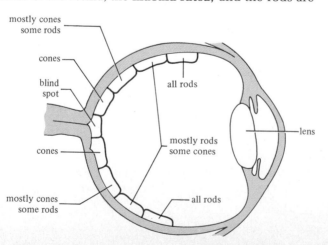

Figure 8-17 Distribution of rods and cones on the retina.

more concentrated at the periphery of the retina. Although both the rods and cones generate impulses in response to light, the function of the two is somewhat different. The rods are the more sensitive of the two, but they only respond to the intensity of light. They are not able to distinguish the color of the incoming light or to form a clearly defined image of the object from which the light is coming. However, because of their great sensitivity the rods are able to respond to very dim light. Thus the primary function of the rods is to enable us to see the general shape of objects when there is very little light, for example, at night or in a dim room.

The rod cells are able to respond to light as a result of the presence in these cells of a special light-sensitive compound called **rhodopsin.** This compound is formed by the combination of a pigment, retinene, and a protein, scotopsin. Retinene is formed in the rods through the action of vitamin A. Light striking the rods causes the chemical breakdown of rhodopsin into its two components, retinene and scotopsin. In an unknown way this chemical breakdown changes the potential across the membrane of the rods. The change in potential initiates action potentials in adjacent neurons. These are then conducted to the brain. Once formed, the retinene and scotopsin recombine into rhodopsin so that the rods can continue to respond to light.

The cones are responsible for distinguishing the colors of the incoming light and for forming a clear image of the object the light is coming from. Like the rods, the cones contain visual pigment which breaks down in the presence of light. Although these compounds have not yet been identified, it is currently thought that there are three types of cones, each of which contains a different type of pigment. The three pigments correspond to the basic colors, blue, green, and red.

### Visual Pathways to the Brain

As illustrated in Figure 8-18, the retina is composed of three layers of cells. The deepest layer is composed of the light-sensitive rods and cones. Adjacent to the rods and cones are short neurons known as bipolar cells. In an unknown manner light stimulation of the rods and cones initiates impulses in the bipolar cells. The bipolar cells synapse with the dendrites of the third layer of cells, the ganglion cells. Thus impulses in the bipolar cells can initiate impulses in the ganglion cells. Axons from the various ganglion cells combine to form the **optic** (second cranial) nerve which carries impulses from the retina to the brain. The area at which the optic nerve leaves the retina is known as the **optic disc.** Since there are no rods or cones at the optic disc, light striking this area does not stimulate vision and the area is known as the blind spot. The optic disc is also the area at which the retinal blood vessels enter and leave the eye.

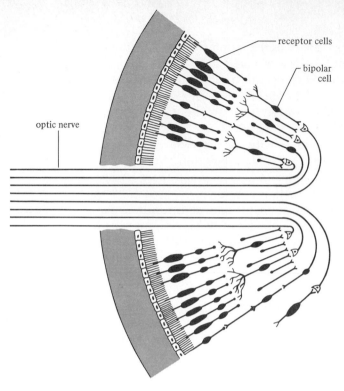

**Figure 8-18** Three layers of the retina.

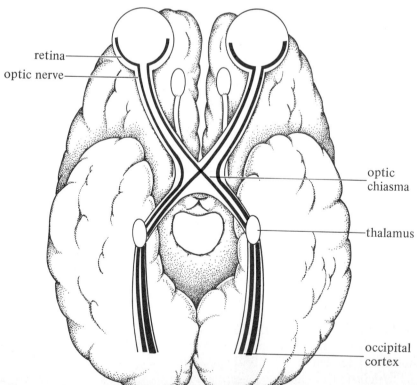

**Figure 8-19** Sensory pathway for visual information.

Figure 8-19 illustrates the pathway from the retina to the brain. The optic nerves from each eye meet at the **optic chiasma** located just in front of the pituitary gland. Neurons from the nasal side of each optic nerve cross at the chiasma to pass to the opposite side of the brain, whereas neurons from the temporal side do not cross and thus pass to the same side of the brain. As a result of this arrangement impulses coming from the right side of each eye pass to the right side of the brain and impulses coming from the left side of each eye pass to the left side of the brain. The right side of each eye receives light from the left visual field and sends this information to the right side of the brain. Likewise the left side of each eye receives information from the right visual field and sends this information to the left side of the brain.

From the optic chiasma the neurons continue, as the optic tracts, to the lateral geniculate bodies of the thalamus. At the thalamus the neurons from the retina synapse with neurons which carry impulses to the occipital lobe of the cerebral cortex. The occipital cortex is responsible for sorting out the incoming visual information and eliciting the conscious sensation of sight.

## THE EAR AND HEARING

### Structure of the Ear

The receptor organs for hearing are the two ears, which convert the pressure waves of sound into impulses that are sent to the brain. Each ear is composed of three sections, the **external ear,** the **middle ear,** and the **inner ear,** as illustrated in Figure 8-20.

The external ear consists of the **auricle** (pinna), a flap of cartilage and skin at the side of the head, and the **external auditory meatus** (auditory canal), a tube leading inward through the temporal bone to the tympanic membrane. Cerumenous glands lining the more distal portion of the external auditory meatus secrete a brownish waxy substance into the canal. This wax and the hairs which line the canal help protect the ear against the entry of infectious agents.

The **tympanic membrane** (eardrum) separates the external ear from the middle ear. It is composed of fibrous connective tissue covered by skin on the side that faces the external ear and by a mucous membrane on the side which faces the middle ear.

The middle ear is an air-filled cavity lined by a mucous membrane. Air enters the middle ear via the **eustachian tube** which extends from the middle ear to the pharynx. The presence of the eustachian tube serves to make the pressure in the middle ear equal to the outside (atmospheric) pressure and thus equal to the pressure in the external ear on the other side of the tympanic membrane.

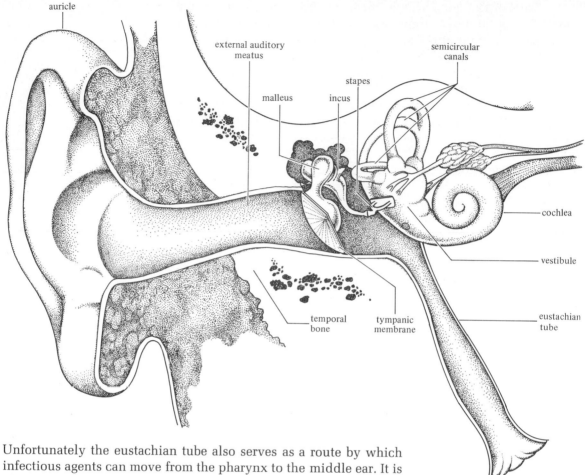

Figure 8-20 Structure of the ear.

Unfortunately the eustachian tube also serves as a route by which infectious agents can move from the pharynx to the middle ear. It is therefore not surprising that infections of the throat often lead to infections of the inner ear. As the mucous membrane that lines the middle ear is continuous with the one lining the air cells of the mastoid process, infection of the middle ear often leads to an infection of the mastoid bone (mastoiditis). To make matters worse the mastoid air cells are only a short distance from the meninges and brain. Thus on occasion infectious agents can spread from the mastoid bone to the meninges (meningitis) and even to the brain itself (encephalitis).

Three small, interconnected bones, the **malleus, incus,** and **stapes,** extend from the tympanic membrane across the middle ear. The stapes is connected to the **oval window,** a membrane that separates the middle ear from the inner ear.

The inner ear consists of a labyrinth of interconnected cavities and canals deep within the temporal bone. As illustrated in Figure 8-21, this labyrinth consists of three main portions: a central, enlarged **vestibule, three semicircular canals** which extend at right

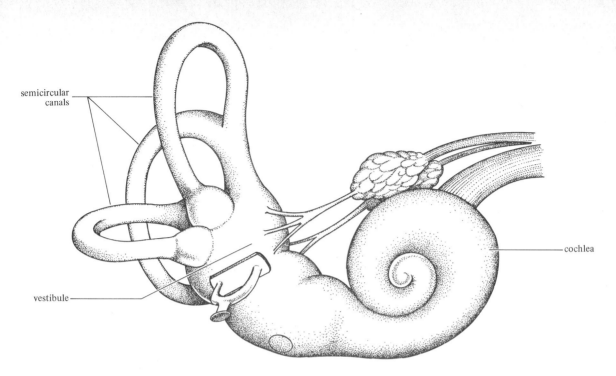

**Figure 8-21** Structure of the inner ear.

angles to each other from the vestibule, and the **cochlea,** a spiral-shaped structure that looks somewhat like a snail. The vestibule and semicircular canals contain the receptor cells for balance whereas the cochlea contains the receptors for hearing.

The internal structure of the cochlea is illustrated in Figure 8-22. It is composed of three canals: the **vestibular canal** which ex-

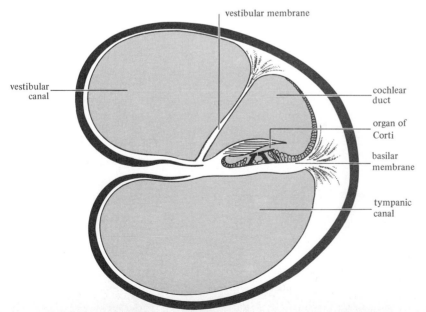

**Figure 8-22** Structure of the cochlea.

tends from the oval window to the tip of the cochlea, the **tympanic canal** which extends from the tip of the cochlea back to the vestibule and ends at the **round window,** and the **cochlear duct** which lies between the vestibular canal and the tympanic canal. The vestibular canal and tympanic canal are continuous and thus filled with the same fluid, the **perilymph;** the cochlear canal is separated from the two and filled with a different fluid, the **endolymph.** The cochlear canal is divided from the vestibular canal by the vestibular membrane and from the tympanic canal by the **basilar membrane.** Projecting from the basilar membrane into the cochlear duct is the **organ of Corti** which contains the receptor cells for hearing.

Figure 8-23 illustrates the structure of the organ of Corti. The cells with tiny projecting hairs, appropriately called **hair cells,** are the receptor cells for hearing. Interspersed among these cells are neuronal endings. It should be noticed that this arrangement of hair cells and neuronal endings is similar to that found in the receptor organs for taste and smell. However, in the organ of Corti the projecting hairs are in contact with a gelatinous structure, the **tectoral membrane.**

### Conversion of Sound Waves to Electrical Impulses

Sound waves are simply physical vibrations of the air which are transmitted from one point to another. As these waves enter the external auditory canal they cause the tympanic membrane to vibrate. Vibrations of the tympanic membrane cause the malleus, incus, and stapes of the middle ear to vibrate. As the stapes vibrates it initiates vibration of the oval window which in turn initiates vibration of the perilymph in the vestibular canal. The three bones are arranged in such a way as to decrease the size but increase the force of the vibrations as they are transmitted from the tympanic membrane to the oval window. This permits the ear to respond to fairly faint sounds.

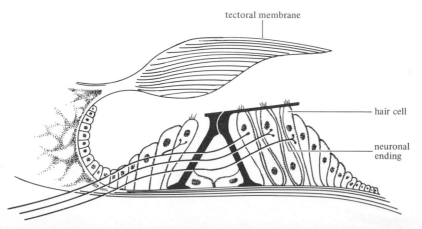

**Figure 8-23** Organ of Corti.

## 205 The ear and hearing

Because fluid can not be compressed, each inward movement of the oval window causes perilymph to move from the vestibular canal to the tympanic canal and push out on the round window. Outward movement of the oval window causes fluid to move back in the opposite direction. This back-and-forth movement of the perilymph causes the basilar membrane to move up and down. As the basilar membrane moves the hair cells are bent against the tectoral membrane. This causes a change in the potential across the hair cell membrane which in some unknown way initiates impulses in the neuronal endings. Figure 8-24 summarizes the sequence of events as vibrations pass through the ear.

The pitch of a sound is determined by the frequency at which the air vibrates. Higher-pitched sounds result from fast vibrations and low-pitched sounds from slower vibrations. It seems that the hair cells at the apex (point) of the cochlea respond to high-frequency vibrations and that the ones at the base (vestibular end) respond to low-frequency vibrations.

### Pathway to the Brain

Impulses are carried from the organ of Corti to the medulla of the brain along the neurons which form the cochlear division of the 8th cranial, or **auditory nerve.** From the medulla impulses can

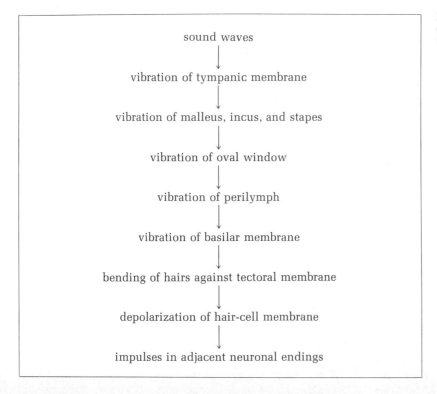

**Figure 8-24** Summary of the conversion of sound waves into electrical impulses.

travel through a number of different pathways to the auditory centers in the temporal portions of each cerebral hemisphere. It seems that impulses from each ear go to both sides of the brain.

## BALANCE

The receptor cells in the vestibule and semicircular canals are responsive to the position of the body as well as its rate and direction of movement. Located inside the vestibule are two membrane-lined sacs, the **utriculus** and **sacculus.** Crystals of calcium carbonate, called otoliths, rest on groups of hair cells which arise from the floor of the sacs. The manner in which these crystals push on the hair cells and initiate impulses depends on the relative position of the head in space. When the impulses are sent to the cerebral cortex and cerebellum the brain is informed of the position of the head.

The three semicircular canals are located at right angles to each other. An expansion at the base of each canal, the **ampulla,** contains groups of hair cells. Movements of the head cause the fluid in the semicircular canals to move and bend the hairs of the hair cell. This initiates impulses which pass along the **vestibular branch** of the 8th cranial nerve to the brain. The direction and rate of head movement will determine which hair cells are stimulated and thus the pattern of impulses which are sent to the brain.

## MOTOR PATHWAYS TO SKELETAL MUSCLE

In order for the muscle cells of a particular muscle to contract they must be stimulated by acetylcholine released from motor neurons. The motor neurons have cell bodies in the ventral horn of the spinal cord or, in the case of the cranial nerves, in special areas of the brainstem. All of the motor neurons which supply a particular muscle are referred to as its final common pathway. The contraction of any muscle ultimately depends upon activation of the final common pathway.

Voluntary, conscious contraction of muscles is initiated by neurons that originate in the motor area of the frontal lobe. There are two major pathways connecting the motor cortex with the final common pathways, as illustrated in Figure 8-25. The pyramidal pathway consists of neurons which extend from the motor cortex to the final pathway with no intermediate synapses. These neurons descend through the internal capsule to the brainstem. At the medulla a majority of the neurons cross and descend through the **lateral corticospinal** tract of the spinal cord to the particular motor

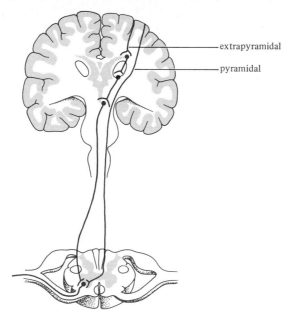

**Figure 8-25** Motor pathways.

neurons they stimulate. Neurons that do not cross descend through the anterior corticospinal tract to the neurons they stimulate.

All of the descending neurons which are not part of the pyramidal pathway are considered part of the extrapyramidal pathway. This pathway involves many synapses in the basal ganglia and brainstem. The interconnections of the extrapyramidal pathway are exceedingly complex and still not completely understood. Some of the neurons which descend as part of the extrapyramidal pathway act to inhibit muscular contraction whereas others stimulate muscular contraction. The function of the extrapyramidal pathway is to regulate and coordinate muscular contraction.

## AUTONOMIC NERVOUS SYSTEM

The autonomic nervous system consists of the efferent neurons which extend from the central nervous system to effector structures other than skeletal muscle. These would include the cardiac muscle of the heart; the smooth muscle of the iris and ciliary body of the eye; the smooth muscle that lines the digestive, respiratory, urinary, and reproductive tracts; the smooth muscle of the blood vessels; and both endocrine and exocrine glands. The basic function of the autonomic system is to help adjust internal function so that it is appropriate to the conditions within the body at a particular time. We will discuss the details of this mechanism in succeeding chapters as they become relevant.

**208** Senses and Motor Pathways

Structurally, and to some degree functionally, the autonomic nervous system can be divided into two divisions, the sympathetic system and the parasympathetic system, as illustrated in Figure 8-26. Each system contains two major neurons, a preganglionic and postganglionic neuron, that connect the central nervous system with the effector. In the **sympathetic nervous system** the preganglionic fibers leave the cord in the thoracic and lumbar regions. These preganglionic fibers leave the cord through the ventral root and then pass through the white ramus to synapse with the postganglionic fibers at the sympathetic ganglia. Postganglionic fibers then extend from the sympathetic ganglion to the effector. There are two sets of sympathetic ganglia. The **lateral ganglia** form two chains, one on either side of the vertebral column. Each chain consists of 22 interconnected ganglia. The colateral ganglia are the **celiac** and **mesenteric ganglia** in the abdominal cavity. Within the sympathetic

**Figure 8-26** Autonomic nervous system.

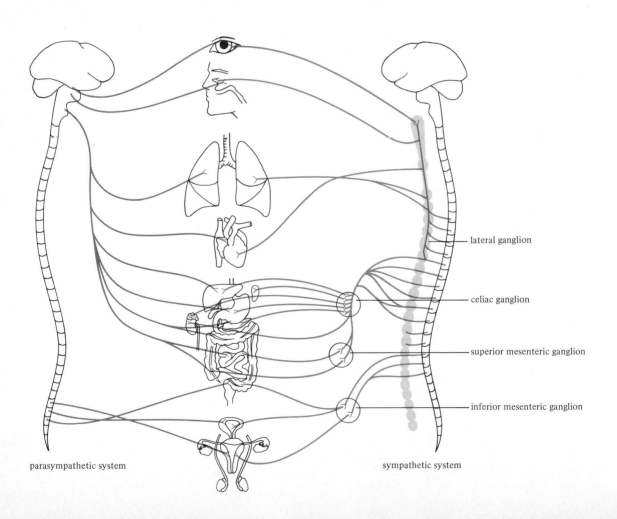

parasympathetic system

sympathetic system

lateral ganglion

celiac ganglion

superior mesenteric ganglion

inferior mesenteric ganglion

ganglia the preganglionic neurons activate the postganglionic neurons through the release of acetyl-choline. Sympathetic postganglionic fibers release **norepinephrine** as a chemical transmitter.

The sympathetic nervous system is activated in stressful or emergency situations. Essentially, it stimulates the internal organs in such a way as to respond to these situations. For example, it stimulates the heart to pump more blood, it dilates the bronchi, it increases the amount of blood flowing to the heart and lungs, and it stimulates the adrenal medulla to release epinephrine.

The **parasympathetic nervous** system differs structurally from the sympathetic system in two respects. In this system the preganglionic fibers originate in the cranial and sacral portions of the central nervous system and the ganglia are located near or in the effectors. The postganglionic nerves are therefore very short.

As in the sympathetic system, the chemical transmitter released by the preganglionic parasympathetic neurons is acetyl-choline. The postganglionic parasympathetic neurons also release acetyl-choline. In general, stimulation of any structure by parasympathetic neurons has the opposite effect of sympathetic stimulation. Table 8-1 lists the effects of sympathetic and parasympathetic stimulation on various structures of the body.

**Table 8-1** Effects of Autonomic Nervous System

| Organ | Action of Sympathetic System | Action of Parasympathetic System |
|---|---|---|
| Iris | Contraction of radial muscle; dilates pupil | Contraction of sphincter muscle; constricts pupil |
| Lacrimal gland | Large amount of secretion | Normal secretion |
| Ciliary muscle | Contraction for close vision | Relaxation for distant vision |
| Heart | Increased pumping of blood and dilation of coronary blood vessels | Decreased pumping of blood |
| Blood vessels | Constriction in skin and visceral organs; Constriction and dilation in skeletal muscle (two types of sympathetics) | Dilation in salivary glands and external genitalia |
| Salivary glands | Secretion of mucus | Secretion of watery saliva |
| Stomach and intestine | Inhibition of motility and secretion; Contraction of sphincters | Stimulation of motility and secretion; Relaxation of sphincters |
| Liver | Stimulation of glycogen breakdown; Inhibition of bile secretion | Stimulation of bile secretion |
| Pancreas | Inhibition of secretion | Stimulation of secretion |
| Spleen | Contraction; adds blood to circulation | |
| Urinary bladder | Inhibition of urination | Stimulation of urination |
| Respiratory system | Dilation of bronchioles and blood vessels | Contraction of bronchioles |
| Sweat glands | Stimulation of secretion | |

## OBJECTIVES FOR THE STUDY OF SENSES AND MOTOR PATHWAYS

At the end of this unit you should be able to:

1. Name three types of receptors that respond to touch.
2. Describe the two pathways by which touch information is transmitted to the cerebral cortex.
3. Name two types of receptors that respond to temperature.
4. Describe the pathway by which temperature and pain information is transmitted to the cerebral cortex.
5. Describe the location and function of proprioceptors.
6. Describe the pathways by which proprioceptive information is transmitted to the brain.
7. Describe the location and structure of taste buds.
8. Describe the afferent pathway by which taste information is transmitted to the cerebral cortex.
9. Describe a smell receptor.
10. Describe the afferent pathway by which olfactory information is transmitted to the brain.
11. Explain how the eye is protected against injury.
12. State the location of the lacrimal gland, lacrimal canals, lacrimal sac, and nasolacrimal duct.
13. Identify from a drawing the following parts of the eye: cornea, sclera, anterior chamber, iris, pupil, lens, suspensory ligaments, ciliary body, ciliary muscles, posterior chamber, canal of Schlemm, vitreous chamber, retina, choroid membrane.
14. State the function of the iris, lens, and retina.
15. Describe the visual problems that exist in astigmatism, myopia, and hyperopia.
16. Distinguish between rods and cones in terms of location and function.
17. Describe the three layers of the retina.
18. Name the area in which the optic nerve leaves the retina.
19. Describe the afferent pathway from the retina to the cerebral cortex.
20. Identify from a drawing the following parts of the ear: external ear, middle ear, inner ear, auricle, external auditory meatus, tympanic membrane, eustachian tube, malleus, incus, stapes, oval window, cochlea, and round window.
21. Describe the internal structure of the cochlea.
22. Describe the structure of the organ of Corti.
23. Describe the sequence of events by which sound waves are converted into impulses by the ear.
24. Describe the afferent pathway from the ear to the brain.
25. State the location of the receptor cells responsible for balance.
26. Distinguish between the pyramidal and extrapyramidal tract in terms of both structure and function.
27. State the anatomical and functional differences between the sympathetic and parasympathetic nervous systems.

# 9 The Endocrine System

INTRODUCTION
PITUITARY GLAND
THYROID GLAND
PARATHYROIDS
ADRENAL GLANDS
PANCREAS
OTHER STRUCTURES WITH ENDOCRINE
   ACTIVITY

# INTRODUCTION

Survival depends on the continuous regulation of the various body processes so that they function in a way that is appropriate to the prevailing conditions in the internal and external environments. The heart must be stimulated to beat considerably faster during exercise than it beats during sleep. Fat cells must be stimulated to store fat after a heavy meal and to release the stored fat during periods of reduced food intake or in starvation. Control over any body function requires both a means of detecting relevant changes in the internal and external environments and a means of directing the appropriate responses to these changes.

In addition, there must be a means of transmitting information from the part of the body detecting the change to the part of the body responding to the change. Chapter 7 described how the nervous system transmits information from one part of the body to another via the propagation of electrochemical impulses along nerve cells. The nervous system is somewhat analogous to a telephone communication system in that the wires (nerves) extend from the point at which the message is sent to the point at which the message is received.

The second major communication system of the body is the endocrine system. In this system, information is sent from one part of the body to another via blood-borne chemical messengers called **hormones.** This system is analogous to a postal communication in which intact messages (hormones) are carried by a transport system (the blood) from one point to another.

Hormones are produced by the endocrine system which is composed of the various **endocrine glands** that manufacture and secrete hormones. As can be seen in Figure 9-1, the endocrine glands are distributed throughout the body. Some of the endocrine glands, such as the pituitary, thyroid, and adrenals, have no function other than the secretion of hormones. Other structures, such as the stomach, kidney, and pancreas, carry out nonendocrine as well as endocrine functions.

## Introduction

The term "endocrine" means internal secretion. It is used to distinguish the endocrine glands, which contain no ducts and release their secretions directly into the blood or lymph, from the **exocrine glands,** which release their secretions into ducts through which they pass into one of the body tubes. The endocrine glands are sometimes referred to as the ductless glands.

Because the endocrine glands release their hormones into the blood, these hormones are carried to all parts of the body. However, only certain cells, known as **target cells,** are capable of responding to a particular hormone. For example, osteoclast bone cells respond to the hormone parathormone by breaking down bone matrix and releasing $Ca^{2+}$ into the blood. Thus osteoclasts would be target cells for parathormone.

### Mechanism of Hormone Action

The exact mechanism by which hormones activate their particular target cells is the subject of a great deal of current research. Clearly the hormone must alter the internal functioning of the cell in some manner. This is most likely accomplished by either ac-

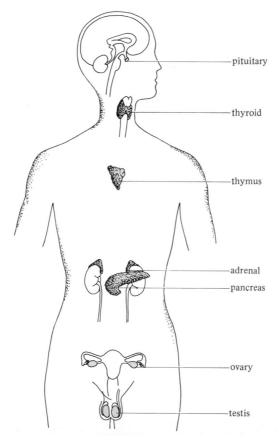

**Figure 9-1** Endocrine glands.

tivating or inhibiting the activity of intracellular enzymes, stimulating the synthesis of new proteins, or altering the permeability of the cellular membrane. A few examples of this would be the stimulation of glucose production in the liver by the hormone epinephrine through the activation of an enzyme which converts stored glycogen to glucose, the stimulation of tissue growth by growth hormone through the stimulation of protein synthesis, and the lowering of blood glucose by insulin through increasing the glucose permeability of cell membranes.

An understanding of the mechanism by which hormones alter the activity of their target cells has been further complicated by the fact that many hormones, particularly those that are proteins or protein derivatives, do not enter cells. One theory of hormone action which takes this fact into account is the cyclic AMP or "second messenger" theory illustrated in Figure 9-2.

According to this theory a protein or protein-derivative hormone acts by combining with the cellular membrane of the target cell in such a way as to activate an enzyme known as adenylcyclase. The adenylcyclase then converts intracellular ATP to a compound known as cyclic adenosine-3′,5′-monophosphate, or simply cyclic AMP. An altered intracellular level of cyclic AMP then leads to the action designated by the hormone, for example, a change in membrane permeability. In recognition of his contribution to the understanding of the role of cyclic AMP in hormone action E. W. Sutherland was awarded the 1971 Nobel Prize in Physiology and Medicine.

**Control of Hormone Secretion**

The rate at which an endocrine gland secretes a particular hormone into the blood must be appropriate to the physiological need

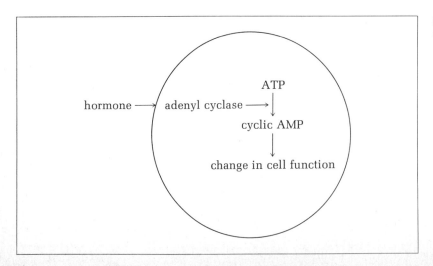

**Figure 9-2** Cyclic AMP theory of hormone action.

for that hormone. There are four basic means by which hormonal secretion is controlled: nerves, other hormones, blood metabolite levels, and negative feedback. Each of these means of control over hormonal secretion will be discussed as they become relevant to particular hormones.

### Endocrine Disease

Endocrine disease can result from either excessive or deficient hormonal secretion. Excessive hormonal secretion is usually the result of a tumorous growth of hormone-secreting cells. The consequences of the excessive secretion are simply exaggerated manifestations of the normal hormone action. For example, the pituitary hormone somatotropin stimulates the protein synthesis necessary for normal growth. A child with a pituitary tumor which produces excessive amounts of growth hormone will be oversized and may even become a giant.

Deficient hormone production results when the endocrine tissue becomes diseased or worn out. Obviously in this case the hormone is unavailable to carry out its normal physiological activity. Thus the high level of blood glucose in diabetes is caused by the lack of insulin and the inability of glucose to enter cells.

## PITUITARY GLAND (HYPOPHYSIS)

The pituitary is a small gland, only about one centimeter in diameter, located at the base of the brain just below the hypothalamus. As can be seen in Figure 9-3 it is contained within a depression of the

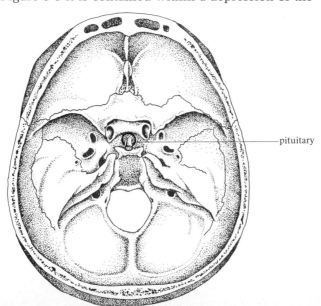

**Figure 9-3** Pituitary gland within the sella turcica.

bony floor of the cranium. This depression, called the **hypophyseal fossa,** is the deepest portion within the sella turcica of the sphenoid bone. The pituitary gland is attached to the hypothalamus by a short stalk, the **hypophyseal stalk.**

This pituitary gland, illustrated in Figure 9-4, contains two major parts: the **anterior pituitary,** or adenohypophysis, and the **posterior pituitary.** The anterior pituitary is composed of glandular tissue which is separated from the overlying hypothalamus by connective tissue. This connective tissue contains a rich supply of blood vessels through which blood flows directly from the hypothalamus to the pituitary. The significance of these vessels will become clear when we discuss the control of anterior pituitary hormonal secretion.

Composed of nervous tissue, the posterior pituitary is actually a continuation of the hypothalamus. The posterior pituitary contains the axons of neurons the cell bodies of which are located within the hypothalamus. These cell bodies manufacture hormones which are then transported intracellularly to the posterior pituitary where they are released into the blood.

The small size of the pituitary belies its enormous functional importance. Hormones secreted by the pituitary play a major role in regulating metabolism, salt and water balance, reproduction, and

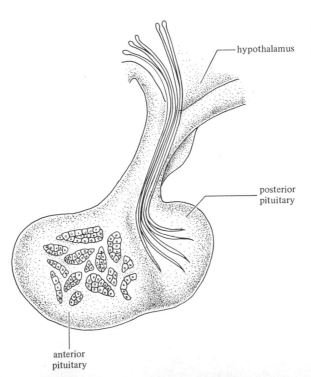

**Figure 9-4** Anterior and posterior portions of the pituitary gland.

the body's response to stressful situations. The secretory activity of the pituitary is controlled by the hypothalamus and in turn controls the secretory activity of a number of other glands. Thus the pituitary serves as a link between nervous activity and endocrine activity. In recognition of the role played by the pituitary in the regulation of other endocrine glands, the pituitary is sometimes referred to as the master gland of the body.

**Anterior Pituitary Hormones**

The anterior portion of the pituitary gland secretes seven basic hormones: thyroid-stimulating hormone (TSH), adrenocorticotropic hormone (ACTH), follicle-stimulating hormone (FSH), luteinizing hormone (LH), lactogenic hormone (LTH), growth hormone (STH), and melanocyte-stimulating hormone (MSH). The first four (TSH, ACTH, FSH, and LH) are referred to as tropic hormones because they stimulate other endocrine glands.

**Thyroid-stimulating hormone (TSH).** As the name implies, thyroid-stimulating hormone acts as a stimulating force on the thyroid gland. TSH seems to stimulate all of the basic activities of the thyroid. That is, it stimulates the increased production and secretion of the thyroid hormone thyroxine by the thyroid cells and the growth of new thyroid cells. In the absence of TSH, the thyroid will atrophy and no longer secrete any thyroid hormone. Current clinical research indicates that a great number of cases of thyroid disease are actually a result of an abnormally high or low production of TSH.

**Adrenocorticotropic hormone (ACTH).** This hormone acts to stimulate the outer portion of the adrenal gland, the adrenal cortex, to produce a group of hormones known as glucocorticoids. As will be discussed later in the chapter, the glucocorticoids play a role in the regulation of metabolism. Cortisol is the predominant glucocorticoid produced in the human adrenal cortex.

Almost any type of stressful situation leads to the release of ACTH by the anterior pituitary and the concomitant release of cortisol by the adrenal cortex. The exact manner in which either or both of these hormones aid in the response to a stressful situation is not at all clear at the present moment.

**Follicle-stimulating hormone (FSH).** Follicle-stimulating hormone is a **gonadotropin:** it stimulates the ovaries in women and the testes in men. In women FSH stimulates the growth and development of an egg and its surrounding follicle during each menstrual cycle. Under the stimulation of FSH, the follicle secretes the hormone estrogen. In men FSH is necessary for spermatogenesis, the production of sperm. The absence of FSH in either men or women would lead to infertility.

**Luteinizing hormone (LH).** Luteinizing hormone is also a

gonadotropin. In women it stimulates ovulation, the release of the egg from the follicle, and the formation of the corpus luteum, the hormone-secreting structure formed by the follicle after the egg is released. In men LH is necessary for the production of the hormone testosterone by the interstitial cells of the testes. The absence of LH would also lead to infertility.

**Latogenic hormone (LTH or prolactin).** Although lactogenic hormone is secreted by the pituitary gland in both men and women it has no known physiological role in men. In women it stimulates milk secretion once the breasts have developed under the influence of the ovarian hormones estrogen and progesterone. It may also play a role in the maintenance of the corpus luteum.

**Growth hormone (STH or somatotropin).** Growth hormone plays an important role in normal growth and development through its stimulation of protein synthesis. In addition, growth hormone is one of a number of hormones involved in the regulation of intermediate metabolism (the pattern by which the body stores and utilizes sources of energy). Specifically, growth hormone acts to mobilize stored fat and block the entry of glucose into most types of cells. Thus in the presence of growth hormone, most cells rely on the breakdown of fatty acids rather than glucose for the formation of ATP. The role of growth hormone in intermediate metabolism will be discussed in greater detail in Chapter 16.

Children whose pituitary gland does not produce adequate amounts of growth hormone become dwarfs. In general, pituitary dwarfs have proportionate features and normal mental development. Very often the inability of the pituitary to produce growth hormone is associated with an inability to produce gonadotropin. Dwarfs who lack gonadotropin are infertile.

As might be expected, oversecretion of growth hormone by the pituitary leads to gigantism. This condition is usually caused by a pituitary tumor in which there is an increase in the number of growth-hormone-producing cells. Figure 9-5 shows a pituitary giant and pituitary dwarf as compared to a normal-sized person.

The excessive amounts of growth hormone in a giant promote all aspects of growth. Hence the skeletal bones become quite long and thick, the internal organs become quite large, and there is an abundance of connective tissue. The external appearance of pituitary giants is characterized not only by their excessive height, but also by the general thickness of their body. This can be seen particularly clearly in their hands, feet, and jaws.

Because growth hormone blocks the entry of glucose into many types of cells, giants have a tendency toward a high level of blood sugar and diabetes. As long as the pancreas is functioning properly, it can release increased amounts of insulin which can overcome the blocking effects of growth hormone. Over the years, this increased production of insulin can wear out the pancreas leading to the development of full-scale diabetes.

In some people a pituitary tumor which secretes excessive amounts of growth hormone arises after the epiphyseal cartilages of the bones are closed and full height has been attained. This condition is referred to as **acromegaly**.

Once the epiphyseal cartilage has closed, the long bones of the arms and legs can no longer grow in length. For this reason a person with acromegaly is usually of normal height. However, the excessive amount of growth hormone does stimulate the widening of bones and the enlargement of the internal organs. Therefore a person with acromegaly has the thickness of a giant without the height. An acromegalic has the same tendency toward diabetes as does a giant.

**Figure 9-5** Pituitary giant and dwarf compared to a normal-sized person. (Photo by Syndication International, Photo Trends.)

**Melanocyte-stimulating hormone (MSH).** This hormone is capable of increasing the pigmentation of the skin. In many animals, MSH is released in varying amounts at different times of the year leading to corresponding changes in skin color. Although it is secreted by the human pituitary, its physiological role is not known at the present time. Human skin has normal pigmentation in the complete absence of MSH. Variations in human skin color are not caused by variations in the amount of MSH present.

### Control of Anterior Pituitary Hormone Secretion

The neurons of the hypothalamus control the secretory activity of the anterior pituitary through the action of a special set of hormones called **hypothalamic releasing factors.** A specific releasing factor is used by the hypothalamus in the control of each of the seven anterior pituitary hormones. Thus **thyroid-stimulating hormone-releasing factor** from the hypothalamus stimulates the secretion of the thyroid-stimulating hormone by the anterior pituitary. **Luteinizing hormone-releasing factor** from the hypothalamus stimulates the release of luteinizing hormone from the anterior pituitary. In a similar manner, there are specific hypothalamic releasing factors that stimulate the release of ACTH, FSH, growth hormone, and MSH by the anterior pituitary. The one exception to this rule is that prolactin secretion by the anterior pituitary is regulated by a hypothalamic inhibitory factor (**prolactin inhibitory factor**) rather than by a stimulating factor.

The manner in which the hypothalamus regulates the anterior pituitary is illustrated in Figure 9-6. A nerve impulse in one of the

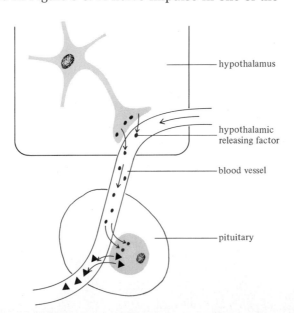

**Figure 9-6** Schematic illustration of the control of the pituitary by the hypothalamus.

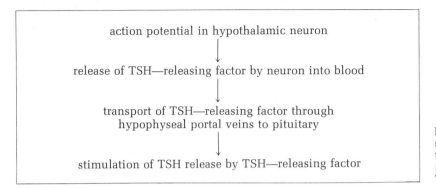

Figure 9-7 Steps in the stimulation of thyroid-stimulating hormone (TSH) secretion by TSH-releasing factor.

neurosecretory cells of the hypothalamus leads to the release of its specific releasing factor. The releasing factor then diffuses into a hypothalamic capillary. Blood flows from the hypothalamic capillaries through a specialized set of blood vessels, the **hypophyseal portal veins,** to capillaries in the anterior pituitary. At the anterior pituitary, the hypothalamic releasing factor diffuses out of the blood and stimulates its specific type of anterior pituitary cell. Figure 9-7 is a flow chart indicating how TSH-releasing factor from the hypothalamus would stimulate the release of TSH by the anterior pituitary.

The importance of the hypothalamic releasing factors is that they serve as a link between the nervous system and the endocrine system. Changes in the internal and external environments can be detected by nervous system receptors and information about these changes transmitted along neurons to the hypothalamus. The hypothalamus can then use both the autonomic nervous system and the endocrine system to direct the appropriate response to these changes.

A second way in which the secretory activity of the anterior pituitary is controlled is through a process known as **negative feedback.** Negative feedback refers to the situation in which a physiological activity is inhibited by the very process it brings about. In the case of the anterior pituitary hormones, the tropic hormones are all inhibited by the hormones whose release they stimulate. That is, TSH is inhibited by thyroxine, ACTH is inhibited by cortisol, and LH is inhibited by progesterone. The negative feedback of thyroxine on TSH is illustrated schematically in Figure 9-8.

### Posterior Pituitary Hormones

**Antidiuretic hormone (ADH).** Antidiuretic hormone, or vasopressin as it is sometimes called, acts on the kidney to cause the retention of water. As a result of this action, the kidneys form a concentrated (hypertonic) urine. ADH is an octapeptide; it is composed

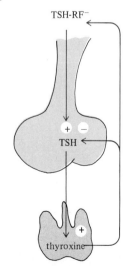

Figure 9-8 Negative feedback of thyroxine on TSH secretion.

of a chain of eight amino acids. The posterior pituitary releases ADH in any situation in which the body must conserve water. Any situation which causes either a drop in blood volume or an increase in blood solute concentration will bring about the release of ADH by the posterior pituitary and the conservation of water by the kidney. Changes in blood concentration are detected by receptors in the hypothalamus.

When the posterior pituitary is incapable of producing ADH, a disease known as **diabetes insipidus** occurs. In the absence of ADH, the ability of the kidney to retain water is severely impaired and a large amount of dilute urine is formed. A person with diabetes insipidis can urinate as much as 25 liters of water a day. This necessitates drinking an equivalent amount of fluid if water balance is to be maintained. The role of ADH in water balance will be discussed in greater detail in Chapter 17.

**Oxytocin.** Oxytocin is an octapeptide quite similar in structure to ADH. This hormone plays an important role in the reproductive process through its ability to stimulate contraction of the smooth muscles of the uterus and the ducts within the breasts. At the time of delivery the posterior pituitary releases oxytocin which then helps stimulate the uterine contractions necessary to expel the fetus.

Oxytocin is also released by the posterior pituitary in response to the newborn infant's suckling of the breast. Once released, oxytocin stimulates contraction of the ducts within the breast and thus the ejection of milk. These processes will be discussed in greater detail in Chapter 18.

## THYROID GLAND

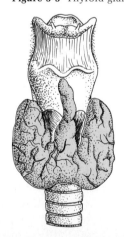

**Figure 9-9** Thyroid gland.

The thyroid gland is located in the neck just below the larynx. As can be seen in Figure 9-9 the gland is composed of three parts: two large **lateral lobes** located on either side of the trachea and a narrow isthmus which passes over the anterior surface of the trachea and connects the two lateral lobes.

Microscopically the thyroid is composed of numerous small follicles. These follicles contain a colloid-filled core surrounded by a layer of cuboidal epithelial tissue.

The epithelial cells lining the thyroid follicles produce the two thyroid hormones, thyroxine and triiodothyronine. Figure 9-10 shows the structure of these two hormones. They are both formed by the combination of the amino acid tyrosine with atoms of iodine. Thyroxine, which is by far the predominant hormone produced by the thyroid, contains four iodine atoms whereas triiodothyronine contains only three iodine atoms. Because the amount of

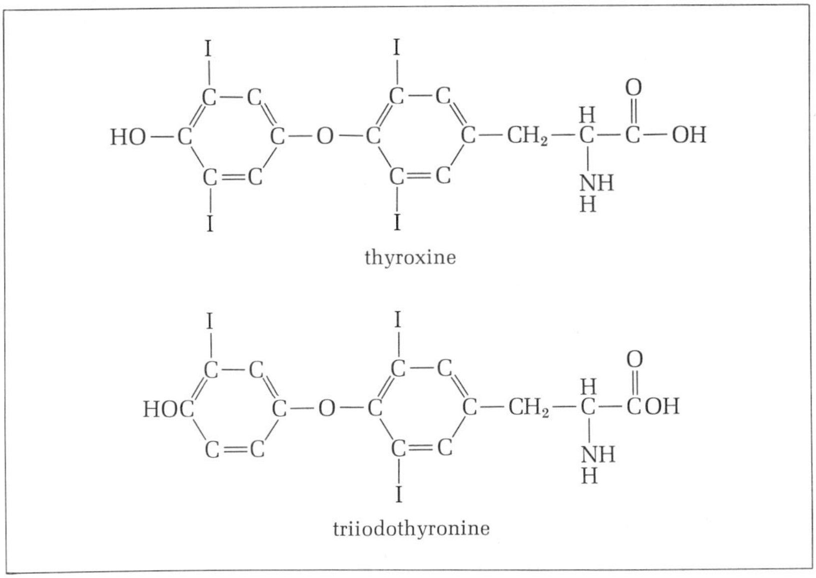

**Figure 9-10** Thyroxine and triiodothyronine.

thyroxine produced by the thyroid is so much larger than the amount of triiodothyronine, the term thyroxine is usually used to designate the thyroid hormones. A lack of dietary iodine will obviously interfere with the ability of the thyroid to manufacture these hormones.

Once formed, thyroxine is stored in the colloidal core of the follicle in combination with a large protein thyroglobulin. Stimulation of the thyroid gland by thyroid-stimulating hormone splits thyroxine from thyroglobulin. Thyroxine then diffuses into the blood from the follicle. Within the blood most of the thyroxine is transported in combination with one of the blood proteins. Because the thyroid hormones are the only molecules in the body that contain iodine, a measurement of the protein-bound iodine (PBI) gives a good indication of the blood level of thyroxine.

The primary effect of thyroxine on the body is to stimulate an increase in metabolic rate, the rate at which oxygen combines with nutrients from ingested food. This increase in metabolic rate leads to a number of secondary effects. In order to meet the demand of an increase in metabolic rate either food consumption must be increased or the body tissue itself must be broken down. The increase in metabolic rate also leads to an increase in heat production. In addition to stimulating metabolic rate, thyroxine also stimulates a number of other physiological processes. Included among these are heart rate, gastrointestinal tract movements, and brain activity. Thyroxine also plays an essential role in growth and development, particularly of the skeletomuscular and nervous system.

Control over thyroxine secretion is exerted primarily through the action of the pituitary hormone TSH. TSH itself is of course controlled by TSH-releasing factor from the hypothalamus. As described previously thyroxine has a negative feedback influence on the pituitary and the hypothalamus. That is, any increase in the circulating level of thyroxine will inhibit the release of TSH-releasing factor and TSH. The subsequent decrease in the blood level of TSH will reduce the rate at which the thyroid secretes thyroxine. In other words if the blood level of thyroxine becomes too high, pituitary stimulation of thyroid secretion will be reduced. This will tend to bring the blood level of thyroxine back to normal.

A major unresolved question in thyroid physiology is the nature of the environmental changes which produce the hypothalamus–pituitary–thyroid response in man. That is to say, we know that TSH-releasing factor stimulates the release of TSH which in turn stimulates the release of thyroxine. However, it is unclear as to what stimulates the hypothalamus to release TSH-releasing factor in the first place. In many mammals, such as dogs and rats, TSH-releasing factor is secreted in response to cold. This is a reasonable response because thyroxine causes increased heat production by increasing the metabolic rate. No one has yet been able to show that this same response takes place in man.

**Thyroid Diseases**

**Hypofunction.** Children born without a functioning thyroid gland develop a disease known as **cretinism.** The absence of thyroxine in these children interferes with normal growth and development, particularly of the skeletomuscular and nervous systems. The children are exceedingly small and mentally retarded. Their metabolic rate is low and most of their physiological function proceeds at a less-than-normal rate. For example, food moves slowly through their digestive tract leading to constipation and the amount of blood pumped by their heart is below normal.

Cretinism can be caused either by a lack of dietary iodine or by a congenital condition in which the thyroid does not develop properly. If caught in time the condition is reversible by the administration of thyroxine. Figure 9-11 shows a cretinous child. If the condition is not treated early enough irreversible brain damage can occur.

Hypofunction of the thyroid in later life leads to a condition known as **myxedema.** The symptoms of myxedema include weight gain, intolerance to cold, fatigue, anemia, decreased resistance to infection, and the accumulation of fluid in the skin. This accumulation of fluid leads to the puffiness characteristic of the external appearance of myxedema. All of these symptoms arise from the low level of circulating thyroxine.

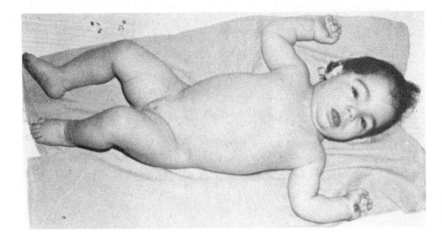

Figure 9-11 Cretinous child (two years old). (Photo by Lester Bergman & Assoc.)

Myxedema can be caused by either a dietary lack of iodine or the destruction of the thyroid by disease. However, in many cases of myxedema the underlying cause of the disease is not known.

**Hyperfunction.** Graves' disease, exophthalmic goiter, and thyrotoxicosis are all terms used to designate the disease in which the thyroid secretes extremely large amounts of thyroxine. The symptoms of this state are the opposite of the symptoms of myxedema. A person with a high level of circulating thyroxine would tend to lose weight in spite of eating large amounts of food, get warm easily, be very nervous, have diarrhea, and have protruding eyes. As can be seen in Figure 9-12, the protruding eyes, known as

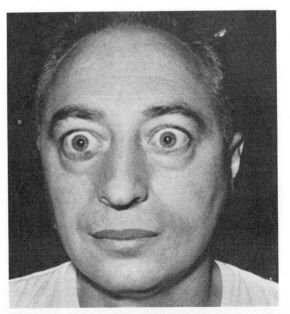

Figure 9-12 Graves' disease. (Photo by Lester Bergman & Assoc.)

$\downarrow I_2 \longrightarrow \downarrow$ thyroxine production $\longrightarrow \uparrow$ TSH $\longrightarrow$ growth of thyroid

**Figure 9-13** Steps by which iodine deficiency can cause the formation of a goiter.

**exopthalamus,** and an enlarged thyroid bulging at the neck are the dominant external manifestations of Graves' disease. This disease is usually caused by a thyroid tumor in which the newly formed cells secrete thyroxine.

**Goiter.** Goiter is the term used to refer to an enlarged thyroid. The enlargement of the gland can come about for one of two reasons: dietary iodine deficiency or a tumor of the gland. In the case of dietary iodine deficiency the gland grows under the stimulation of an increased release of TSH by the pituitary. The steps leading to the goiterous enlargement of the thyroid in iodine deficiency are described in Figure 9-13. A lack of iodine in the diet interferes with the ability of the thyroid to manufacture thyroxine. As the circulating level of thyroxine falls the secretion of TSH-releasing factor and TSH are no longer inhibited by negative feedback. Thus the pituitary begins to secrete larger amounts of TSH. This TSH stimulates the growth of the thyroid and the ability of the gland to manufacture thyroxine. In some cases the enlarged thyroid is able to produce a normal level of thyroxine even though there is not a normal amount of iodine in the diet. However, if dietary iodine is very low or if it is completely absent thyroxine secretion can never be brought to normal. Figure 9-14 shows a person with iodine-deficiency goiter.

**Figure 9-14** Iodine-deficiency goiter. (Photo by Paul Almasy, WHO.)

In Graves' disease the thyroid also becomes goiterous. However, in this case the enlarged thyroid secretes a larger-than-normal amount of thyroxine. Graves' disease can be due to either a tumor of the glandular tissue of the thyroid itself or a tumor of the hypothalamus or pituitary in which an excessive amount of TSH is secreted.

### Thyrocalcitonin

In addition to secreting thyroxine and triiodothyroxine the thyroid secretes a third hormone, **thyrocalcitonin.** This hormone tends to lower blood calcium concentration by inhibiting the breakdown of bone. The overall importance of thyrocalcitonin in calcium balance is debated. Patients whose thyroid is removed are still able to regulate calcium adequately. This is accomplished through the action of **parathormone,** a hormone produced by the parathyroid glands. Parathormone seems to play the dominant role in calcium regulation, which will be discussed in the next section.

## PARATHYROIDS

The parathyroids are four tiny glands embedded within the substance of the thyroid gland. Each gland is about the size of a small pea. As can be seen in Figure 9-15, there are two parathyroid glands on the posterior surface of each lateral lobe of the thyroid.

The hormone produced by the parathyroid glands is called parathormone, a polypeptide composed of 84 amino acids. This hormone plays a crucial role in the maintenance of a stable level of blood calcium. Whenever blood calcium begins to fall the parathyroids release parathormone into the blood. Parathormone then stimulates a number of processes that tend to raise blood calcium back to normal.

The most immediate way in which parathormone acts is to stimulate the breakdown of bone matrix by osteoclast cells. This breakdown leads to the transfer of calcium from the bones to the blood. It should be pointed out that as regards homeostasis blood calcium is more important than bone calcium. When the total amount of calcium in the body is low an attempt is made to maintain an adequate level of blood calcium, even if this means that the bones become calcium deficient.

Parathormone is also able to raise blood calcium in two other ways. It increases the rate at which calcium is absorbed from the digestive tract and it decreases the rate at which calcium is excreted by the kidneys. Figure 9-16 is a schematic illustration of the ways in which parathormone brings about an increase in blood calcium.

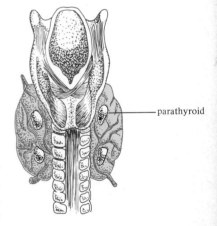

**Figure 9-15** Parathyroid glands.

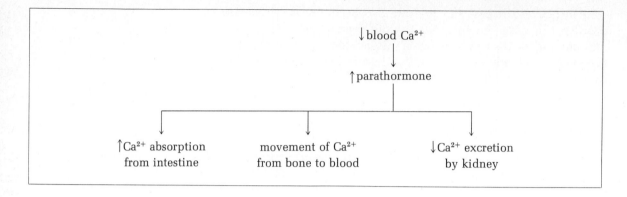

Figure 9-16 Mechanisms by which parathormone raises blood calcium.

It should be kept in mind that if blood calcium becomes too high the parathyroids will secrete only small amounts of parathormone. The consequences of this would be a low rate of bone matrix breakdown, an inhibition of calcium absorption from the digestive tract, and an increased rate of calcium excretion by the kidneys.

**Hypoparathyroidism** refers to an inadequate blood level of parathormone. This condition can be caused either by disease of the parathyroids or by the accidental removal of the parathyroids during thyroid surgery. In the absence of parathormone blood calcium begins to fall. This causes spasm and tetany of the skeletal muscles. It seems that a drop in blood calcium makes nerve and muscle cells more permeable to sodium and thus more excitable. Hypoparathyroidism can be fatal if the blood level of calcium falls low enough to initiate spasm of the diaphragm. However, the consequences of hypoparathyroidism can be avoided by the regular administration of parathormone and calcium.

**Hyperparathyroidism** is the condition in which the parathyroids secrete excessive amounts of parathormone. It is caused by a tumor of the parathyroids. Just as hypoparathyroidism leads to a low blood level of calcium, hyperparathyroidism leads to a high blood level of calcium. All of the symptoms of hyperparathyroidism are due to the high blood calcium. These symptoms include a depression of the central nervous system and muscular weakness. The most dangerous aspect of high blood calcium is its depression of cardiac function. If blood calcium becomes high enough the heart will go into fibrillation. This is a spasm of the heart in which it no longer contracts and relaxes in an orderly manner.

The high blood level of calcium in hyperparathyroidism is in part a result of the movement of calcium from the bone to the blood. In addition to depressing nervous and muscular activity this situation obviously leaves the bones weak. The only cure for hyperparathyroidism is the removal of the tumorous gland.

## ADRENAL GLANDS

The two adrenal glands are small structures located at the top of each kidney, as illustrated in Figure 9-1. Each adrenal gland is composed of two functionally unrelated parts, an outer adrenal cortex and an inner adrenal medulla. The adrenal cortex produces a number of different hormones all of which have a steroid (see Chapter 2) structure. The adrenal medulla, which is functionally part of the sympathetic nervous system, produces the hormone epinephrine.

### Hormones of the Adrenal Cortex

The hormones produced by the adrenal cortex can be divided into these categories: **mineralcorticoids, glucocorticoids,** and **androgens.** Aldosterone, the predominant mineralcorticoid, plays an essential role in regulating sodium, potassium, and water balance. Specifically aldosterone acts on the kidney to reduce sodium and water excretion and to increase potassium excretion. The role of aldosterone in salt and water balance will be discussed in detail in Chapter 17.

As the name implies the glucocorticoids are adrenocortical hormones which play a role in regulating the blood level of glucose. The most important glucocorticoid produced by the adrenal cortex is cortisol. Cortisol acts to increase the blood level of glucose in two ways. It stimulates **gluconeogenesis,** the production of glucose by the liver, and it inhibits the utilization of glucose by a number of different tissues.

In addition to raising blood glucose, cortisol also acts to mobilize stored fat and protein. The mobilization of stored protein leads to an increase in the amount of amino acid circulating in the blood. Because the liver uses amino acids to manufacture glucose, the stimulation of gluconeogenesis by cortisol may simply be a result of the increased availability of amino acids. The role played by cortisol in regulating the pattern by which the body stores and uses glucose, amino acids, and fats will be discussed in detail in Chapter 16.

Cortisol seems to play an important role in the body response to stressful situations. In any stressful situation, for example, infection or tissue injury, the hypothalamus secretes adrenocorticotropic hormone-releasing factor which then stimulates the pituitary to secrete ACTH. The ACTH in turn stimulates the adrenal cortex to secrete cortisol. Unfortunately, the way in which cortisol helps the body respond to stressful situations is not at all clear at the present time. One possibility is that the mobilization of protein brought about by cortisol makes amino acids available for the rebuilding of damaged tissues.

In addition to producing aldosterone and cortisol the adrenal cortex also produces hormones known as **adrenal androgens.** These androgens act in a manner similar to that of the male reproductive hormone testosterone, which is produced by the testes. They promote the growth and development of the male reproductive system as well as the development of the male secondary sexual characteristics such as facial hair and body shape.

The role of the adrenal androgens in normal physiology is still unclear. They are produced in both men and women. In women their masculinizing effects are masked by the female reproductive hormones, estrogen and progesterone. Even in males the adrenal androgens are not able to maintain normal reproductive function in the absence of testosterone.

The adrenal androgens can alter normal reproductive function if they are produced in excess. Occasionally young boys develop an androgen-producing tumor of the adrenal cortex. This will lead to precocious puberty (the early onset of puberty).

**Adrenocortical Diseases**

Hypoadrenalism, the inadequate production of adrenocortical hormones, is known as Addison's disease. This disease can be fatal owing to the excessive loss of salt and water that occurs in the absence of aldosterone. However, treatment with mineralcorticoids can reverse this process and prevent death.

The lack of cortisol in Addison's disease can cause a number of problems. The patient has a difficult time maintaining an adequate level of blood glucose because amino acids are not readily available for glucose synthesis by the liver. In addition the absence of cortisol interferes with the patient's ability to respond fully to stressful situations. Cortisol must also be necessary for normal cerebral function as patients with Addison's disease tend to show abnormal psychological behavior. Full treatment of Addison's disease requires glucocorticoid as well as mineralcorticoid replacement.

Hyperadrenalism, the overproduction of adrenocortical hormones, is known as **Cushing's disease,** as illustrated in Figure 9-17. This condition is brought about by a tumor of cortical tissue. In some cases the growth is caused by an abnormality of the adrenal cortex itself. In other cases it is secondary to overproduction of ACTH by the anterior pituitary. The problems of a patient with Cushing's disease are in a sense the reverse of those of a patient with Addison's disease. There is a moderate retention of salt and water leading to an expansion of the body fluids and hypertension. The blood glucose level tends to be elevated. This leads to an increased production of insulin, a hormone that will be discussed in a subsequent section of this chapter. However, the elevated blood glucose may eventually wear out the ability of the pancreas

## 231  Adrenal glands

to produce insulin, in which cases diabetes mellitus develops.

Cushing's disease can lead to widespread tissue damage owing to the protein breakdown caused by the excessive amounts of cortisol. A person with Cushing's disease characteristically has a puffy face and an unusual distribution of body fat. The abdomen tends to be thick with large fat deposits whereas the limbs are thin with relatively little fat.

### Adrenal Medulla

The inner portion of the adrenal gland, the **adrenal medulla,** is essentially a specialized portion of the sympathetic nervous system. Sympathetic neurons which leave the spinal cord in the lower thoracic region reach the adrenal medulla via the splanchnic nerve. At the adrenal medulla these neurons synapse with the secretory cells of the gland. Activation of the sympathetic nervous system stimulates the secretory cells to release their hormones into the blood.

**Epinephrine** is the principal hormone secreted by the adrenal medulla. However, a small amount of **norepinephrine,** which is

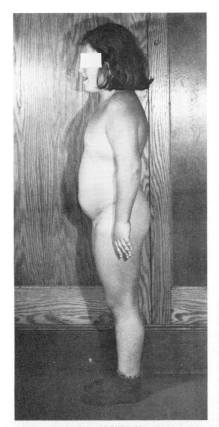
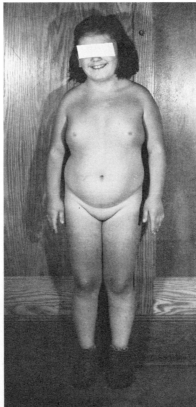

**Figure 9-17** Cushing's disease. (Photo by Lester Bergman & Assoc.)

similar in both structure and function to epinephrine, is also secreted. Epinephrine and norepinephrine are referred to as **catecholamines**.

The physiological effects of epinephrine are similar to those of the sympathetic nervous system in general. As was discussed in Chapter 8 many of the internal organs are innervated by sympathetic nerves. These nerves stimulate the structures they innervate to function in such a way as to help the body to handle emergency (fight or flight) situations. Epinephrine seems to both prolong and intensify the various responses to direct sympathetic nerve stimulation.

Epinephrine affects a large number of internal processes. It stimulates the heart to pump more blood by increasing the heart rate and the force of cardiac contraction. It reduces the blood flow to structures such as the digestive tract which are not needed in the response to emergency situations. This is accomplished by causing the small arteries within these structures to constrict. However, epinephrine produces an increase in the amount of blood flowing to the heart, brain, and skeletal muscles by dilating the small arteries in these structures. Epinephrine increases the availability of cellular energy sources by mobilizing stored carbohydrate and fat. It does this by stimulating the liver to convert glycogen to glucose and by stimulating the adipose cells to convert fat to glycerol and fatty acids. Epinephrine aids respiration by producing dilation of the bronchioles. This provides a wider air passage and makes it easier for air to enter and leave the lungs. Finally, epinephrine acts on the central nervous system to increase alertness and reduce reaction time.

There are a number of situations in which epinephrine is released by the adrenal medulla. It is released at times of fear. It is also released whenever there is a fall in blood pressure (hypotension). By stimulating the heart to pump more blood epinephrine helps restore blood pressure to normal. The control of blood pressure will be discussed in detail in Chapter 13. A fall in blood sugar (hypoglycemia) is another stimulus that elicits epinephrine secretion. The epinephrine helps elevate blood sugar through its stimulation of glycogen and fat breakdown. In Chapter 16 we will discuss the control of blood sugar.

**Adrenal Medullary Disease**

Tumors of the adrenal medulla can secrete excessive amounts of epinephrine and norepinephrine into the blood. The physiological response to very high levels of these hormones is simply an exaggerated version of the response to lower levels.

Complete absence of the adrenal medullary hormones does not cause any serious problems. This is because their effects can all be produced by sympathetic nervous stimulation.

## PANCREAS

The pancreas is located in the upper posterior portion of the abdomen at the level of the second and third lumbar vertabrae, just below the stomach. The head of the pancreas fits into the concavity of the first portion of the small intestine. Its body and tail extend to the left toward the spleen.

The pancreas is both an exocrine and an endocrine gland. As an endocrine gland the pancreas secretes two hormones, insulin and glucagon.

The exocrine and endocrine functions of the pancreas are carried out by different types of cells. The exocrine **acinar cells** are grouped around a system of pancreatic tubules. These tubules empty into the pancreatic duct, which carries the exocrine secretions to the digestive tract. The endocrine-producing cells of the pancreas are located in the **islets of Langerhans,** interspersed between the exocrine acini throughout the pancreas. The islets of Langerhans contain two different types of cells: alpha cells and beta cells.

The beta cells of the islets of Langerhans produce the hormone **insulin.** Insulin is a protein composed of two chains of amino acids. It is probably the most important regulator of the pattern in which the body uses and stores glucose, fat, and protein. This pattern and its regulation by insulin and other hormones will be discussed in detail in Chapter 16. In this chapter we are going to describe only briefly some of the effects of insulin.

The principal action of insulin is to increase the rate at which glucose enters most cells. In the absence of insulin most, but not all, cells are quite impermeable to glucose. On the other hand, in the presence of insulin glucose enters the cells of the body quite easily. The exact manner in which insulin increases cellular permeability to glucose is not understood at the present time.

There are a number of physiological consequences of this basic action of insulin. When the blood level of insulin is high glucose is converted to glycogen for storage by the liver and skeletal muscle. Glucose is also converted to fat for storage by adipose tissue. In most of the remaining tissues of the body glucose is used as a source of ATP production. The movement of glucose into the various cells serves to lower the blood level of glucose.

The principal stimulus for the release of insulin by the pancreas is an elevation of blood glucose. As more insulin is released glucose moves out of the blood and into the cells. This lowers the blood level of glucose back toward normal. Thus insulin helps maintain homeostasis for blood glucose. When the pancreas cannot produce an adequate amount of insulin the blood level of glucose becomes very high. This condition is known as diabetes mellitus and will be discussed in Chapter 16.

In addition to producing insulin the pancreas produces a second hormone called glucagon. This hormone is produced in the alpha cells of the islets of Langerhans. Essentially the effects of glucagon on the body are the opposite of those of insulin. Glucagon acts to raise the blood level of sugar whereas insulin acts to lower it. The increase in blood sugar brought about by glucagon results primarily from the effects of glucagon on the liver. Glucagon stimulates the liver to break down glycogen to glucose and to synthesize new glucose (gluconeogenesis). The glucose produced by both of these processes is then secreted into the blood.

Just as the stimulus for insulin release by the pancreas is an elevation in blood sugar, the stimulus for glucagon release is a drop in blood sugar; thus insulin acts to lower blood sugar when it gets too high and glucagon acts to raise blood sugar when it gets too low.

## OTHER STRUCTURES WITH ENDOCRINE ACTIVITY

In addition to the glands already discussed, a number of other structures produce hormones. Included among these are the ovaries, testes, gastrointestinal tract, kidneys, and thymus gland. The hormones produced by these structures will be discussed in subsequent chapters as they become relevant.

## OBJECTIVES FOR THE STUDY OF THE ENDOCRINE SYSTEM

At the end of this unit you should be able to:

1. State the function of the endocrine system.
2. Distinguish between endocrine glands and exocrine glands.
3. Name the four basic means by which hormonal secretion is controlled.
4. State the location of the pituitary gland.
5. Distinguish between the anterior and posterior portion of the pituitary on the basis of structure.
6. Name the seven hormones secreted by the anterior pituitary gland and state the function of each.
7. Name three diseases caused by abnormal amounts of growth hormone and state the distinguishing characteristics of each.
8. Describe the manner in which hypothalamic releasing factors control the secretion of anterior pituitary hormones.
9. Describe the negative feedback control over anterior pituitary secretion.
10. Name the two hormones secreted by the posterior pituitary and state the function of each.
11. Name the disease caused by a lack of antidiuretic hormone and explain the major problem that occurs.
12. State the location of the thyroid gland.
13. Name the three hormones produced by the thyroid gland and state the function of each.

14. Describe the distinguishing characteristics of a person with each of the following disorders: cretinism, myxedema, and Graves' disease.
15. Describe two different situations that can lead to goiter formation.
16. Describe the location of the parathyroid glands.
17. State the function of parathormone.
18. Describe the distinguishing characteristics of hypoparathyroidism and hyperparathyroidism.
19. State the location of the adrenal glands.
20. Name the three types of hormones secreted by the adrenal cortex and state the function of each.
21. Describe the distinguishing characteristics of Addison's disease and Cushing's disease.
22. Describe the physiological actions of epinephrine and norepinephrine.
23. Name the two hormones secreted by the pancreas and state the function of each.

# 10
# Circulation: Introduction to the Circulatory System and the Blood

INTRODUCTION
BLOOD
WHITE BLOOD CELLS AND BODY DEFENSE
HEMOSTASIS
BLOOD TYPES AND BLOOD TRANSFUSIONS

# INTRODUCTION

The circulatory system carries out a number of essential functions: it transports material from one part of the body to another; it aids in defending the body against infectious organisms and in repairing damaged tissue; it plays an important role in acid-base balance; and it helps regulate body temperature. Every cell in the body depends on the circulatory system to provide it with oxygen and essential nutrients and to remove wastes. If the circulation to any region of the body is cut off, all of the cells in that region will soon die.

The circulatory system is composed of three basic components: the blood, the heart, and the blood vessels. **Blood** is the material that flows through the circulatory system. It serves as the carrier by which material is transported from one part of the body to another. It also contains phagocytic cells and antibodies which help defend against infectious organisms. The **heart** is a pump which provides the force necessary to move the blood. **Blood vessels** are the tubes through which the blood flows as it circulates around the body. There are three types of blood vessels. **Arteries** transport blood from the heart to capillaries within the body tissues. **Capillaries** are very thin-walled vessels in which materials can be exchanged between the circulating blood and the interstitial fluid that surrounds the cells. **Veins** are the vessels through which blood flows from the capillaries back to the heart. As we shall see in Chapter 11, the circulatory system is designed in such a way that blood brought back to the heart from the tissue capillaries is pumped by the heart to the lungs, where it picks up oxygen and loses carbon dioxide, before it is pumped back to the tissue capillaries.

In this chapter we are going to look in detail at the composition and function of the blood. In the next three chapters, we will discuss the heart, the blood vessels, and the mechanisms by which blood pressure and blood flow are controlled.

# BLOOD

Blood is a thick, viscous material composed of cells, protein, water, and dissolved solutes. Normally the blood constitutes about 8% of the total body weight. In a 70-kilogram (150 lb) man this amounts to about 5600 ml of blood.

### Composition of Blood

There are three basic types of cells, or formed elements as they are sometimes called, found in the blood. Figure 10-1 is a blood smear illustrating the different types of cells. **Erythrocytes** (red blood cells) are the most abundant type of cell. They contain large amounts of the pigmented protein hemoglobin which functions to transport oxygen and carbon dioxide around the body. Hemoglobin also plays a role in regulating acid-base balance. **Leukocytes** (white blood cells) act to help defend the body against infectious organisms such as bacteria. They do this by phagocytosis and by the production of antibodies. **Thrombocytes** are cell fragments that play an important role in hemostasis, the process by which blood is prevented from leaving an injured blood vessel.

When blood is placed in a special tube, called a Wintrobe tube, and is centrifuged so that the cells become packed at the bottom, one can determine the **hematocrit**. This is the percentage of the total blood volume occupied by the red blood cells. In a normal person the hematocrit value is in the range of 42–47. Men usually have a higher hematocrit than women. Any change in the hemaocrit is an indication of an excess or deficiency in erythrocytes.

The extracellular fluid of the blood is called the **blood plasma**.

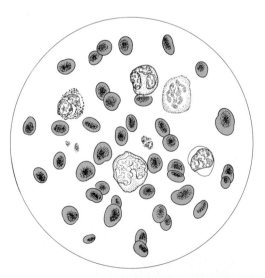

**Figure 10-1** Blood cells.

This represents all of the fluid outside of the cells; therefore its percentage of the blood volume is equal to 100 minus the hematocrit value. Thus, if the hematocrit is 45, the plasma will be 55% of the total blood volume. Approximately 6–8% of the plasma is composed of proteins. By far, the most abundant protein found in the blood is **albumin.** The primary function of albumin appears to be its role in causing the osmotic movement of water from the interstitial fluid to the blood. The fact that protein, primarily albumin, is found in the blood but not in the interstitial fluid means that the total solute concentration, or osmolarity, of the blood exceeds that of the interstitial fluid. This sets up an osmotic gradient which acts as a force, moving water from the interstitial fluid into the blood. The concentration difference caused by the presence of protein in blood is called the **colloid osmotic pressure.** In Chapter 13 we will discuss the movement of water between the blood and interstitial fluid in detail.

**Alpha globulins** and **beta globulins** are blood proteins that function primarily as carrier molecules within the blood. Large molecules, such as hormones, which are not soluble in the blood, attach to these proteins so that they can be transported by the blood from one part of the body to another.

**Gamma globulins** are the protein, or humeral, antibodies which aid in the defense against infectious organisms. The production and action of these substances will be discussed in a later section of this chapter.

**Fibrinogen** and **prothrombin** are two blood proteins which play an important role in blood clotting. The mechanism by which a blood clot is formed will also be discussed later in this chapter.

All of the blood proteins, except the gamma globulins, are produced by the cells of the liver. Gamma globulins are produced by the plasma cells located in the lymph nodes.

About 92% of the blood plasma is composed of water. The major solutes dissolved in the plasma water are glucose, amino acids, the inorganic salts (or electrolytes), waste products such as urea, and small amounts of oxygen and carbon dioxide.

### Erythrocytes

**Shape.** As can be seen in Figure 10-2, the erythrocytes, or red blood cells, are flat, biconcave discs. They have a diameter of about 6–9 microns (6–9/25,000 of an inch). Erythrocytes are quite flexible. This permits them to change shape as they squeeze through tiny capillaries that have a diameter that is approximately the same as their own. Mature red blood cells have no nucleus.

**Function.** The function of erythrocytes is to transport oxygen and carbon dioxide and to help regulate acid-base balance. All of these functions are carried out by the hemoglobin which is con-

**Figure 10-2** Erythrocytes.

tained in these cells. Hemoglobin is composed of an iron-containing pigment, heme, and protein, globin. Normal blood contains about 15 milligrams (mg) of hemoglobin per hundred milliliters. The value is somewhat higher in men than in women.

**Production.** Red blood cells are produced by the red bone marrow. The only exception to this is in the fetus in which red blood cells are produced by the liver until a few months before birth. Figure 10-3 lists the various stages in erythropoiesis, the formation of a mature red blood cell from a primitive stem cell in the red marrow. The primitive stem cell, the **hemocytoblast,** is first converted into an **erythroblast.** The erythroblast begins to produce hemoglobin and is then converted to a **normoblast.** The normoblast is in turn converted into a **reticulocyte** which begins to lose its nucleus. The mature red blood cell, the **erythrocyte,** has no nucleus as it enters the bloodstream. Normally a certain number of immature reticulocytes enter the blood; thus about 5% of the circulating red blood cells are reticulocytes. An increase in the number of reticulocytes or the presence of normoblasts in the blood is a sign that the body has an increased demand for red cells and that these cells are being released by the red marrow before they are fully mature.

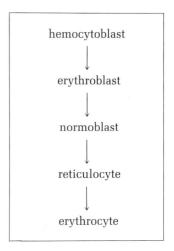

**Figure 10-3** Erythropoiesis.

hemocytoblast
↓
erythroblast
↓
normoblast
↓
reticulocyte
↓
erythrocyte

The rate at which the red marrow produces red blood cells seems to depend on the oxygen needs of the body tissues. Control over this rate is mediated primarily by the hormone erythropoietin which is produced by the kidneys and possibly by other tissues as well. The relationships between oxygen lack, erythropoietin production, and red blood cell formation are illustrated in Figure 10-4. Whenever the blood level of oxygen falls the kidney responds by secreting erythropoietin. The erythropoietin in turn stimulates the red bone marrow to produce red blood cells at a faster rate. Situations that would lead to an increased demand for red blood cells include: hemorrhage, movement to a higher altitude where the oxygen content of the air is low, and an increased level of physical activity.

In order to produce an adequate number of red blood cells certain materials are needed by the red bone marrow either directly or indirectly. Iron is essential for the production of hemoglobin in the

↓ blood $O_2$
↓
↑ secretion of erythropoietin
↓
↑ production of red blood cells

**Figure 10-4** Role of erythropoietin in red blood cell production.

cells. Because not all of the iron taken in with food is absorbed from the digestive tract, more iron must be consumed than is actually needed to make red blood cells. In general, adult men require about 10 mg of iron per day and adult women require about 18 mg of iron per day. The difference is at least partly caused by the iron lost by a women during menstruation.

**Vitamin $B_{12}$** (cyanocobalamin), also known as **extrinsic factor,** is necessary for DNA synthesis. A lack of vitamin $B_{12}$ leads to a reduction in the rate of red blood cell formation. Vitamin $B_{12}$ is almost always present in adequate amounts in the diet. However, the ability to use dietary vitamin $B_{12}$ depends on a substance known as **intrinsic factor.** This substance, which is secreted by the lining of the stomach, is necessary for the absorption of $B_{12}$ from the digestive tract to the blood. Apparently, intrinsic factor combines with vitamin $B_{12}$ in the intestinal tract and prevents it from being digested by the gastrointestinal enzymes. Certain people lack the ability to make intrinsic factor, a condition known as **pernicious anemia.** These people must be treated with injections of vitamin $B_{12}$ because they cannot absorb any vitamin $B_{12}$ from their digestive tract.

Other substances necessary for red blood cell formation include: folic acid, vitamin C, riboflavin, and nicotinic acid.

**Number.** The average red blood cell count for an adult is usually in the range of $4\frac{1}{2}$–$5\frac{1}{2}$ million per cubic millimeter of blood. It should be kept in mind that a cubic millimeter is only 1/1000 of a milliliter (ml). Thus, the total number of red blood cells in an adult is enormous—somewhere around 25 trillion.

A number of factors influence the red blood cell count in any individual. Men tend to have a red blood cell count about 20% higher than that of women at the same age and level of physical activity. As the level of physical activity increases over a period of time, the red blood cell count increases accordingly. A well-trained athlete may have a red blood cell count of 6–7 million per cubic millimeter, whereas a bedridden hospital patient may have a count of only 3–4 million per cubic millimeter. The red blood cell count increases with increasing altitude as a result of the lower density of atmospheric oxygen. Any condition that interferes with the production of red blood cells, or leads to an increased level of red blood cell destruction, will obviously lower the red blood cell count. These conditions are known as **anemias.**

**Red blood cell destruction.** Normally, red blood cells live for about 100–120 days within the blood. Eventually they become quite fragile and are destroyed by phagocytic cells lining the blood vessels of the spleen, liver, and red marrow itself. As the red blood cells are destroyed, the iron is removed from the hemoglobin and is saved for reuse in the formation of new red blood cells. The protein is broken down and the constituent amino acids become part of the pool of amino acids circulating around the body. These amino acids can be used by cells in the production of protein. The heme portion of the hemoglobin minus the iron which has been removed is converted to a substance known as **bilirubin.** Once formed, bilirubin is secreted into the digestive tract by the liver as one of the bile pigments. It then eventually leaves the body as part of the feces. If there is liver disease, or if the bile ducts are obstructed, bilirubin will build up in the blood. This tends to give the tissues a yellowish tint, a condition known as **jaundice.** A sudden increase in the rate at which red blood cells are destroyed can also lead to jaundice if the liver cannot secrete bilirubin as fast as it is being formed.

## WHITE BLOOD CELLS AND BODY DEFENSE

There are a number of means by which the body is protected against the harmful effects of infectious organisms, such as bacteria and viruses. The first line of defense is the epithelial tissue which lines both the outer surface of the body and the inner surface of all the body tubes that open to the outside. As described in Chapter 4, epithelial tissue is composed of cells that are packed very close together. This serves as a barrier which foreign organisms have a hard time crossing.

A second line of defense against invading organisms consists of the phagocytic histiocytes contained in the connective tissue that underlies the epithelial tissue. If any foreign organisms break through the epithelial barrier, these histiocytes attempt to destroy them by phagocytosis.

### Reticuloendothelial System

The **reticuloendothelial system** is composed of phagocytic cells that are located within blood and lymph vessels, particularly in the liver, spleen, red bone marrow, and lymph nodes. These phagocytic cells attack any foreign organisms that gain access to the blood or lymph. They are also responsible for destroying worn-out blood cells in the circulation.

### Leukocytes

Leukocytes (white blood cells) are circulating cells that help defend the body against foreign organisms. As illustrated in Figure 10-5, there are five different types of leukocytes. Three of these, the neutrophils, basophils, and eosinophils, are known as **granular,** or **polymorphonuclear,** leukocytes because they contain small granules and have nuclei that contain a number of different lobes.

**Figure 10-5** Types of leukocytes.

agranular leukocytes

monocyte

lymphocyte

granular leukocytes

basophil

neutrophil

eosinophil

The **neutrophils** are the most abundant of the white blood cells; they constitute about 60–70% of the total number. Neutrophils have a nucleus that contains three to five lobes. They seem to act primarily as phagocytic cells.

**Basophils** are named for the fact that they stain with basic dyes. They have an S-shaped nucleus. Basophils are similar to the mast cells of connective tissue in that they contain large amounts of histamine and heparin. The exact role played by these cells in body defense is not clear at the present time.

**Eosinophils** stain with acid dyes and have a two-lobed nucleus. These cells become quite abundant when foreign proteins or parasites enter the body. However, it is not clear what role they play in fighting these agents.

The two types of **agranular leukocytes,** the monocytes and lymphocytes, are characterized by the fact that they do not contain granules in their cytoplasm.

**Monocytes** are the largest white blood cells. They function to remove large foreign organisms or dead tissue cells by phagocytosis.

**Lymphocytes** are the second most abundant type of white blood cells, constituting about 20–25% of the total. These cells contain a relatively large nucleus surrounded by a small amount of cytoplasm. They play an important role in the production of antibodies.

All of the granular leukocytes are produced in the red bone marrow. The agranular leukocytes are produced in lymphatic structures such as the lymph nodes, tonsils, spleen, and thymus gland. It is possible that the cells in the lymphatic structures that produce agranular leukocytes are themselves cells that originated in the red bone marrow.

The normal red blood cell count is usually in the range of 5000–10,000 cells per cubic millimeter of blood. This number increases tremendously in the presence of infectious organisms. Table 10-1 lists the percentage composition of various types of leukocytes.

**Table 10-1** Percentage of White Blood Cells

| Type of Leukocyte | Composition (%) |
| --- | --- |
| Neutrophils | 62 |
| Eosinophils | 2.3 |
| Basophils | 0.4 |
| Monocytes | 5.3 |
| Lymphocytes | 30 |

### Inflammation

If tissue is damaged for any reason, for example, by infectious organisms, physical injury, or toxic chemicals, the mast cells of the damaged tissue release the chemical histamine. This chemical acts to dilate the local blood vessels and thereby increases the flow of blood to the area. It also increases capillary permeability so that fluid and protein can move out of the blood and into the interstitial space of the damaged area. As this happens, white blood cells, particularly neutrophils and monocytes, move out of the blood and into the damaged area by squeezing through the space between the endothelial cells of the capillaries, a process known as **diapedesis.** At the site of the damaged tissue the neutrophils and monocytes, along with the tissue histiocytes, attack any foreign organisms by phagocytosis. These same cells also remove any dead tissue cells by phagocytosis. Fibrinogen that leaks into the same area tends to clot and thus helps form a seal around the area. This prevents any foreign organisms present from spreading to other parts of the body. The pus found at the site of an infection consists of damaged tissue, dead bacteria, and dead phagocytes.

**Antibody Formation**

One of the principal means the body has for protecting itself against foreign organisms is through the production of antibodies. There are two basic types of antibodies; humoral antibodies and cellular antibodies. **Humoral antibodies** are gamma globulin proteins produced by the plasma cells of the lymph nodes. **Cellular antibodies** are small lymphocytes that function as antibodies. Both humoral and cellular antibodies are able to combine with and inactivate specific foreign molecules known as **antigens.**

Antigens are usually molecules present in the cell membrane of an invading foreign organism. As illustrated in Figure 10-6, the presence of an antigen stimulates the production of an antibody specific to that antigen. Once the antibody is produced, it combines with the antigen in such a way as to cause the death of the organism that contains the antigen. This can be brought about in a number of ways. In some cases the combination of the antibody

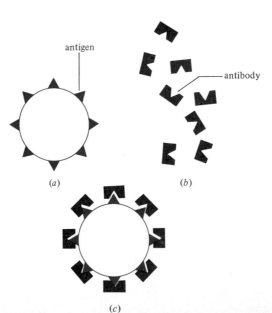

**Figure 10-6** Relationship between antigen and antibody. (a) Presence of cell containing antigen. (b) Production of antibody in response to antigen. (c) Combination of antibody with antigen.

with the antigen causes the cell membrane to rupture. In other situations the one antibody combines with antigens in two different cells so that the cells clump, or agglutinate. The combination of antibody and antigen can also make the organism more susceptible to phagocytosis. Specific antibodies that are produced against toxins secreted by bacteria are known as antitoxins. These antibodies combine with the toxins and neutralize them.

Antibodies are produced against any type of foreign material that enters the body. This complicates transplanting tissues from one person to another because the person who receives the transplanted tissue will produce antibodies that tend to destroy (reject) it.

Occasionally the body will start to produce antibodies against some of its own tissue. This condition, known as **autoimmune disease,** can lead to serious damage to the tissue against which the antibodies are produced. A number of common diseases are now thought to be caused by autoimmunity. These include rheumatoid arthritis, some forms of valvular heart disease, and possibly glomerulonephritis.

### Diseases Involving White Blood Cells

**Leukemia** is a disease in which the production of leukocytes by either the red bone marrow or the lymph nodes becomes out of control. Large numbers of immature white blood cells are released into the blood. These cells tend to use up the oxygen and nutrients required by other cells.

**Agranulocytosis** is a condition in which the red bone marrow stops producing leukocytes due to its destruction by drug poisoning or radiation. Left untreated, this condition is rapidly fatal.

## HEMOSTASIS

**Hemostasis** is the process by which blood is prevented from leaving a cut or injured vessel. This process involves three stages: the local constriction of the blood vessel; the clumping together of thrombocytes to form a plug of cells; and the formation of a blood clot.

When a vessel is injured, chemicals are released which cause the vessel to constrict. This reduces the amount of blood flowing into the injured vessel.

**Thrombocytes** (blood platelets) are round or oval discs approximately 2–5 microns in diameter. They are formed in the red bone marrow as fragments of larger cells called megakaryocytes. There are about 150,000–300,000 thrombocytes per cubic millimeter of blood.

Any damage to the wall of a blood vessel causes the platelets to stick together and form a plug which blocks off the hole in the wall, as illustrated in Figure 10-7. Once the platelets clump together, they release a chemical called platelet factor which initiates the formation of a blood clot.

The mechanism by which blood clots involves a large number of steps and is not very well understood at the present time. It is known that the initial event involves either the release of platelet factor by the platelets as they clump together or the release of a substance known as thromboplastin by injured tissue. As illustrated in Figure 10-8, the release of platelet factor, or thromboplastin, leads to the conversion of the blood protein prothrombin into the protein **thrombin.** Once formed, thrombin causes the conversion of the blood protein fibrinogen into the protein **fibrin.** Fibrinogen molecules are globular whereas fibrin molecules are long threadlike strands. When fibrin molecules are formed, they wrap around each other and form a type of screen that can trap blood cells. The fibrin molecules and the cells trapped within them constitute a blood clot. After a blood clot is formed, chemicals released by the blood platelets cause the clot to shrink and become firm. Once an injured blood vessel has healed, the clot is removed either chemically by fibrinolysin or heparin circulating in the blood or by phagocytosis.

In order for blood to clot a number of accessory factors must be present. **Antihemophilic factor** is a blood protein that is necessary for the conversion of prothrombin to thrombin. As the name implies, this factor is absent in hemophiliacs who have a tendency to bleed excessively. **Vitamin K** is necessary for the production of prothrombin in the liver. Calcium is necessary for a number of steps in the clotting process; however, blood calcium never falls low enough to interfere with the clotting process.

Any interference with the normal smoothness of the inner surface of blood vessel walls can initiate the clumping of platelets and the formation of a clot. Degenerative vascular disease such as arteriosclerosis or atherosclerosis can lead to the formation of a stationary clot, known as a **thrombus.** Once formed, a thrombus can interfere with the flow of blood through the vessel. Thrombus formation can also be brought about by bacterial-induced inflamma-

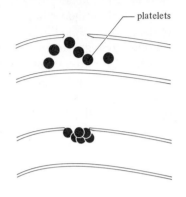

**Figure 10-7** Clumping of platelets to block off a hole in a blood vessel.

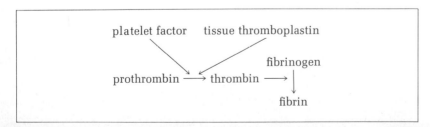

**Figure 10-8** Major steps in the formation of a blood clot.

tion, as in phlebitis, or by an abnormally slow blood flow, as occurs when a patient must stay in bed for a long time. If a thrombus breaks loose from a vessel wall, it will be carried upstream by the flowing blood. Such a moving clot, known as an **embolus,** is particularly dangerous because it can completely obstruct a smaller vessel. For example, when a thrombus breaks loose from one of the large veins in the leg, it can flow through the heart and then block off one of the arteries to the lungs. This condition, known as **pulmonary embolism,** can be rapidly fatal.

# BLOOD TYPES AND BLOOD TRANSFUSIONS

Whenever blood is given from one person to another, great care must be taken that there are no antibodies present in one of the bloods that can combine with antigens present in the other blood. Blood is typed according to the presence or absence of certain antigens in the red blood cell membrane. There are two major systems of antigens which must be considered in giving a transfusion: the ABO system and the Rh system.

### ABO System

Figure 10-9 illustrates the four different types of blood found in the ABO system. Type A blood contains the A antigen, or agglutinogen, in the red blood membranes. Likewise type B blood contains the B antigen, or agglutinogen. Type AB blood has both the A antigen and the type B antigen whereas type O blood has neither

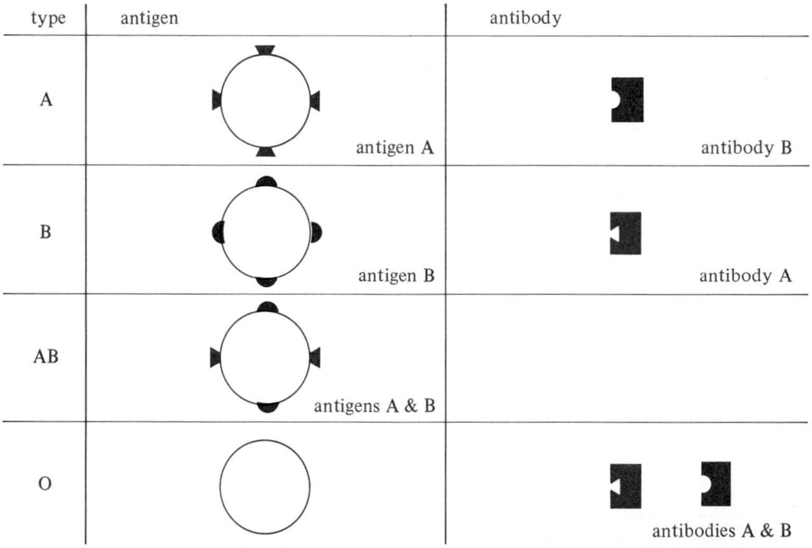

| type | antigen | antibody |
|------|---------|----------|
| A | antigen A | antibody B |
| B | antigen B | antibody A |
| AB | antigens A & B | |
| O | | antibodies A & B |

**Figure 10-9** ABO blood types.

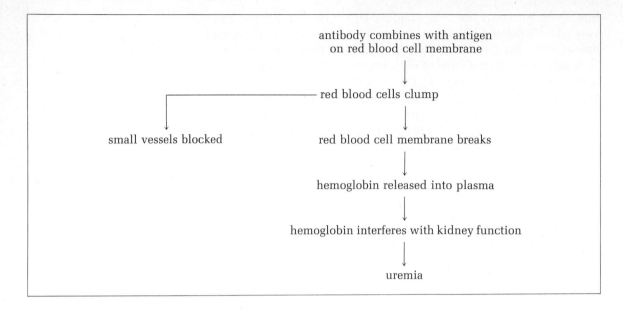

Figure 10-10 Transfusion reaction.

the A nor the B antigen. Each of the four types of blood in the ABO system contains antibodies, or agglutinins, for those antigens it lacks. Thus, type A blood has the antibody to B, type B blood has the antibody to A, type AB blood has neither antibody, and type O blood has both the antibody to A and the antibody to B.

If bloods are mismatched so that agglutinins from one blood can attach to agglutinogens in the red blood cells of the other blood, the red blood cells will agglutinate or clump together. These agglutinated cells are attacked and destroyed by phagocytosis and their hemoglobin is released into the blood. Hemoglobin can deposit in the kidney leading to serious kidney damage. Figure 10-10 sums up the events of a transfusion reaction that occurs if bloods are mismatched.

When only small quantities of blood are given to a person, one normally only has to worry about the recipient having agglutinins against the donor's agglutinogens. This is because the donor's agglutinins become so diluted by the recipient's blood that they are no longer effective. Figure 10-11 shows the various types of donor

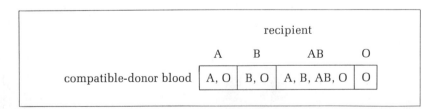

Figure 10-11 Compatible ABO blood types.

blood that can be given to each type of recipient if only small amounts of blood are given. Because type O blood has no agglutinogens, it can be given in small amounts to any recipient. For this reason, type O is sometimes referred to as a universal donor. Likewise type AB blood has no agglutinins and thus a type AB person can receive a small amount of any of the ABO types. A type AB person is sometimes referred to as a universal recipient. If relatively large amounts of blood are given, there is a danger that agglutinins in the donor blood can combine with agglutinogens in the red blood cells of the recipient. In this situation, a person can only receive their own type blood.

**Rh System**

The second major group of antigens by which blood must be classified is the Rh system. Figure 10-12 shows the agglutinogens and agglutinins present in two different Rh types. Rh-positive blood contains the Rh agglutinogen in its red blood cells and Rh-negative blood does not. An Rh-negative person will form an agglutinin against Rh-positive blood if he is exposed to it. This is different from the ABO system in which the agglutinin is present even if there has been no exposure to the agglutinogen.

Rh incompatibility between a mother and fetus can cause serious complications. If a mother is Rh-negative and the fetus is Rh-positive, the mother will produce Rh agglutinins if any of the fetal red blood cells enter her circulation. This is usually not a serious problem with the first Rh-positive child of an Rh-negative mother because fetal cells rarely enter the mother's circulation except at birth. However, the second time an Rh-negative mother is carrying an Rh-positive child, the agglutinins formed by the mother can enter the child's blood and destroy its red blood cells. This condition is known as **erythroblastosis fetalis.** Administration of Rh agglutins to an Rh-negative mother just after she delivers her first Rh-positive child seems to suppress her own production of Rh agglutins and thus prevent erythroblastosis fetalis with the second child.

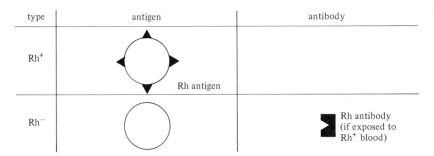

**Figure 10-12** Rh blood types.

# OBJECTIVES FOR THE STUDY OF CIRCULATION

At the end of this unit you should be able to:

1. State four functions of the circulatory system.
2. Name the three major types of blood vessels.
3. Name which type of blood vessel takes blood away from the heart and which type returns blood to the heart.
4. Name the three types of blood cells and state the major function of each.
5. Define the term hematocrit.
6. State the normal numerical range of the hematocrit valve.
7. Define the term blood plasma.
8. State what percentage of the plasma is composed of protein.
9. State the function and site of production of the following blood proteins: albumin, alpha globulins, beta globulins, gamma globulins, fibrinogen, and prothrombin.
10. Describe the structure of red blood cells.
11. Describe the chemical composition of hemoglobin.
12. State the site of red blood cell production.
13. Name the hormone that regulates red blood cell production and state the stimulus for its release.
14. Distinguish between the role of extrinsic factor and intrinsic factor in red blood cell production.
15. State the normal range for the red blood cell count of an adult.
16. Describe where in the body red blood cells are destroyed and the process by which this takes place.
17. Describe the portion of the hemoglobin molecule from which bilirubin is formed and how bilirubin is eliminated from the body.
18. Name the condition that results from the buildup of bilirubin in the blood.
19. Describe the composition of the reticuloendothelial system.
20. Name the five different types of leukocytes and state which are granular and which are agranular.
21. State the normal range for the white blood cell count of an adult.
22. Describe the process of inflammation.
23. Distinguish between humoral antibodies and cellular antibodies.
24. Define the term antigen.
25. Explain how antibodies can destroy an organism that contains an antigen.
26. Explain what is meant by autoimmune disease.
27. Define the term hemostasis.
28. State the site of thrombocyte formation.
29. State the three ways in which blood platelets are involved in hemostasis.
31. Name the vitamin necessary for prothrombin production by the liver.
32. Define the terms thrombus and embolus.
33. State the basis by which blood is typed.
34. Name the antigens and antibodies found in the following blood types: A, B, AB, O, $Rh^+$, $Rh^-$.
35. Determine if the donor's blood and recipient's blood are compatible once the blood types of the donor and the recipient are known.
36. Describe the type of reaction that takes place if bloods are mismatched.
37. Describe under what circumstances Rh incompatibility between mother and father can occur and the consequences of such incompatibility.

# 11
# The Heart

INTRODUCTION
CHAMBERS OF THE HEART
HEART VALVES
CONTRACTION OF THE HEART
CARDIAC CYCLE
HEART SOUNDS
CARDIAC OUTPUT

## INTRODUCTION

The heart is a muscular pump that develops the pressure necessary to propel blood through the blood vessels. Contracting at a rate of 70 times a minute, the heart beats over 3 billion times in a 70-year life-span. During this time it pumps about 50 million gallons of blood.

**Location of the heart.** As illustrated in Figure 11-1, the heart is located within the thoracic cavity in the space between the two lungs. This space is known as the mediastinum. The heart is not centered in the middle of the thorax. Rather, it is positioned at an angle with about two-thirds of the heart located to the left of the

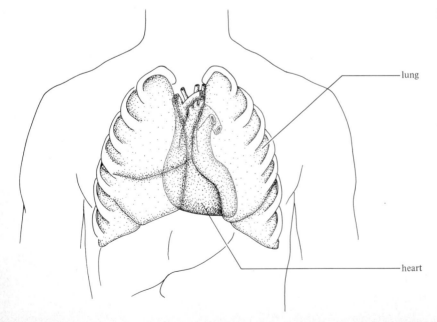

**Figure 11-1** Heart within the thoracic cage.

midline. The apex, or point, of the heart faces to the left and rests on the diaphragm. It is situated at the level of the space between the 5th and 6th ribs. An apical pulse is taken by placing a stethoscope over the space between these ribs on the left side and then listening to the heart. The base of the heart faces toward the right and is located at the level of the 2nd rib.

The heart is surrounded by a fluid-filled sac known as the pericardium. As illustrated in Figure 11-2, the pericardium is a double-layered sac that folds back on itself. The outer layer adheres loosely to the heart and the inner layer forms the outer portion of the heart wall. The fluid occupies the space between the two layers.

**Layers of the heart.** As illustrated in Figure 11-3, the wall of the heart itself can be divided into three layers. The outer layer, known as the **epicardium,** is simply the visceral infolding of the pericardium—it is a serosal membrane with an outer layer of epithelial cells that secrete fluid into the pericardial space.

Situated below the epicardium is the heart muscle tissue, the **myocardium,** which makes up the bulk of the heart wall. Cardiac muscle is similar to skeletal muscle in that the actin and myosin filaments are arranged in an orderly fashion giving the muscle cells a

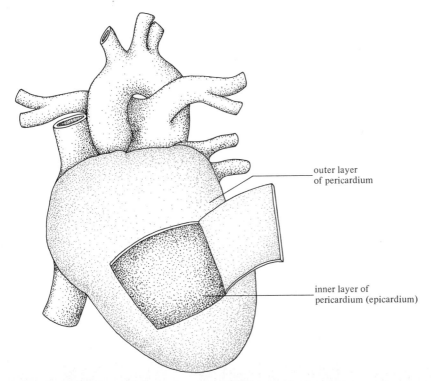

**Figure 11-2** Pericardium.

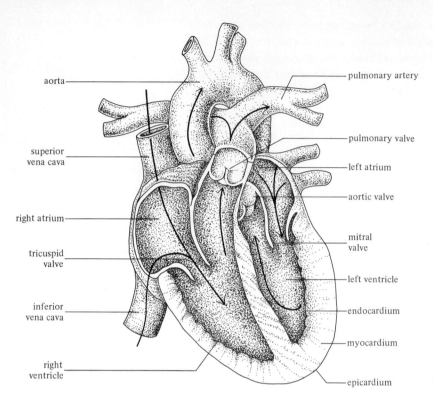

**Figure 11-3** The heart.

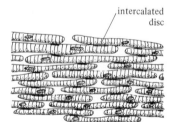

**Figure 11-4** Cardiac muscle.

striated appearance. However, as illustrated in Figure 11-4, at certain points the individual cells are held exceedingly close together by specialized structures known as intercalated discs. The discs are areas of very low electrical resistance that allow impulses to pass freely from one muscle cell to the next. This means that once a single muscle cell of the heart depolarizes, the pulse will travel to all the other muscle cells of the heart. The significance of this for normal cardiac function and for disease will be seen in later sections of this chapter.

The inner layer of the heart wall is the **endocardium.** This layer consists of the connective tissue next to the myocardium and of the epithelial tissue that lines the chambers of the heart.

## CHAMBERS OF THE HEART

Internally the heart is divided into four chambers, two upper atria and two lower ventricles. These are illustrated in Figure 11-3. The upper atria have relatively thin walls, whereas the lower ventricles have thicker walls. The partition between the right and left atria is called the **interatrial septum** and the partition between the right and left ventricles is called the **interventricular septum.** There are two small appendages, shaped somewhat like a dog's ear, one attached to each atrium. These are called the auricular appendages.

The veins which bring blood back to the heart empty into the two atria. Blood returning to the heart from all tissue except the lungs enters the right atrium through either the **superior vena cava, inferior vena cava,** or **coronary sinus.** Blood returning from the lungs enters the left atrium through one of the four **pulmonary veins.**

The ventricles receive blood from the atria and pump it out of the heart into the arteries. Blood from the right ventricle flows into the pulmonary artery and then to the lungs. Blood from the left ventricle flows into the aorta and then to all of the other tissues of the body.

Figure 11-5 is a schematic view of the flow of blood through

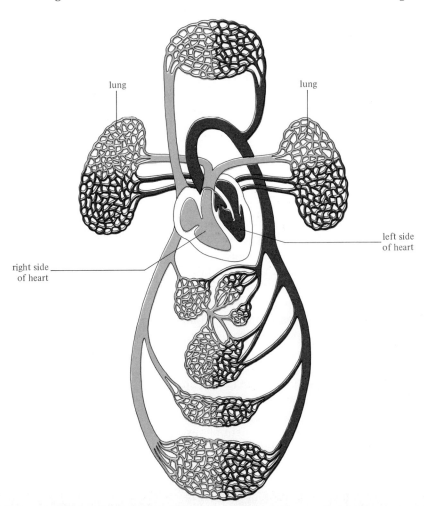

**Figure 11-5** Schematic illustration of the pattern of blood flow.

the heart. Blood returning from the body tissues is carried into the right atrium through the superior and inferior venae cavae (blood from the tissues of the heart itself enters the right atrium through the coronary sinus). The blood then flows from the right atrium into the right ventricle. From the right ventricle blood flows into the pulmonary artery and then to the capillaries of the lungs where it picks up $O_2$ and eliminates $CO_2$. The pulmonary veins bring blood from the capillaries of the lungs back to the left atrium. From the left atrium blood flows into the left ventricle. The left ventricle then pumps blood into the aorta. From the aorta blood flows into smaller arteries which carry it to capillaries in all the body tissues. Blood from the tissues above the heart flows into the superior vena cava and then into the right atrium. Blood from tissues below the heart flows into the right atrium through the inferior vena cava.

Functionally the circulatory system consists of two halves. The flow of blood from the right side of the heart through the lungs to the left side of the heart is called the **pulmonary circulation.** The flow of blood from the left side of the heart through the body tissues to the right side of the heart is called the **systemic circulation.** This design of the circulatory system permits all of the blood that is returned to the right side of the heart from the body tissues to be sent to the lungs, where it can pick up oxygen and eliminate carbon dioxide, before being pumped back to the body tissues by the left side of the heart. Figure 11-6 sums up the flow of blood through the heart.

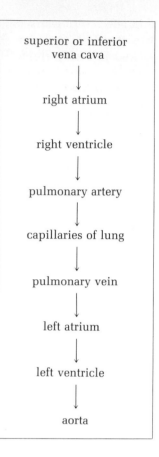

**Figure 11-6** Flow of blood through the heart.

## HEART VALVES

To ensure that blood flows through the heart in the proper direction various parts of the heart are separated by valves. Figure 11-7 is a schematic illustration of how a valve works. A valve contains flaps of tissue that are cupped in one direction. As seen in Figure 11-7a, a force from behind the cup opens the valve. A force in the opposite direction, as seen in Figure 11-7b, closes the valve. Thus the valve shown in Figure 11-7 would allow movement from left to right but not from right to left.

There are four valves within the heart. Their location is shown in Figure 11-3 and their structure is illustrated in Figure 11-8. The **tricuspid valve** is composed of three leaflets or cusps. It is attached to a fibrous ring that surrounds the opening between the right atrium and the right ventricle. The valve between the left atrium and left ventricle is called the **bicuspid valve (mitral valve).** The mitral valve is somewhat similar in structure to the tricuspid valve except that it contains only two cusps. In addition the opening of the bicuspid valve is somewhat smaller than the opening of the tricuspid. The tricuspid and mitral valves are referred to as the atrioventricular or A-V valves. Both A-V valves are attached by fibrous strands, the cordae tendonae, to small muscular projections of the inner ventricular wall. These muscles, which look somewhat like nipples, are called the **papillary muscles.** The papillary muscles contract at the same time as the ventricles and prevent the A-V valves from buckling backward.

The valve between the right ventricle and the pulmonary artery is called the **pulmonary valve.** The valve between the left ventricle and the aorta is called the **aortic valve.** Both of these valves have three flaps shaped somewhat like a half-moon, which is why these valves are known as the **semilunar valves.**

Unfortunately, for various reasons, such as a bacterial infection, the valves of the heart can become damaged. In some cases the valves become damaged in such a way that they do not open completely. This is called **stenosis.** Thus in mitral stenosis the valve between the left atrium and left ventricle does not open completely. This interferes with the flow of blood into the left ventricle causing the blood to back up into the pulmonary circulation. In other cases the valves are damaged in such a way that they do not close completely. This is called valvular insufficiency. In tricuspid insufficiency the valve between the right atrium and the right ventricle does not close completely. This causes some of the blood to flow from the right ventricle back to the right atrium when the right ventricle contracts.

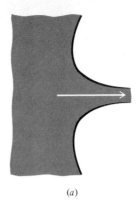

(a)

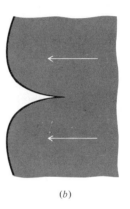

(b)

## CONTRACTION OF THE HEART

### Conduction System

The heart is able to pump the blood as a result of the contraction and relaxation of the myocardium, the muscular wall of the

**Figure 11-7** Mechanism of valvular function.

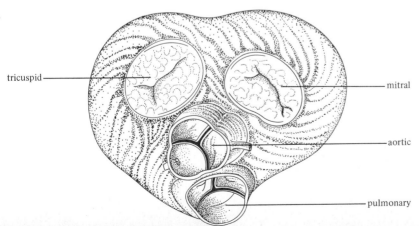

**Figure 11-8** Valves of the heart.

heart chambers. Like any other muscle cells, the heart muscle cells must first be depolarized before they can contract. The function of the conduction system of the heart is to depolarize the cells of the myocardium so that they contract in an orderly way. Figure 11-9 illustrates the structure of the conduction system.

Depolarization of the heart is initiated at the **sinoatrial** or S-A node located in the wall of the right atrium near the entrance of the superior vena cava. Like all cardiac muscle the S-A node has the property of autorhythmicity—it is able to depolarize and repolarize with its own inherent rhythm, free of external stimuli. However, the rate at which the S-A node rhythmically depolarizes is faster than that of any other part of the heart. For this reason the S-A node serves as the **pacemaker** of the heart. Each time the S-A node depolarizes, the depolarization spreads over the rest of the heart. Depolarization of the heart leads to contraction. Thus the rate at which the S-A node depolarizes determines the rate at which the heart contracts.

A second important structure of the cardiac conduction system is the **atrioventricular, or A-V node.** The A-V node is located at the base of the right atrium near the interatrial septum. It is connected to a group of cells known as the **atrioventricular bundle,**

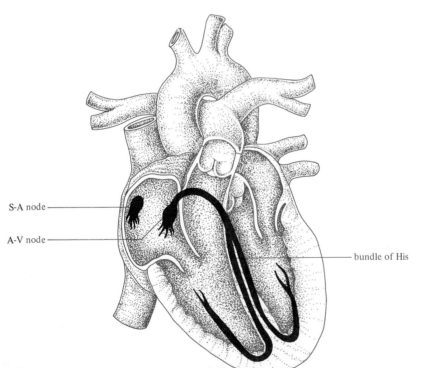

**Figure 11-9** Conduction system of the heart.

**259** Contraction of the heart

or the bundle of His. The bundle of His extends from the A-V node to the top of the interventricular septum where it divides into right and left bundles that pass along the right and left sides of the interventricular septum. At the bottom of the interventricular septum the bundles of His are connected to the **Purkinje fibers** which extend up into the myocardium of the ventricles.

The pattern in which depolarization spreads over the heart is illustrated in Figure 11-10. The sequence of events is as follows: Depolarization starts at the S-A node, as illustrated in Figure 11-10a. It then spreads down over both the right and left atria to the A-V node, as shown in Figure 11-10b. From the A-V node the depolarization continues down the bundle of His to the base of the ventricles, as shown in Figure 11-10c. The depolarization then spreads up the Purkinje fibers into the myocardium of the ventricles, as shown in Figure 11-10d.

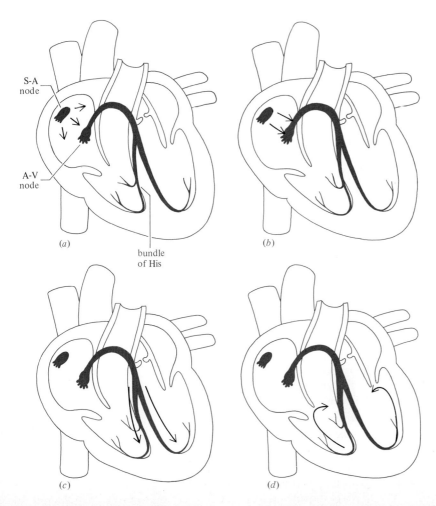

**Figure 11-10** Spread of depolarization over the heart.

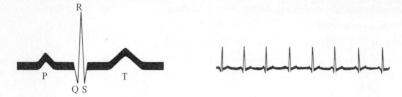

**Figure 11-11** Normal electrocardiogram.

This pattern of depolarization guarantees that the different parts of the heart contract in the right order. The fact that the depolarization spreads over the atria before it spreads over the ventricles guarantees that the atria contract before the ventricles do. Because depolarization spreads downward over the atria, the atria contract from top to bottom. This enables the atria to push blood downward into the ventricles. The bundle of His and the Purkinje fibers cause depolarization to spread upward along the walls of the ventricles. Thus the bases of the ventricles contract before the tops. This enables the ventricles to push blood upward into the pulmonary artery and aorta.

The electrical activity of the heart can be measured by a special instrument called an electrocardiograph. This instrument is connected to the skin by wires that respond to electrical changes caused by depolarization and repolarization of the various parts of the heart. The tracings of these changes constitute an electrocardiogram, as illustrated in Figure 11-11. The first wave is called a P wave and is associated with depolarization of the atria. The second wave is called the QRS complex and is associated with the depolarization of the ventricles. The third wave is called the T wave and is associated with the repolarization of the ventricles. Repolarization of the atria occurs during the time that the ventricles depolarize and does not show up on the electrocardiogram.

A diseased heart will produce an electrocardiogram that is different from normal. Analysis of the electrocardiogram is used as a tool in pinpointing what is wrong with a diseased or damaged heart.

## CARDIAC CYCLE

The cardiac cycle refers to the series of events that takes place during each contraction of the heart. This cycle is divided into two main stages: **systole** and **diastole.** Systole refers to the time during which the ventricles are contracting while diastole refers to the time during which the ventricles are relaxed. Thus one complete cardiac cycle occurs from the beginning of one diastole through systole to the beginning of the next diastole.

Figure 11-12 shows the heart at the beginning of diastole. At this point both the atria and ventricles are relaxed. Blood is flowing into the right atrium from the superior and inferior venae cavae and is entering the left atrium from the pulmonary veins. Because the pressure of the blood in the atria is greater than the pressure of the blood in the ventricles the tricuspid and mitral valves are open. Thus much of the blood that enters the atria passes through the A-V openings and into the ventricles. Approximately 70% of ventricular filling occurs before the atria contract. Because the semilunar valves guarding the arterial exits are closed the blood that enters the ventricles stays there.

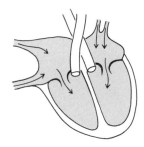

Figure 11-12 Heart in diastole.

Once the S-A node depolarizes, the impulse spreads over the atria and the atria contract. Figure 11-13 illustrates atrial contraction. As the atria contract they push the blood they contain down into the ventricles. This blood is added to the blood that entered the ventricles prior to atrial contraction.

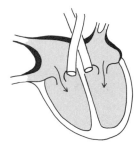

Figure 11-13 Atrial contraction.

By the time the atria have contracted, depolarization has spread over the ventricles and they begin to contract. Contraction of the ventricles is illustrated in Figure 11-14. As the ventricles contract, the pressure of the blood in the ventricles is increased. As soon as the pressure of the blood in the ventricles exceeds the pressure of the blood in the atria the tricuspid and mitral valves close. This prevents blood from flowing back into the atria. As the ventricles continue to contract, the pressure of the blood within them rises even higher. Soon the pressure in the ventricles is great enough to force open the aortic and pulmonary valves. Blood then flows from the right ventricle into the pulmonary artery and from the left ventricle into the aorta.

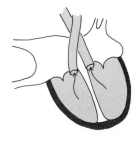

Figure 11-14 Ventricular contraction.

Figure 11-15 illustrates what happens once blood has left the ventricles and the ventricles have begun to relax. The pressure in the pulmonary artery is now greater than the pressure in the right ventricle; the pressure in the aorta is greater than the pressure in the left ventricle. This causes the pulmonary and aortic valves to close, preventing the backflow of blood into the ventricles.

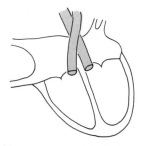

Figure 11-15 Ventricular relaxation.

Further relaxation of the ventricles causes the pressure within them to fall below the pressure in the atria, which have been filling up with blood during ventricular systole. Once the pressure in the atria exceeds the pressure in the ventricles the A-V valves open and a new cycle begins.

The sequence of events of the cardiac cycle takes place simultaneously in the right and left sides of the heart. That is, the right and left atria contract at the same time, the right and left ventricles contract at the same time, and the tricuspid and pulmonary valves open and close at the same time as the mitral and aortic valves. The only difference between the cardiac cycle on the two sides of the heart is in the amount of pressure. Pulmonary artery pressure varies from about 10 millimeters of mercury (mm Hg) during diastole to about 25 mm Hg at the peak of systole. In contrast aortic pressure varies from about 80 mm Hg during diastole to about 120 mm Hg during systole. Thus, the pressure in the right ventricle and pulmonary artery is considerably lower than the pressure in the left ventricle and aorta.

## HEART SOUNDS

If one listens to the heart through a stethoscope two distinct sounds can be heard. The first sound is louder, lower-pitched, and of longer duration than the second and sounds like "lubb." It is caused by the closing of the tricuspid and mitral valves at the beginning of ventricular contraction. The second sound is softer, higher-pitched, and shorter than the first and sounds something like "dubb." This sound is caused by the sudden closing of the pulmonary and aortic valves at the beginning of diastole (relaxation of the ventricles). Various diseases of the heart can lead to abnormal sounds known as heart murmurs. For example, in mitral stenosis the blood rushing through the narrow mitral valve produces a sound that can be heard through a stethoscope.

## CARDIAC OUTPUT

The amount of blood pumped by each ventricle per minute is known as the **cardiac output,** or minute volume. In a normal, resting adult the cardiac output is about 5 liters per minute. During strenuous exercise the cardiac output can be increased to as much as 20–25 liters per minute.

The amount of blood pumped by the heart per minute depends on two factors: the **heart rate** and the **stroke volume.** The heart rate is the number of times the heart beats per minute. Normally the heart rate of a resting adult is about 72 beats per minute. The stroke volume is the amount of blood pumped by the heart during each contraction. At rest stroke volume is about 70 ml of blood per beat.

The cardiac output is equal to the stroke volume multiplied by the heart rate. That is, the amount of blood pumped by the heart per minute is equal to the amount of blood pumped per beat multiplied by the number of times the heart beats per minute. For example, if the stroke volume is 70 ml per beat and the heart rate is 72 beats per minute, the cardiac output will equal $72 \times 70$ or 5,040 ml per minute.

Cardiac output is not constant. Rather it is continuously being altered to meet changing conditions. For example, your heart must pump much more blood when you are running than when you are sitting in a chair. Because cardiac output depends on the heart rate and the stroke volume, the only way it can be altered is by a variation in one or both of these factors. In other words the amount of blood pumped by the heart is altered by changing either the rate at which it contracts, or the strength with which it contracts, or both.

### Control of Heart Rate

The heart rate is set by the number of times the S-A node depolarizes per minute. Although the S-A node has an inherent rate of depolarization, this rate can be altered by the action of nerves and hormones. The S-A node receives nerve fibers from both the sympathetic and parasympathetic nervous systems. The sympathetic fibers are postganglionic fibers from the cervical and upper thoracic ganglia. These fibers release the chemical transmitter norepinephrine that speeds up the heart. Thus as the number of impulses traveling along the sympathetic nerves to the S-A node increases, more norepinephrine is released, and the heart rate increases.

Parasympathetic fibers reach the S-A node as part of the cardiac branch of the vagus nerve. These fibers release acetylcholine, a chemical transmitter that slows down the heart. When parasympathetic activity is reduced this also acts to speed up the heart.

The rate at which the S-A node depolarizes is also influenced by a number of hormones. Physiologically the most important of these is epinephrine which speeds up the heart rate. The higher the blood level of epinephrine the faster the heart will beat.

In combination, the sympathetic nerves, parasympathetic nerves, and epinephrine can be used to change the heart rate from a resting level of 72 beats per minute to as high as 200 beats per minute. As the heart rate increases over this range the cardiac output also increases. Once the heart rate exceeds 200 beats per minute the cardiac output begins to fall. This is because at this rate the ventricles do not have time to fill adequately between contractions.

### Control of Stroke Volume

The stroke volume is determined by the strength with which the myocardium contracts. The strength of myocardial contraction is influenced by a number of factors. One of these is the amount of blood in the ventricles at the time they begin to contract. This is illustrated in Figure 11-16. Up to a certain point, the more blood there is in the ventricles at the time they begin to contract, the greater the stroke volume. This is because the strength with which myocardial muscle cells contract increases as their length increases. Thus the more blood there is in the ventricles at the time

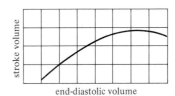

**Figure 11-16** Starling's law.

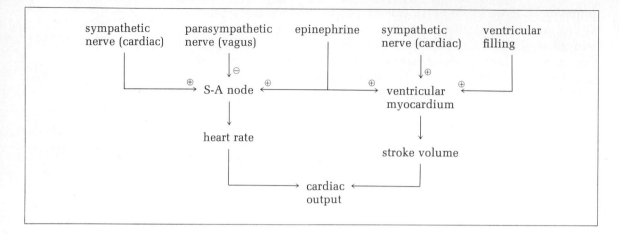

**Figure 11-17** Factors that control cardiac output.

they begin to contract, the more stretched the myocardial muscle cells will be, and the stronger they will contract. This relationship is known as **Starling's law of the heart** after the British physiologist.

The stroke volume is also influenced by neural and hormonal factors. Sympathetic fibers go to the ventricular myocardium as well as to the S-A node. These fibers release norepinephrine, which causes the myocardial muscle cells to contract more forcefully, thereby increasing stroke volume. In addition the hormone epinephrine increases stroke volume. In a normal person the various factors mentioned can increase stroke volume from a resting level of 70 ml per beat to 140 ml per beat. In a well-trained athlete the stroke volume can go as high as 200 ml per beat.

Figure 11-17 sums up the various factors that influence how much blood is pumped by the heart, the cardiac output. The cardiac output depends on the heart rate and stroke volume. As mentioned previously heart rate is determined by the combined action of sympathetic nerves, parasympathetic nerves, and epinephrine. Stroke volume is determined by the amount of blood in the ventricles at the time they start to contract, by the action of sympathetic nerves, and by epinephrine. Any change in the amount of blood pumped by the heart is brought about through the action of some or all of these factors. In Chapter 13 we will discuss some of the situations in which these factors are used to alter cardiac output.

# OBJECTIVES FOR THE STUDY OF THE HEART

At the end of this unit you should be able to:

1. Describe the location of the heart.
2. Name the sac that surrounds the heart.
3. Name the three layers of the heart wall.
4. Name the four chambers of the heart and state the relative location of each.
5. Name the blood vessels that are connected to each of the chambers of the heart.
6. Name the wall separating the two atria and the wall separating the two ventricles.
7. Describe the flow of blood from the venae cavae to the aorta.
8. Distinguish between the pulmonary circulation and the systemic circulation.
9. Name the four valves found in the heart and state the location of each.
10. Explain the importance of depolarization of the heart.
11. State the location of the S-A node and describe its function.
12. State the location of the A-V node and the bundle of His.
13. Describe the sequence of events in the spread of depolarization from the S-A node to the rest of the heart.
14. State the cardiac events associated with the P wave, QRS complex, and T wave of an electrocardiogram.
15. Distinguish between the terms systole and diastole.
16. Describe one heart beat in terms of the sequence of depolarization, contraction, pressure changes, the flow of blood, and the opening and closing of the various valves.
17. State the causes of the two primary heart sounds.
18. Define the terms cardiac output, heart rate, and stroke volume.
19. Give the approximate normal, resting adult values for cardiac output, heart rate, and stroke volume.
20. State the equation relating cardiac output, heart rate, and stroke volume.
21. Describe the ways in which heart rate can be changed.
22. Describe the ways in which stroke volume can be changed.

# 12

## The Blood Vessels and Lymphatics

ARTERIES
CAPILLARIES
VEINS
LYMPHATIC SYSTEM
SPLEEN
THYMUS

The blood, which transports material from one part of the body to another, flows through a system of tubes known collectively as the blood vessels. These tubes form a network which enables the blood to flow within a short distance of every cell in the body. It has been estimated that no cell is more than 0.005 inch from a capillary. Over this short distance material can move back and forth between the blood and cells simply by diffusion.

As mentioned previously there are three major types of blood vessels: arteries, capillaries, and veins. The arteries are a branching system of smaller and smaller tubes through which blood flows from the heart to the tiny capillaries located in all of the body tissues. At the capillaries there is an exchange of material between the blood and the interstitial fluid. Thus the capillaries are the site at which material enters and leaves the bloodstream. The veins are in a sense the reverse of the arteries in that they are a system of larger and larger tubes that join together. It is through this system of tubes that blood flows from the capillaries back to the heart.

## ARTERIES

### Structure

The walls of the arteries are designed to withstand the relatively high pressure of the blood that is leaving the heart. They are composed of three basic layers: the tunica intima, the tunica media, and the tunica externa. Although the walls of all arteries contain these three basic layers the exact composition of each layer depends on the size of the artery. The inner layer, the **tunica intima,** of all arteries is composed primarily of endothelial tissue. However, in larger arteries there is a thin layer of elastic tissue within the tunica intima. The middle layer of the arterial wall, the **tunica media,** is the thickest of the three layers. In very large arteries this layer is composed almost entirely of elastic tissue. As the arteries get smaller and smaller there is a transition from elastic tissue to

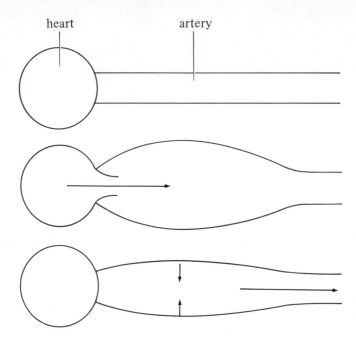

**Figure 12-1** Expansion and contraction of the large arteries during the cardiac cycle.

smooth muscle tissue within the tunica media. In all arteries the outer layer, the **tunica externa,** is composed of fibrous connective tissue. This tissue helps give strength to the arterial wall.

The elastic tissue within the walls of the large arteries enables them to stretch as they absorb the blood which rapidly leaves the heart with each contraction. Upon completion of contraction the large arteries return to normal size forcing blood on to the capillaries and veins. This process is illustrated in Figure 12-1.

The smooth muscle found in the smaller arteries allows the diameter of these vessels to be adjusted. As illustrated in Figure 12-2, contraction of this muscle narrows the vessels and makes it more difficult for blood to flow through them. This process is known as **vasoconstriction.** The reverse process, **vasodilation,** is a relaxation of the arterial smooth muscle. Upon vasodilation the small arteries become wider and blood flows through them more readily. In Chapter 13 we will discuss how vasoconstriction and vasodilation are used to regulate both the blood pressure and the amount of blood flowing to the various tissues.

As a person grows older the arterial walls undergo various degenerative changes. These changes are responsible either directly or indirectly for approximately 60% of all deaths in this country. Three major types of changes occur. **Arteriosclerosis,** or hardening of the arteries, refers to a loss of elasticity in the arterial wall. As a result of this the arteries cannot swell as readily to accommodate the large amount of entering blood. A larger volume of blood in a smaller space leads to an increase in blood pressure. **Atherosclerosis** is a degenerative disease involving fat deposits in the

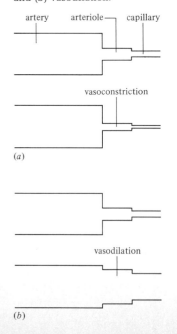

**Figure 12-2**
(a) Vasoconstriction and (b) vasodilation.

arterial wall. The exact cause of this disease is not known; however, it does seem to have some relationship to a diet high in animal fat and cholesterol. The danger of atherosclerosis is that the fat deposits can block off the vessel they form in, or break loose and block off smaller vessels further downstream. The fat deposits also can initiate the formation of a **thrombus,** an internal blood clot. An **aneurysm** is a weakening of the arterial wall so that it bulges out. If the wall becomes too weak it is no longer able to withstand the pressure of the blood and it bursts. This leads to a hemorrhage which can be fatal.

## Major Arteries and Their Distribution

### Pulmonary Arteries

As mentioned in the previous chapter blood leaves the heart through two large arteries: the **pulmonary artery** and the **aorta.** Shortly after leaving the right ventricle the pulmonary artery branches into the right pulmonary artery which carries blood to the right lung and the left pulmonary artery which carries blood to the left lung. Within each lung these arteries branch into smaller and smaller arteries which carry blood to the capillaries adjacent to the alveoli. It is at these capillaries that oxygen enters and carbon dioxide leaves the blood.

### Aorta

The oxygenated blood that returns from the lungs is pumped by the left side of the heart into the aorta. Blood then flows from the aorta into branching arteries which eventually carry the oxygenated blood to all the body tissues. Figure 12-3 is a schematic drawing which shows the major large arteries. The aorta itself is divided into four parts: the ascending aorta, the aortic arch, the thoracic aorta, and the abdominal aorta. Arising from the base of the left ventricle the **ascending aorta** extends upward for a short distance above the heart. The **aortic arch** then curves backward and to the left so that by the end of the arch, at the level of the 4th thoracic vertebra, the aorta is extending downward along the back. Arteries that transport blood to the head and upper limbs arise from the aortic arch. The thoracic aorta extends from the 4th to the 12th thoracic vertebrae. It gives rise to the arteries which supply the viscera and walls of the thoracic cage. The final section of the aorta is the **abdominal aorta.** It extends to the level of the 4th lumbar vertebra where it divides into the left and right common iliac arteries. Branches of the abdominal aorta carry blood to the organs of the abdominal cavity and the walls of the abdominal cavity. The common ilac arteries carry blood to the lower limbs.

**Coronary arteries.** The right and left coronary arteries branch off the ascending aorta just above the aortic valve (see Figure 12-6).

These blood vessels and their smaller branches carry blood to the capillaries of the myocardium. Blockage of one of the smaller coronary arteries by an atherosclerotic plaque, a thrombus, or a combination of the two is the most common cause of heart attacks.

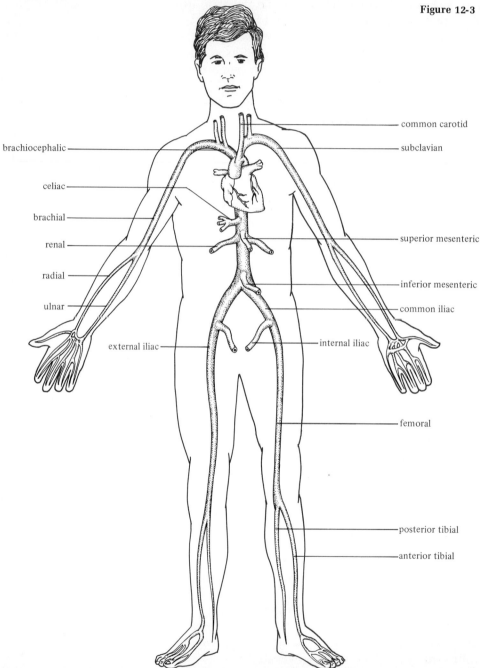

Figure 12-3 Major arteries.

# Arteries

**Brachiocephalic artery.** The brachiocephalic artery is the first artery that branches off the aortic arch. It extends upward and to the right for a short distance before branching into the right common carotid and right subclavian arteries. The left common carotid and left subclavian arteries arise directly from the aortic arch.

**Common carotid arteries.** The common carotid arteries transport blood to the head and neck. The common carotids pass up through the neck on either side of the trachea (see Figure 12-4). At the upper border of the thyroid cartilage (Adam's apple) they each branch into internal and external carotid arteries. As mentioned above, the right common carotid arises from the brachiocephalis artery whereas the left common carotid arises directly from the aorta.

**External carotids.** The external carotids arise from the common carotids and pass up along the neck to the level of the top of the mandible. At this point the external carotid branches into **superficial temporal** and **internal maxillary** arteries. The superficial temporal passes up the side of the head and the internal maxillary supplies the deep structures of the face. The pulse can be felt by placing two fingers on the superficial temporal artery just in front of the ear. Figure 12-4 shows the external carotid and some of

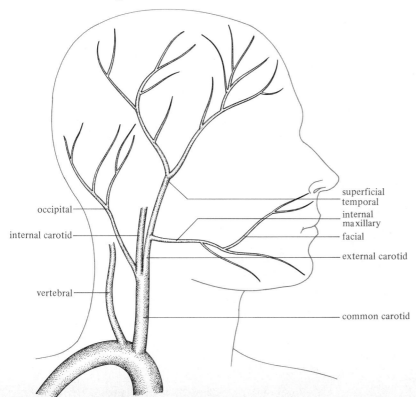

**Figure 12-4** External carotid artery and its major branches.

its more important branches. The major structures supplied by this artery are the thyroid, the muscles of the neck, the pharynx, the structures of the mouth (tongue, teeth, salivary glands), the muscles and skin of the face, the nose, the ear, and the dura mater. In other words the external carotid and its branches supply the structures of the upper neck and most of the skull other than the brain.

**Internal carotids.** The internal carotid arteries enter the cranial cavity through a hole in the temporal bone and then pass to the base of the brain. At the base of the brain the internal carotid gives rise to four major arteries: the **ophthalmic artery,** the **middle cerebral artery,** the **anterior cerebral artery,** and the **posterior communicating artery.** The ophthalmic artery supplies the orbital cavity and the eyeball. It also sends blood to the nose and forehead. The middle cerebral artery supplies the basal ganglia, the internal capsule, and the temporal surface of the brain. The anterior cerebral artery supplies the basal ganglia as well as the corpus callosum and the frontal lobes of the brain. The posterior communicating artery along with the anterior cerebral artery enters into the formation of (the **circle of Willis**) an interconnecting system. Figure 12-5 illustrates the circle of Willis.

**Circle of Willis.** At the base of the brain blood is brought into the circle of Willis by three major arteries: the basilar artery and the two internal carotids. The basilar artery is formed by the union of the two vertebral arteries which are in turn branches of the subclavian arteries. The circle itself is formed by the two anterior cerebral arteries, the anterior communicating artery, the two posterior

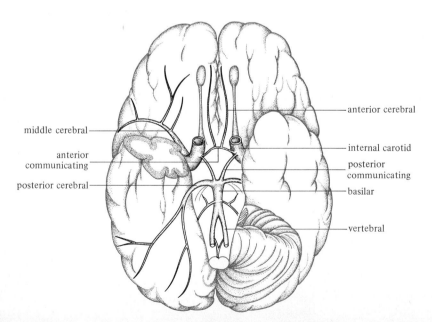

Figure 12-5 Circle of Willis (inferior view).

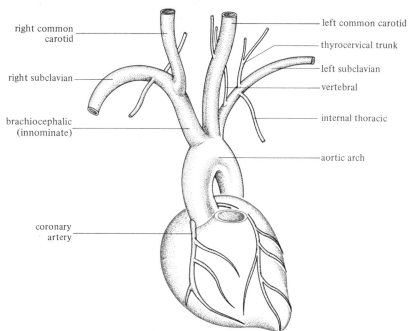

**Figure 12-6** Branches of the aortic arch.

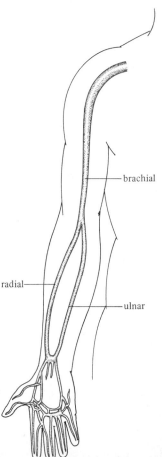

**Figure 12-7** Major arteries of the arm.

communicating arteries, and the two posterior cerebral arteries. As is best understood at the present time the function of the circle is to permit an alternate route for blood flow if a major artery to one side of the brain is obstructed.

**Subclavian.** The subclavian arteries transport blood to the shoulder region. The left subclavian arises directly from the aorta whereas the right subclavian is a branch of the brachiocephalic artery. There are three main branches off each subclavian artery: the **vertebral artery** which passes to the brain and then forms the basilar artery; the **thyrocervical trunk** which supplies the thyroid, larynx, and trachea; and the **internal throacic,** or mammary artery, which supplies the skin and muscles of the anterior thorax. Figure 12-6 illustrates the two subclavian arteries and their branches.

**Axillary.** The axillary artery is simply the continuation of the subclavian artery as it passes through the axillary region. This artery turns into the brachial artery as it enters the upper arm.

**Brachial.** The brachial artery passes through the upper arm. At the elbow it passes in front of the humerus and then divides to form the **radial artery,** which supplies the thumb side of the lower arm, and the **ulnar artery,** which supplies the little finger side of the lower arm. Figure 12-7 shows the arteries of the arm.

**Branches of the abdominal aorta.** The abdominal aorta extends from the end of the thoracic aorta to the level of the 4th lumbar vertebra. For the purpose of identification its many branches can be divided into visceral branches, which carry blood to the internal organs of the abdomen, and parietal branches, which carry blood to the skin and muscles of the abdominal walls. The most important visceral branches are the celiac, superior mesenteric, middle suprarenal, renal, spermatic (or ovarian), and inferior mesenteric arteries. Figure 12-8 illustrates the abdominal aorta and its main branches.

**Celiac artery.** The celiac artery arises from the anterior surface of the abdominal aorta just below the diaphragm. It passes a short distance and then divides into three branches: the **left gastric artery** which supplies the bottom of the esophagus and top of the stomach; the **hepatic artery** which supplies the liver; and the **splenic artery** which supplies the spleen. The hepatic artery has two important branches: the **right gastric artery** which supplies the stom-

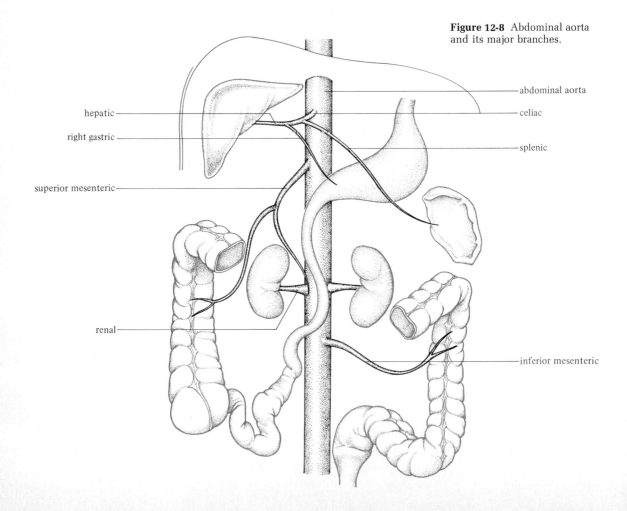

**Figure 12-8** Abdominal aorta and its major branches.

ach and the **gastroduodenal artery** which supplies the lower part of the stomach, the duodenum, and the pancreas.

**Superior mesenteric artery.** The superior mesenteric artery arises from the anterior aorta just below the celiac artery. It supplies all of the small intestine except the duodenum. It also supplies the right side of the large intestine, that is, the cecum, appendix, ascending colon, and the right half of the transverse colon.

**Middle suprarenal arteries.** These two arteries arise from the side of the aorta at the level of the superior mesenteric artery. Each supplies the adrenal gland on its side.

**Renal arteries.** The two renal arteries branch off the abdominal aorta just below the two suprarenal arteries. They carry blood to the kidneys on their side.

**Spermatic or ovarian arteries.** In the male the two spermatic arteries branch off the aorta and descend through the inguinal canal to the testes. The corresponding arteries in the female are the **ovarian arteries.** These arteries pass from the abdominal aorta to the ovaries.

**Inferior mesenteric artery.** The inferior mesenteric artery arises from the base of the abdominal aorta just above the point where it divides to form the common iliac arteries. It supplies the left portion of the large intestine (the left side of the transverse colon, the descending colon, the sigmoid colon, and most of the rectum).

The parietal branches of the abdominal aorta include the inferior phrenic arteries, the lumbar arteries, and the middle sacral artery.

**Inferior phrenic arteries.** The inferior phrenic arteries arise from the abdominal aorta just above the celiac artery. They supply the undersurface of the diaphragm.

**Lumbar arteries.** The four pairs of lumbar arteries supply the muscles of the abdominal wall, the abdominal portion of the spinal cord, the lumbar vertebrae, and the muscles and skin of the lower back.

**Middle sacral artery.** The middle sacral artery extends from the back of the base of the aorta. It supplies blood to the sacrum and coccyx.

**Common iliac arteries.** The two common iliac arteries are formed from the terminal division of the abdominal aorta at the fourth lumbar vertebra. They extend to the level of the sacroiliac joint on their corresponding side at which point they divide into external and internal iliac arteries. Branches of the common iliac arteries supply blood to the peritoneum and to the ureters. Figure 12-9 illustrates the major arteries of the hips and legs.

**Internal iliac arteries (hypogastric arteries).** Each internal iliac artery branches off the common iliac on its side and enters the

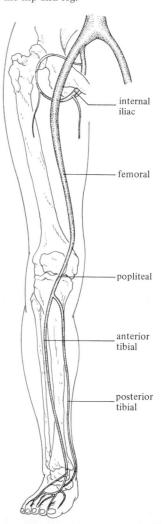

**Figure 12-9** Major arteries of the hip and leg.

pelvic region. These arteries supply the urinary bladder, rectum, reproductive structures other than the testes and ovaries, buttock, and the medial portion of the thigh.

**External iliac artery.** The external iliac arteries are continuations of the common iliac arteries. They extend to the level of the inguinal ligament where they enter the thighs as the femoral arteries.

**Femoral artery.** Each femoral artery passes in front of the hip joint and then descends along the anterior, medial surface of the thigh. Branches of the femoral artery supply blood to the skin, muscles, and bone of the thigh.

**Popliteal artery.** The popliteal artery is a continuation of the femoral artery beginning in the lower third of the thigh. It passes behind the knee and then divides to form the anterior and posterior tibial arteries.

**Anterior tibial artery.** The anterior tibial artery passes between the tibia and fibula along the lateral, anterior surface of the lower leg. It supplies the muscles and skin of the front of the lower leg.

**Posterior tibial artery.** The posterior tibial artery passes down the back of the lower leg. It supplies the skin and muscles of this region.

# CAPILLARIES

It is at the capillaries that materials can be exchanged between the flowing blood and the interstitial fluid bathing the cells. This exchange can take place because of the exceedingly thin walls of these tiny vessels. A capillary wall is composed of a layer of endothelial tissue only one cell thick. At the present time it is not clear whether materials cross the capillary wall by passing through the endothelial cells, between the cells, or by a combination of both of these.

In general the capillary walls seem to be quite permeable to most substances smaller than proteins. However, the degree of permeability differs from one tissue to another. The capillaries in the liver are the most permeable whereas those in the brain are the least permeable.

Capillaries are distributed in great abundance throughout all tissues of the body. It has been estimated that if all the capillaries were put end to end they would stretch about 60,000 miles, even though individual capillaries are only about 1 millimeter long. This abundance of capillaries guarantees that every cell is close enough to a capillary to permit the ready diffusion of material between the blood and the immediate surroundings of the cell.

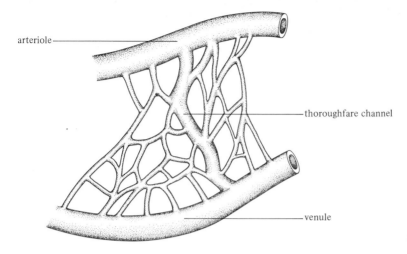

**Figure 12-10** Capillary bed.

As illustrated in Figure 12-10, microscopic examination shows each capillary bed to be made up of two types of capillaries. **Arteriovenous capillaries,** or thoroughfare channels, are direct connections between the arterioles and venules. Blood flowing through these channels does not exchange material with the surrounding extracellular fluid. These channels permit blood to flow rapidly through a tissue that is not metabolically active at the moment. The second type of capillaries are the **true capillaries** along which materials are exchanged with the extracellular fluid. These capillaries form a winding network between the arterioles and the venules.

Rings of smooth muscle known as precapillary sphincters are found at the entrances to the true capillaries. These sphincters regulate the amount of blood flowing into each part of the capillary bed according to local metabolic conditions.

**Sinusoids** are specialized capillary-type vessels found in the liver, spleen, and red bone marrow. These vessels have irregularly structured walls which are much more permeable than normal capillaries. For example, protein can usually pass freely through the walls of a sinusoid. The walls of a sinusoid usually contain phagocytic cells.

## VEINS

The walls of veins are composed of the same three basic layers as are those of arteries: a tunica intima, tunica media, and tunica externa. However, the venous walls contain considerably less elastic connective tissue than do the arterial walls and are therefore much thinner. This is possible because the blood in veins is under less pressure than the blood in arteries. The tunica media and tunica externa of the veins contain smooth muscle which can be used to adjust the venous diameter. Normally about 60% of the total blood volume is within the veins. In times of stress the smooth muscles of the veins constrict. This increases the rate at which blood is returned to the heart thereby increasing cardiac output. The increase in cardiac output in turn increases the amount of blood in the arteries and capillaries. In other words constriction of the veins leads to a decrease in the amount of blood in the veins and an increase in the amount of blood in the arteries and capillaries. Venous constriction is brought about as a result of stimulation by the sympathetic nervous system.

Many veins, particularly those in the extremities, contain valves which prevent the backflow of blood. Figure 12-11 illustrates the structure of a venous valve. These valves are composed of two pocketlike flaps of epithelial and connective tissue. If the pressure behind the valve exceeds the pressure in front of the valve, the valve will open and blood will flow toward the heart. If the pressure in front of the valve exceeds the pressure behind the valve, the pockets will fill with blood and the valves will close. This prevents blood from flowing backward in the veins, away from the heart.

For a variety of reasons the valves in the veins of the legs occasionally wear out. When this happens the veins become very dilated, a condition known as **varicose veins.**

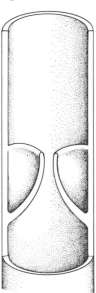

Figure 12-11 Venous valve.

## Major Veins

### Drainage of Heart
**Coronary sinus.** Situated in the groove between the left atrium and left ventricle on the posterior surface of the heart, the coronary sinus receives blood from the smaller coronary veins of the heart. The coronary sinus empties blood into the right atrium just below the opening of the inferior vena cava. Some of the coronary veins open directly into the right atrium.

### Drainage of the Head and Neck
The cranial venous sinuses are spaces formed between the two layers of dura mater, as illustrated in Figure 12-12. Blood from veins in the cranial and orbital cavities eventually flows into these sinuses. The major sinuses include the superior sagittal, inferior sagittal, straight, occipital, and transverse sinuses. These sinuses are illustrated in Figure 12-13. Blood from all these sinuses eventually enters the internal jugular veins via the sigmoid portions of the transverse sinuses on each side of the head.

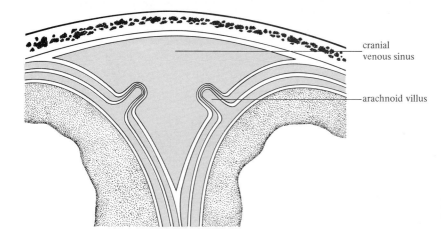

Figure 12-12 Structure of a cranial venous sinus.

### Internal Jugular Veins

The internal jugular veins start at the base of the skull and descend down through the neck just lateral to the common carotids. They join the subclavian veins at the level of the sternoclavicular joint. The internal jugulars serve to drain blood from the brain, eyes, and superficial parts of the face and neck.

### External Jugular Veins

Blood from the exterior portion of the cranium and from the deep parts of the neck and face is drained via the external jugular

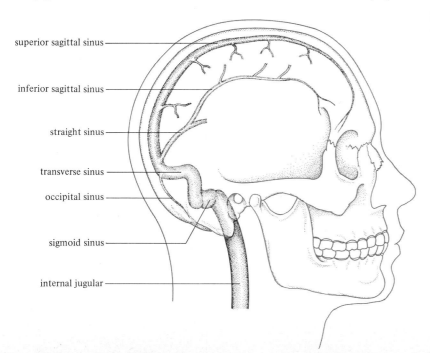

Figure 12-13 Cranial venous sinuses.

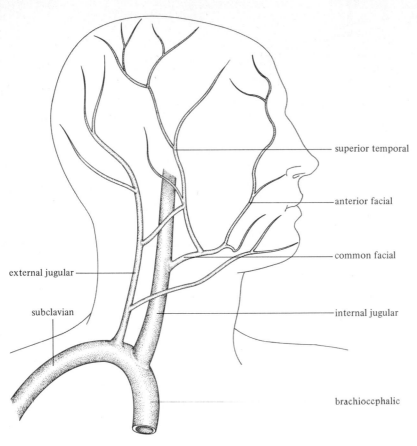

**Figure 12-14** Veins that drain the head and neck.

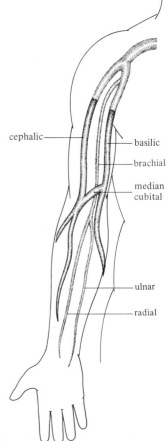

**Figure 12-15** Major veins of the arm.

veins. As illustrated in Figure 12-14, these veins start at the level of the angle of the mandible and descend to the subclavian vein.

### Veins of the Upper Extremities

The veins of the arm fall into two major categories: superficial veins and deep veins. As the names imply, the superficial veins are near the surface whereas the deep veins are deeper in the tissues. However, the two sets of veins are connected by many anastomoses so that blood can move from one set to the other. Figure 12-15 shows some of the main veins of the arm.

#### Superficial Veins of the Upper Extremities

**Cephalic vein.** The cephalic vein starts on the thumb side of the wrists. It then ascends along the lateral surface of the forearm and upper arm. It then passes between the deltoid and pectoralis muscles to join the axillary vein in the shoulder just below the clavicle.

**Basilic vein.** The basilic vein starts at the little finger side of the wrist and ascends along the medial surface of the arm. At the upper end of the arm it joins the deep brachial vein to form the axillary vein.

**Median cubital vein.** The median cubital vein connects the cephalic vein and the basilic vein at the front of the elbow; it is frequently used as a site for intravenous injection.

### Deep Veins of the Upper Extremity

In general the deep veins of the arm follow the same path as the arteries and are, in fact, often enclosed in the same sheath. Often the deep veins are found in pairs, one on either side of the corresponding artery.

**Radial veins.** The radial veins pass through the deep portion of the lateral side of the forearm.

**Ulnar veins.** The ulnar veins are in the deep portion of the medial side of the forearm.

**Brachial veins.** The brachial veins are formed at the elbow by the joining of the radial and ulnar veins. They ascend through the upper arm to the bottom border of the pectoralis major where they join with the basilic vein to form the axillary vein.

**Axillary vein.** The axillary vein passes through the shoulder and becomes the subclavian vein at the outer border of the first rib.

**Subclavian vein.** The subclavian vein drains blood from the arm and neck. At the junction of the sternum and clavicle the subclavian vein joins with the internal jugular vein to form the brachiocephalic vein.

### Veins of the Thorax

**Brachiocephalic veins.** As illustrated in Figure 12-16, the

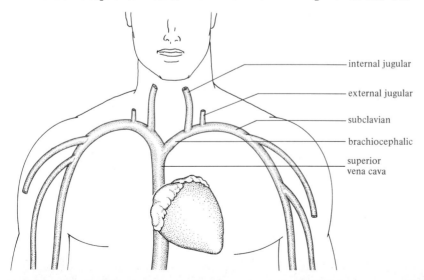

**Figure 12-16** Major veins that drain into the superior vena cava.

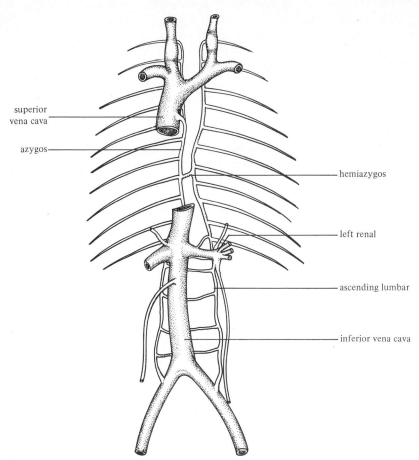

Figure 12-17 Azygous system of veins.

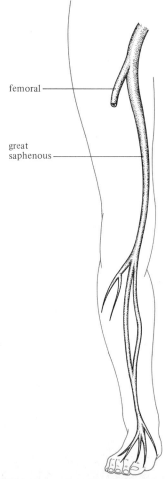

Figure 12-18 Saphenous vein.

brachiocephalic veins are formed by the union of the internal jugular veins and subclavian veins on each side. The two brachiocephalic veins are joined together to form the superior vena cava at the level of the first costal cartilage. The brachiocephalic veins receive blood from the internal thoracic, vertebral, and inferior thyroid veins in addition to the blood they receive from the subclavian and internal jugular veins.

**Azygos veins.** As illustrated in Figure 12-17, the azygos vein passes in front of the vertebral column slightly to the right of the midline. This vein is a continuation of the right ascending lumbar vein. As it ascends along the vertebral column it receives blood from the intercostal, esophageal, mediastinal, pericardial, right bronchial, and hemiazygos veins. Because the right lumbar vein has connections with the right common iliac vein and the inferior vena cava, blood from these veins can also enter the azygos vein. The azygos vein enters the superior vena cava at the level of the 4th thoracic vertebra.

**Hemiazygos veins.** The hemiazygos vein is a continuation of the left lumbar vein. It ascends along the left side of the vertebral column to the level of the 9th thoracic vertebra where it crosses over to join the azygos vein. The hemiazygos receives blood from the lowest four or five intercostal veins as well as from some esophageal and mediastinal veins. In addition a communicating vein which connects the left renal vein with the hemiazygos veins serves as a channel connecting the inferior vena cava and other veins in the lower part of the body with the superior vena cava. Thus blood from the lower part of the body can be returned to the heart via the superior vena cava if the inferior vena cava is obstructed.

## Veins of the Lower Extremities

Like those of the upper extremities, the veins of the lower extremities can be divided into deep veins and superficial veins. The superficial veins of the legs can become varicosed and are therefore of considerable clinical importance.

### Superficial Veins of the Lower Extremities

**Great saphenous vein.** The great saphenous vein begins at the medial side of the top of the foot. As illustrated in Figure 12-18, it passes in front of the medial malleolus and then ascends along the medial side of the thigh and leg. At the level of the groin in the top of the leg it empties into the femoral vein.

**Small saphenous vein.** The small saphenous vein begins at the back of the ankle and passes behind the lateral malleolus. It then ascends along the lateral, posterior portion of the leg to the back of the knee where it empties into the popliteal vein.

### Deep Veins of the Lower Leg

**Posterior tibial veins.** The posterior tibial veins begin at the back of the foot, behind the medial malleolus, as illustrated in Figure 12-19. They ascend behind the tibia to just below the knee where they join the anterior tibial veins to form the popliteal vein.

**Anterior tibial veins.** The anterior tibial veins begin at the top of the foot in front of the medial malleolus. They ascend along the anterior surface of the leg in front of the tibia. Just below the knee the anterior tibial veins pass behind the tibia to join the posterior tibial veins in the formation of the popliteal vein.

**Popliteal vein.** Formed by the junction of the anterior and posterior tibial veins, the popliteal vein passes behind the knee joint. In the thigh the popliteal vein becomes the femoral vein.

**Femoral vein.** The femoral vein is a continuation of the popliteal vein. It passes through the thigh medial to the femur. In the upper portion of the thigh it receives blood from the great sa-

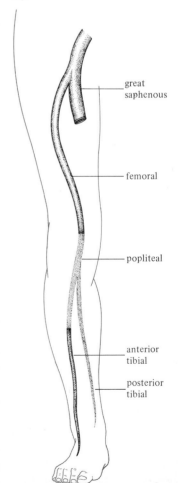

**Figure 12-19** Deep veins of the leg.

phenous vein. The femoral vein then passes under the inguinal ligament into the pelvic region where it continues as the external iliac vein.

**External iliac vein.** The external iliac is the continuation of the femoral vein in the pelvic region. It ascends in a superior and medial direction to the sacroiliac joint where it connects with the internal iliac vein to form the common iliac vein.

**Internal iliac vein (hypogastric).** The internal iliacs drain blood from the pelvic structures, the buttocks, and the medial surface of the thigh. They join with the external iliacs on each side to form the common iliacs.

**Common iliac vein.** The common iliac veins are formed by the union of the external and internal iliacs. They ascend toward the midline of the body where they join to form the inferior vena cava in front of the 5th lumbar vertebra.

## Veins of the Abdomen

**Inferior vena cava.** The inferior vena cava is formed by the union of the two common iliac veins. It ascends along the front of the vertebral column to the right of the aorta. The inferior vena cava passes along a groove in the posterior surface of the liver and then through the diaphragm before entering the right atrium of the heart. As it ascends through the abdominal cavity the inferior vena cava receives blood from the abdominal viscera and the abdominal body walls. Figure 12-20 illustrates the inferior vena cava and its major tributaries.

**Lumbar veins.** There are four lumbar veins on each side of the body. They drain blood from the skin and muscles of the lower back as well as the lumbar region of the vertebral column and spinal cord. The lumbar veins empty blood into the inferior vena cava. In addition they are interconnected by the ascending lumbar veins on either side of the vertebral column. As mentioned previously the right ascending lumbar vein continues as the azygos vein and the left lumbar vein continues as the hemiazygos vein.

**Spermatic (testicular) veins.** The spermatic veins in the male drain blood from the testes. They pass into the abdominal cavity through the inguinal canal. The right spermatic vein empties directly into the inferior vena cava and the left spermatic vein empties into the left renal vein.

**Ovarian veins.** The ovarian veins in the female drain blood from the ovaries. As is the case with spermatic veins, the right ovarian vein empties into the inferior vena cava whereas the left ovarian vein empties into the left renal.

**Renal veins.** The renal veins drain blood from the kidneys and ureters. They join the inferior vena cava above the level of the lumbar veins. The left renal vein receives blood from the left sper-

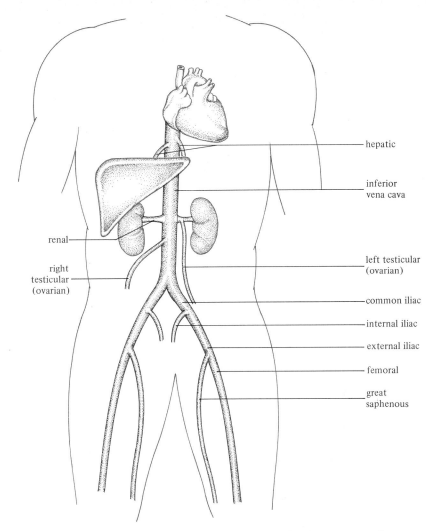

Figure 12-20 Major veins that drain into the inferior vena cava.

matic (ovarian) vein, left inferior phrenic, and left suprarenal veins in addition to blood from the left kidney and ureter.

**Suprarenal veins.** The suprarenal veins drain blood from the adrenal glands. The right suprarenal empties directly into the inferior vena cava and the left suprarenal empties into the left renal vein.

**Hepatic veins.** The hepatic veins drain blood from the sinusoids of the liver. They enter the anterior surface of the inferior vena cava just below the diaphragm.

## Portal System of Veins

The portal system of veins transports blood from the capillaries of the digestive tract, pancreas, spleen, and gallbladder to the sinusoids of the liver. Thus, blood flowing through this system flows through two sets of capillaries from the time it leaves the heart until it returns to the heart. Figure 12-21 illustrates the portal system of veins.

**Hepatic portal vein.** The hepatic portal vein is formed by the union of the superior mesenteric vein and the splenic vein. It ascends ventral to the inferior cava and enters the lower surface of the liver where it divides into right and left branches. Within the liver the hepatic portal vein forms smaller and smaller branches which eventually empty into the liver sinusoids. As it ascends through the abdominal cavity the hepatic portal vein receives blood from the following veins: the left gastric which drains the lower esophagus and part of the stomach, the paraumbilical which drains the anterior abdominal wall, and the cystic which drains the gallbladder.

**Superior mesenteric vein.** The superior mesenteric vein drains the small intestine and the first half of the large intestine. It ascends from the pelvic region to about the level of the duodenum

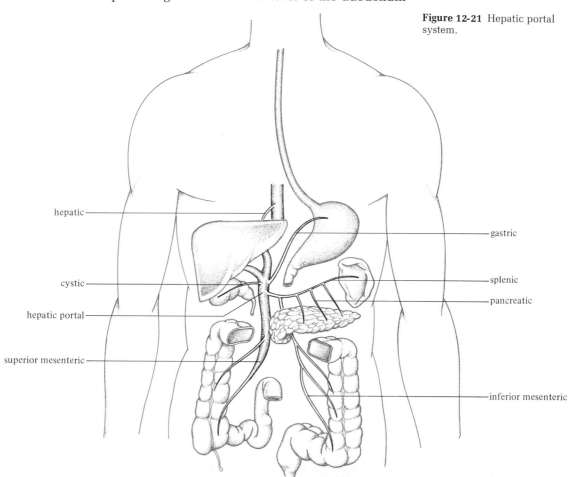

**Figure 12-21** Hepatic portal system.

where it joins with the splenic vein to form the hepatic portal vein.

**Splenic vein.** The splenic vein drains the spleen and then passes from left to right until it joins with the superior mesenteric to form the hepatic portal vein. Along its course the splenic vein receives blood from the following veins: the gastric veins from the stomach, the pancreatic veins from the pancreas, and the inferior mesenteric vein which drains the second half of the large intestine.

### Portal Hypertension

Any obstruction of blood flow through the liver can cause a buildup of pressure in the portal system. This is known as portal hypertension. It may be caused by liver disease such as cirrhosis or by a hepatic tumor. It may also be caused by heart disease in which the pressure in the inferior vena cava and hepatic vein builds up.

The buildup of pressure in the portal system causes the veins that compose this system to dilate. Dilation of the network of veins in the esophagus leads to a condition known as esophageal varices. These esophageal varices can rupture leading to a fatal hemorrhage.

When portal hypertension occurs some of the blood in the portal system is drained back to the heart through a collateral circulation. There are communications through which blood can flow from the inferior mesenteric and rectal vein to the common iliac veins. There are also communications from the esophageal and gastric veins to the azygous veins. If portal hypertension is severe an operation is performed in which a portacaval shunt is established. This consists of a connection between the hepatic portal vein and the inferior vena cava which bypasses the liver.

## LYMPHATIC SYSTEM

The lymphatic system functions along with the circulatory system to transport material through the body. In addition the lymphatic system plays a major role in the protection of the body. Specialized structures known as lymph nodes filter the fluid moving through the lymphatic system and remove harmful agents. The lymphatic system also plays a major role in the production of antibodies.

As illustrated in Figure 12-22, the lymphatic system is composed of lymphatic vessels and lymph nodes. The smallest lymphatic vessels are the lymph capillaries. These capillaries differ from blood capillaries in that they are dead-ended and considerably more permeable. In general slightly more fluid leaves the blood to enter the interstitial space than leaves the interstitial space to enter the blood. This excess fluid enters the lymphatic capillaries and is eventually returned to the blood circulation via larger lymphatic vessels. Also a small amount of protein tends to leave the blood

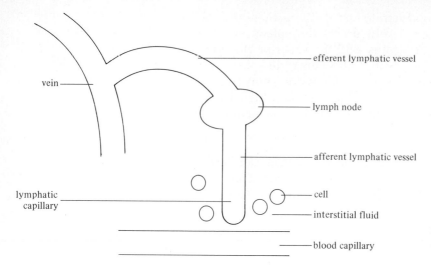

Figure 12-22 Schematic illustration of the lymphatic system.

capillaries and enter the interstitial fluid. This protein then moves from the interstitial fluid into the lymph capillaries. It is also eventually returned to the circulation by the larger lymph vessels. Figure 12-23 shows the movement of fluid and protein from the blood capillaries to the interstitial fluid to the lymphatic capillaries. Thus one function of the lymphatic system is to return excess filtered material back to the circulatory system.

The large lymphatic vessels are composed of a three-layered wall of endothelium, smooth muscle, and collagen. It is felt that the smooth muscle in the walls of these larger vessels helps move the lymph through the lymphatic vessels. Valves that prevent the backflow of lymph are found along the course of the lymphatic vessels. These valves are similar to those found in veins; they consist of two flaps on opposite sides of the vessel.

### Lymph Nodes

Figure 12-24 illustrates a typical lymph node. The capsule, or

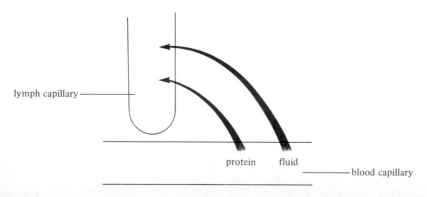

Figure 12-23 Schematic illustration of the movement of fluid and protein from blood capillaries to lymphatic capillaries.

# 289  Lymphatic system

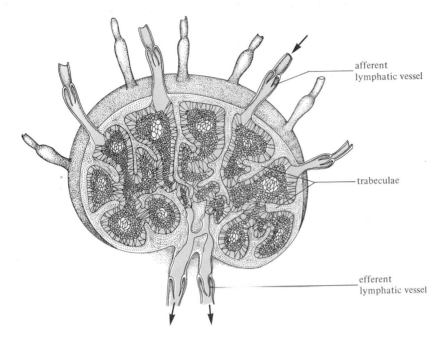

**Figure 12-24** Lymph nodes.

covering, consists of dense, collagenous connective tissue. Extensions of connective tissue called trabeculae subdivide the capsule into nodes. Very thin reticular fibers extend from the trabeculae into the sinuses through which lymph flows. The reticular fibers form a mesh that can filter the flowing lymph. There are macrophages attached to these fibers which can remove foreign matter by phagocytosis.

Lymph which is formed in the lymph capillaries is transported to the lymph nodes by afferent lymph vessels. Within the lymph nodes the lymph flows through the sinuses and is filtered by the meshwork of reticular fibers and macrophages. Bacteria in the lymph are destroyed at this point. Because the lymph capillaries are more permeable than blood capillaries, bacteria enter the lymph much more readily than the blood. Once in the lymph the bacteria are transported to the lymph nodes where they can be destroyed.

Lymph flows out of the lymph nodes through efferent lymphatic vessels which leave the nodes at the hilus. The hilus is also the site where blood vessels enter and leave the nodes. The efferent vessels from various nodes join together to form the large lymphatic trunks which eventually return the lymph to the blood.

Figure 12-25 illustrates the major large lymphatic-collecting trunks. The thoracic duct receives lymph from all parts of the body below the diaphragm and from the left side above the diaphragm. The thoracic duct starts at the level of the 2nd lumbar vertebra, passes upward along the spinal cord, and finally empties into the

**290** THE BLOOD VESSELS AND LYMPHATICS

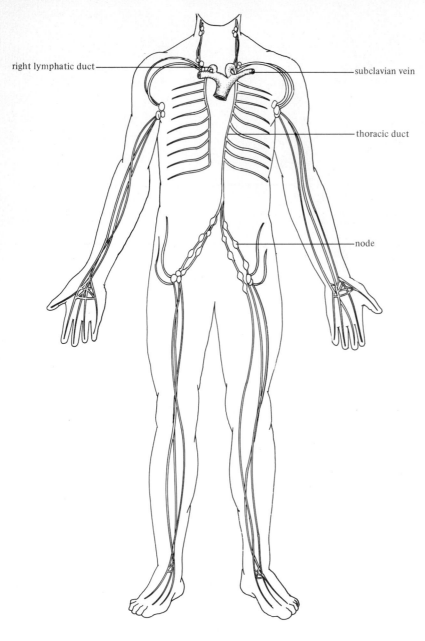

**Figure 12-25** Major lymphatic collecting trunks.

venous system at the junction of the left subclavian and left internal jugular vein. Valves at the entrance prevent blood from flowing backward into the lymphatic system.

The right lymphatic duct drains blood from the right side of the body above the diaphragm. It empties into the right brachiocephalic vein at the union of the right internal jugular vein and right internal carotid.

In summary, the lymphatic system carries out two major functions. It returns fluid and protein from the interstitial space to the blood and it filters out harmful material. Lymph which is formed in the lymph capillaries flows through afferent lymph vessels to the lymph nodes where it is filtered. From the lymph nodes lymph flows through efferent lymph vessels to one of the larger lymph-collecting ducts: the thoracic duct and the right lymphatic duct. These ducts empty lymph into the large blood vessels just above the heart.

## SPLEEN

The spleen is a large organ which contains lymphatic tissue. It is designed to produce lymphatic cells and to filter the blood. The spleen is located on the left side of the abdominal cavity below the diaphragm. It lies posterior and lateral to the stomach.

The spleen is similar in structure to a lymph node. It contains a collagenous capsule that sends trabeculae into the interior. However, the capsule of the spleen also contains smooth muscle. The interior of the spleen contains two major types of structures: the white pulp and the red pulp.

The white pulp consists of lymphatic tissue. This tissue produces circulating lymphocytes and plasma cells which produce antibodies. The white pulp is scattered throughout the red pulp.

The red pulp of the spleen consists of a network of blood sinusoids which contain a mesh of reticular fibers. Along these fibers are many macrophages. As blood flows through the splenic sinusoids it is filtered by the reticular fibers and the macrophages. Old red blood cells and bacteria can be removed at this point. Thus the spleen filters the blood in much the same way as the lymph nodes filter the lymph. As mentioned earlier in this chapter blood enters the spleen through the splenic artery and leaves through the splenic vein, which is part of the portal system.

The sinusoids of the spleen contain a fairly large amount of blood. In stressful situations in which there is a demand for a larger volume of blood the smooth muscle of the splenic capsule contracts. This forces blood out of the spleen and adds it to the general circulation.

## THYMUS

The thymus is a mass of lymphatic tissue in the neck and behind the sternum in the upper thorax. Structurally it is similar to lymph nodes with a connective tissue capsule and trabeculae.

The thymus seems to play an important role in the development of the body's immune ability. Certain types of lymphocytes, the circulating T cells, will only develop if a child has an intact thymus. These cells combat viruses, tumor cells, and tissue transplants, by an antigen-antibody type response. Once a child reaches puberty the thymus seems to atrophy but the T cells continue to function. The plasma cells which produce circulating protein antibodies do not appear to depend on the thymus. The circulating protein antibodies are important in fighting bacteria.

## OBJECTIVES FOR THE STUDY OF THE BLOOD VESSELS AND LYMPHATICS

At the end of this unit you should be able to:

1. Name the three layers of an arterial wall.
2. Distinguish between the structure of large arteries and the structure of small arteries.
3. Explain the function of the smooth muscle in the walls of arterioles.
4. Name the four portions of the aorta.
5. Name the area of the body that receives blood from the following arteries: pulmonary, coronary, brachiocephalic, common carotid, external carotids, internal carotids, subclavian, vertebral, axillary, brachial, radial, ulnar, celiac, left gastric, hepatic, splenic, right gastric, superior mesenteric, middle suprarenal, renal, spermatic, ovarian, inferior mesenteric, lumbar, common iliac, internal iliac, external iliac, femoral, anterior tibial, and posterior tibial.
6. State the function of capillaries.
7. State how a sinusoid differs from a capillary.
8. Describe how the wall of a vein differs from the wall of an artery.
9. Describe the physiological effects of venous constriction.
10. Describe the function of venous valves.
11. State the location of the coronary sinus.
12. State the location of the cranial venous sinuses.
13. Name the area of the body from which blood is drained into the following veins: internal jugular, external jugular, cephalic, basilic, radial, ulnar, brachial, axillary, subclavian, brachiocephalic, superior vena cava, azygos, hemiazygos, great saphenous, small saphenous, posterior tibial, anterior tibial, femoral, external iliac, internal iliac, common iliac, inferior vena cava, lumbar, spermatic, ovarium, renal, suprarenal, and hepatic.
14. Name the major veins that form the hepatic portal vein.
15. State the function of the hepatic portal vein.
16. State the functions of the lymphatic system.
17. Describe how lymphatic capillaries differ from blood capillaries.
18. Describe the structure and function of a lymph node.
19. State the parts of the body drained by the thoracic duct and by the right lymphatic duct.
20. Describe the location of the spleen.
21. Explain the major functions of the spleen.
22. Distinguish between the function of the red pulp and white pulp of the spleen.
23. Describe the function of the spleen.

# 13

# Blood Pressure, Blood Flow, and Capillary Exchange

INTRODUCTION
MAINTENANCE OF STABLE ARTERIAL
 PRESSURE
CAPILLARY BLOOD FLOW
CAPILLARY EXCHANGE
HEMORRHAGE

# INTRODUCTION

As the ventricles contract, blood is pumped under pressure into the pulmonary artery and aorta. This pressure provides the force that causes blood to flow through the blood vessels. In this chapter we are going to discuss how the body controls blood pressure so that it is neither too high nor too low and how the body controls blood flow so that all of the tissues receive an amount of blood that is appropriate to their metabolic needs. In addition, we are going to discuss the mechanisms by which materials are exchanged between the blood and interstitial fluid.

Pressure is defined as the force exerted by a substance, divided by the area against which this force is exerted. This is called the force per unit area. Thus, the pressure of the blood at any point within the blood vessels is the force exerted by the blood per unit area of the blood vessel wall. As illustrated in Figure 13-1, the pressure at any point is equal in all directions. This means that the blood not only exerts a force against the walls of the vessel but that it also exerts a force that tends to move blood along the vessels. Because any fluid flows from an area of higher pressure to an area of lower pressure, within the circulatory system blood pressure decreases from arteries to capillaries to veins.

Within any blood vessel the pressure depends on the amount of blood in the vessel and the size of the vessel. If the amount of blood in the vessel is increased, the pressure will also increase. For this reason, the arterial blood pressure is not constant; rather, it varies with the cardiac cycle. When the ventricles contract and pump blood into the arteries, the arterial pressure increases to its peak value. This is known as the **systolic pressure.** While the ventricles are relaxed and the semilunar valves are closed, blood flows out of the arteries and into the capillaries. As the amount of blood in the arteries decreases, the arterial blood pressure decreases. It reaches its lowest value just before the next contraction of the heart. This is known as the **diastolic pressure.** Blood pressure is

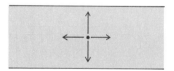

**Figure 13-1** Schematic illustration showing that the pressure at any point is equal in all directions.

usually measured in millimeters of mercury (mm Hg) with the arterial blood pressure expressed as the systolic value over the diastolic value. The normal systemic arterial blood pressure of a young adult is usually about 120/80. This means a systolic pressure of 120 mm Hg and a diastolic pressure of 80 mm Hg. The difference between the systolic and diastolic pressures is known as the **pulse pressure.** As mentioned previously, capillary pressure is less than arterial pressure and venous pressure is less than capillary pressure. Systemic capillary pressure is usually about 17 mm Hg and systemic venous pressure averages about 7 mm Hg. Figure 13-2 shows the blood pressures in different areas of the systemic circulation.

The pressures in the pulmonary portion of the circulation are considerably less than those of the systemic portion. Pulmonary artery pressure averages a systolic value of about 22 mm Hg and a diastolic value of about 8 mm Hg. In the pulmonary veins the pressure falls to about 4 mm Hg.

## MAINTENANCE OF STABLE ARTERIAL PRESSURE

### Importance

Although the arterial pressure varies with each cardiac cycle, the average, or mean, arterial pressure is kept quite constant. If this pressure were to drop too low, there would not be an adequate force to drive blood into the capillaries and the metabolic needs of the cells could not be met. On the other hand, if the arterial pressure becomes too high, there is a danger that one of the blood vessels will not be able to withstand the high pressure and will burst, leading to a severe hemorrhage and the loss of blood flow to

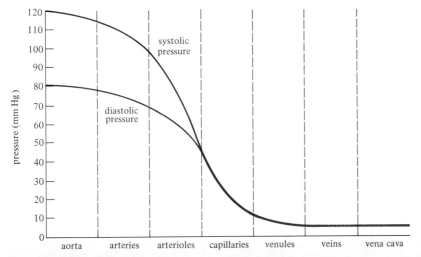

**Figure 13-2** Blood pressures in different areas of the circulatory system.

the area supplied by the vessel. This is particularly dangerous when a cerebral vessel bursts, a condition known as a **stroke.** High blood pressure also presents a strain on the heart. When the ventricles contract, they must develop a pressure that is greater than the arterial pressure in order for blood to flow out of the ventricles and into the arteries. The higher the arterial pressure, particularly the diastolic pressure, the more forcefully the ventricles must contract. **Hypertension,** or high blood pressure, is said to exist when the diastolic pressure exceeds 90 mm Hg.

### Determinants of Blood Pressure

In order to understand the mechanisms by which the body maintains a stable arterial pressure, we must first look at the factors that determine the arterial blood pressure. The arterial blood pressure depends on the volume of the arteries and the amount of blood in the arteries. Normally, as blood flows into the arteries during systole, the arteries expand to accommodate the entering blood. During diastole the elastic tissue in the arteries causes them to return to their initial size as blood flows along into the capillaries. As one gets older, the arteries often undergo degenerative changes and become less distensible, a process known as **arteriosclerosis.** During this process the systolic blood pressure will increase because the arteries no longer expand as readily when blood enters them. Except during the disease process of arteriosclerosis, the volume of the arteries does not play an important role in determining the average blood pressure. The normal mean blood pressure is determined by the amount of blood in the arteries. As illustrated in Figure 13-3, the amount of blood in the arteries depends on the amount of blood entering the arteries from the heart and the amount of blood flowing out of the arteries through the arterioles.

The amount of blood entering the arteries from the heart depends on the cardiac output. If the cardiac output is increased, more blood will enter the arteries and blood pressure will rise. Likewise if the cardiac output falls, less blood will enter the arteries and there will be a drop in arterial blood pressure.

The amount of blood leaving the arteries through the arterioles depends on the diameter of the arterioles and the thickness, or **viscosity,** of the blood. As discussed in Chapter 12, the arterioles contain smooth muscle which can contract or relax. Contraction of the

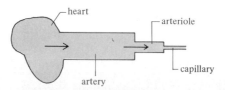

**Figure 13-3** Factors determining the amount of blood in the arteries.

arteriolar smooth muscle (vasoconstriction) narrows the arterioles and makes it more difficult for blood to flow out of the arteries. This serves to increase blood pressure. During vasodilation, when the arteriolar smooth muscle relaxes, it is easier for blood to flow out of the arteries and, thus, blood pressure falls.

The ease with which blood flows out of the arteries also depends on the viscosity of the blood. As blood becomes thicker it offers more resistance to flow. The viscosity of the blood depends primarily on the hematocrit. If the percentage of red blood cells increases, the blood becomes thicker and more resistant to flow; and if the percentage decreases, it becomes thinner and less resistant to flow.

### Control of Arterial Blood Pressure

The control of arterial blood pressure is basically a reflex process. As described in Chapter 7, any reflex involves receptors, afferent pathways, an integrating center in the central nervous system, efferent pathways, and effectors.

**Receptors.** The most important receptors for detecting changes in blood pressure are the **baroreceptors** in the carotid sinuses and the aortic arch. The carotid sinuses are located in the internal carotid arteries just past the point where they branch off the common carotid arteries. Essentially, these baroreceptors are stretch receptors that generate impulses at a rate proportionate to the degree to which the arterial wall is stretched. Because the amount of stretch depends on the arterial blood pressure, the baroreceptors detect any change in this pressure. An increase in arterial blood pressure will increase the rate at which the receptors generate impulses and a decrease in blood pressure will lower the rate at which they generate impulses. Although the baroreceptors in the internal carotid arteries and aortic arch seem to be the most important receptors for blood pressure, there are also receptors in the heart and venae cavae which can detect changes in blood pressure.

**Afferent pathways.** Impulses generated in the carotid sinus baroreceptors are conducted to the vasomotor center in the medulla along the glossopharyngeal nerves, as illustrated in Figure 13-4. The vagus nerve conducts impulses from the aortic arch baroreceptors to the vasomotor center.

**Vasomotor center.** The vasomotor center in the medulla integrates the information it receives about the blood pressure from the baroreceptors and directs any adjustments which must be made. Thus, if the vasomotor center receives information that the arterial blood pressure is less than it should be, it directs responses that will bring about an increase in blood pressure. Likewise, if the vasomotor center finds out that the blood pressure is too high, it must direct responses that will lower blood pressure.

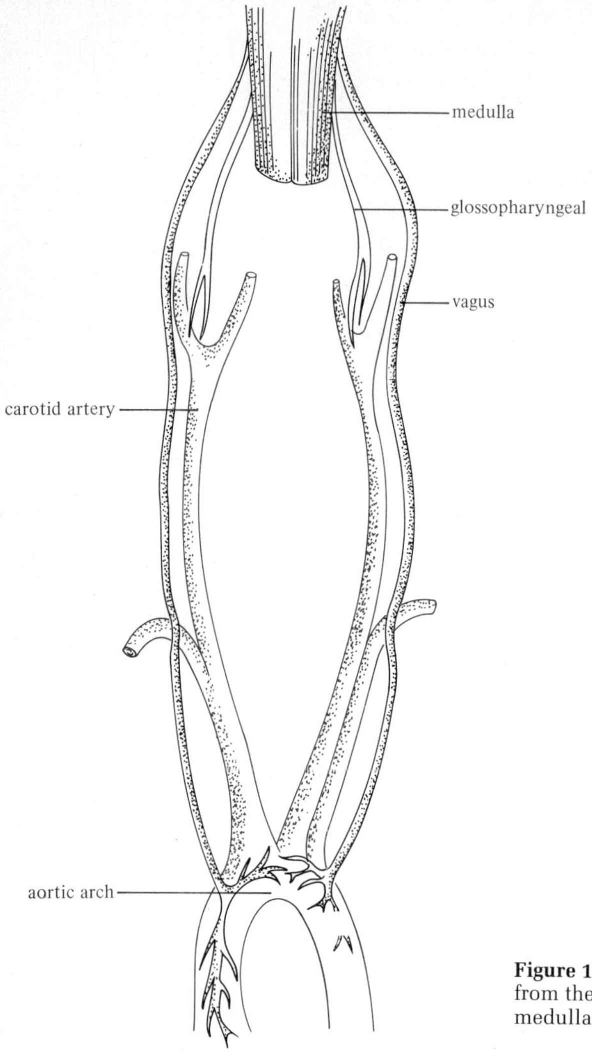

Figure 13-4 Afferent pathways from the baroreceptors to the medulla.

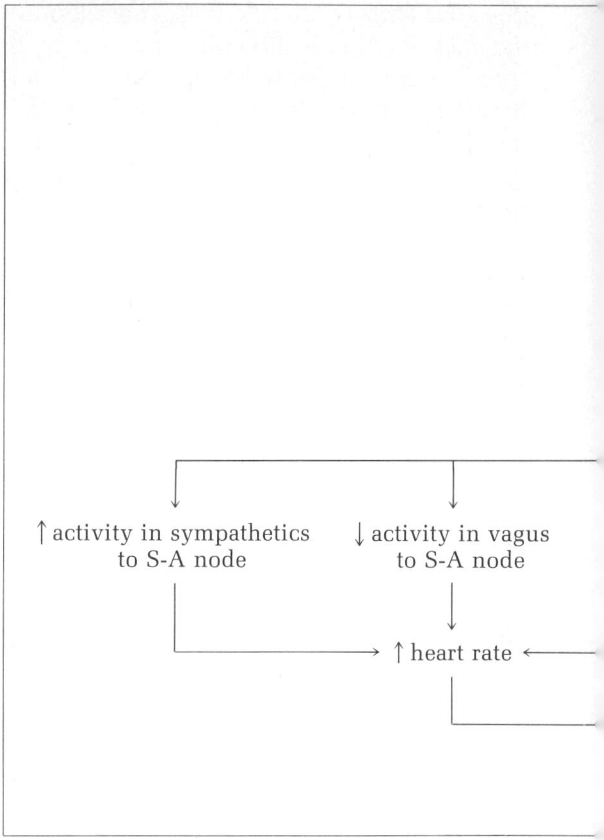

**Efferent pathways.** The efferent pathways conduct impulses from the vasomotor center to those structures that can affect blood pressure. They consist of pathways to the heart: the arterioles, the veins, and the adrenal medulla. The efferent pathways to the heart are the sympathetic nerves to the S-A node which speed up the heart, the parasympathetic vagus to the S-A node which slows down the heart, and the sympathetic nerves to the ventricular myocardium which can stimulate the heart to contract with more force. Because the sympathetic nerves to the arterioles are tonically active, these nerves can be used to either constrict or dilate the arterioles. If these nerves are made more active, there will be vasoconstriction; if they are made less active, there will be vasodilation.

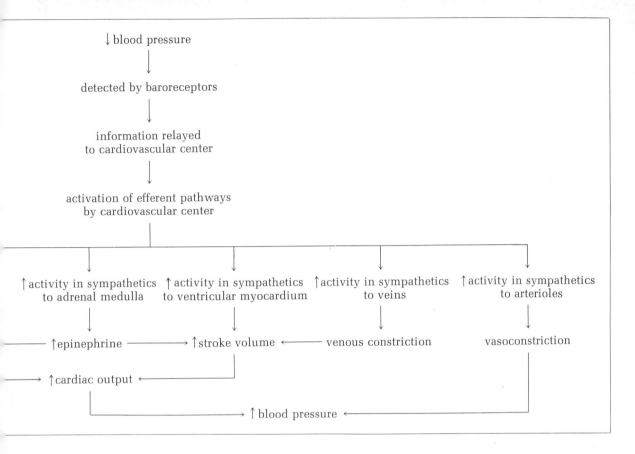

**Figure 13-5** Flow chart of the reflex response to a drop in blood pressure.

The sympathetic nerves to the veins can stimulate the veins to constrict and, thus, increase the rate at which blood returns to the heart. This will, in turn, lead to an increase in stroke volume, hence cardiac output. The sympathetic nerves to the adrenal medulla stimulate it to secrete norepinephrine and epinephrine. These hormones stimulate the heart to beat faster and more forcefully, thereby increasing cardiac output. They can also cause overall vasoconstriction or vasodilation depending on their relative concentrations.

Figure 13-5 traces the steps involved in the body's response to a drop in blood pressure. When the blood pressure drops, the baroreceptors detect the drop and relay this information along the glossopharyngeal and vagus nerves to the vasomotor center in the medulla. Once the vasomotor center discovers that the blood pressure has dropped, it activates efferent pathways that can raise the blood pressure back toward normal. The sympathetic nerves to the S-A node are made more active and the vagus to the S-A node is inhibited. This leads to an increase in heart rate and, thus, in cardiac output. The sympathetic nerves to the ventricular myocardium stimulate the heart to contract more forcefully so that stroke volume is increased. This adds further to the increase in cardiac output. Increased activity in the sympathetic nerves to the adrenal medulla stimulate it to release epinephrine and norepinephrine. These hormones increase both heart rate and stroke volume over and above the increase caused by the nerves to the heart. Thus, as a result of stimulation by both nerves and hormones, cardiac output is increased. The increased cardiac output helps raise the blood pressure back to its normal level.

In addition to directing the increase in cardiac output, the vasomotor center increases activity in the sympathetic nerves to the arterioles. These nerves stimulate the smooth muscle of the arterioles to contract. As vasoconstriction takes place, blood has a harder time flowing out of the arteries and the blood pressure is increased. The vasomotor center also activates sympathetic nerves to the veins causing them to constrict. This increases the rate at which blood returns to the heart and contributes to the increase in cardiac output.

In summary, the vasomotor center responds to a drop in blood pressure by directing an increase in cardiac output and vasoconstriction. The increase in cardiac output is brought about by both an increase in heart rate and stroke volume. Constriction of the veins contributes to the increase in cardiac output.

**Other Factors that Affect Blood Pressure**

**Chemoreceptors.** Just adjacent to the baroreceptors in the carotid arteries and aortic arch, there are chemoreceptors that detect changes in the oxygen and carbon dioxide concentrations of the blood. If the oxygen concentration of the blood falls, these chemoreceptors initiate a reflex attempt to raise blood pressure. The point of this reflex is to increase the amount of blood flowing to the brain so that it will receive sufficient oxygen. However, under normal circumstances the chemoreceptors seem to play only a minor role in the regulation of blood pressure. Their primary function is to aid in the regulation of respiration, as will be described in Chapter 14.

**Aldosterone.** Aldosterone, a hormone secreted by the adrenal cortex, acts on the kidney to cause the retention of salt and water. This hormone is secreted in most situations in which blood pressure falls. As a result of aldosterone stimulation, the retention of salt and water by the kidney tends to increase the blood volume and thereby helps return blood pressure to its normal level. In Chapter 17 we will discuss the relationship between aldosterone and fluid volume in detail.

# CAPILLARY BLOOD FLOW

The amount of blood flowing into the capillaries of any tissue depends primarily on the arterial blood pressure and the extent to which the arterioles in that region are constricted or dilated. This is illustrated in Figure 13-6. If the diameter of the arterioles is kept constant, an increase in arterial blood pressure will lead to an increase in the amount of blood flowing into the capillaries. On the other hand, if the arterial blood pressure is kept constant, vasoconstriction will reduce the amount of blood flowing into the capillaries and vasodilation will increase the blood flow. Under normal conditions the arterial blood pressure is kept constant and the amount of blood flowing into the capillaries is adjusted by changes in the diameter of the arterioles.

To a large degree, the amount of blood flowing into any particular area of the body is related to the level of metabolic activity in that area. Whenever a particular tissue increases its level of metabolic activity, blood flow into that tissue is increased through vasodilation. For example, the amount of blood flowing into a resting skeletal muscle is quite small. When the muscle is contracting vigorously, the amount of blood flowing into the muscle may increase 20-fold.

The exact means by which arteriole size is adjusted to the local metabolic needs of a tissue is not completely understood. It is felt that vasodilation is probably brought about by some change in the composition of the local extracellular fluid. This change could be a drop in oxygen concentration or an increase in hydrogen, carbon dioxide, or lactate concentration, all of which occur when a tissue increases its level of metabolic activity. Whether one or all of these factors is the immediate cause of vasodilation remains to be discovered.

When a particular tissue increases its level of metabolic activity over a long period of time, the number of blood vessels in the

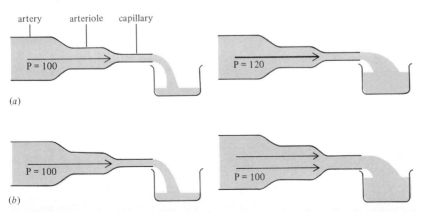

**Figure 13-6** Effects of (a) blood pressure and (b) arteriole size on blood flow into the capillaries.

area is also increased. This growth of blood vessels into the area provides for a larger blood flow. One of the reasons it is felt that exercise helps prevent heart attacks is that a well-exercised heart has a more extensive blood supply than does a poorly exercised heart.

In addition to being influenced by local metabolic factors, the blood flow to different areas of the body is regulated by the sympathetic nerves to the arterioles. These nerves can be used to bring about vasoconstriction or vasodilation in such a way as to redistribute blood flow during different situations. For example, during exercise the sympathetic nerves initiate vasodilation in the exercising skeletal muscles and vasoconstriction in most of the abdominal organs. Thus, blood flow to the muscles is increased and blood flow to organs such as the stomach, intestines, and kidneys is reduced. Sympathetic regulation over local blood flow is also one of the mechanisms by which the body temperature is regulated. When heat must be conserved within the body, the sympathetic nerves cause vasoconstriction of the arterioles in the skin. This effects a decrease in the amount of blood flowing to the skin and a reduction in the amount of heat lost to the external environment.

The hormones norepinephrine and epinephrine also play a role in regulating arteriole size and, thus, blood flow. Norepinephrine is a vasoconstrictor and, therefore, tends to reduce blood flow to most tissues. Epinephrine is a vasoconstrictor in most tissues but a vasodilator in the heart and skeletal muscles.

The arterioles in the blood vessels to the brain are not under nervous or hormonal control, although they are influenced to a certain extent by local metabolic factors. Blood flow to the brain depends primarily on the arterial blood pressure. This is one of the main reasons the body must make all attempts not to allow the arterial blood pressure to fall. When arterial pressure falls, the amount of blood flowing to the brain is reduced. The nerve cells of the brain are exceedingly sensitive and can only live a very short time without an adequate supply of nutrients and oxygen.

## CAPILLARY EXCHANGE

The capillaries are the site at which materials are exchanged between the blood and interstitial fluid. Transportation of material across the capillary wall is brought about by two basic processes: diffusion and filtration. In most tissue the capillary walls are quite permeable to all molecules smaller than proteins. Those molecules, such as glucose and oxygen, which are more concentrated in the blood than the interstitial fluid will diffuse out of the blood along their concentration gradient. Likewise, molecules, such as carbon dioxide, which are more concentrated in the interstitial fluid, will diffuse into the blood.

**303**  Capillary exchange

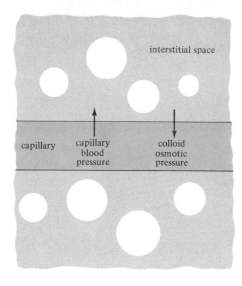

**Figure 13-7** Forces acting to regulate fluid movement across the capillary wall.

Filtration is brought about by a pressure difference between the blood and interstitial fluid. As illustrated in Figure 13-7, the blood pressure is about 32 mm Hg at the arterial end of the capillary and falls to about 15 mm Hg at the venous end. Interstitial fluid has a pressure of about 2 mm Hg. The difference between the capillary pressure and the interstitial pressure acts as a force that tends to drive fluid out of the blood. However, this force is balanced by the colloid osmotic pressure (oncotic pressure) brought about by the presence of large amounts of protein in the blood and only very small amounts of protein in the interstitial fluid. This difference in protein concentration acts as an osmotic force causing fluid to move from the interstitial fluid to the blood. At the arterial end of the capillary, the hydrostatic (fluid) pressure difference between the blood and the interstitial fluid is larger than the osmotic pressure difference and there is a net movement of fluid out of the capillaries. At the venous end, the osmotic pressure difference is greater than the hydrostatic difference and there is a net movement of fluid into the capillaries. Under normal circumstances the amount of fluid entering the venous end of the capillary is almost equal to the amount leaving at the arterial end. Any fluid that is filtered out of the capillaries but not back into them is eventually returned to the circulation via the lymphatics.

**Edema**

**Edema** refers to a buildup in the amount of fluid in the interstitial space. This will occur whenever the amount of fluid moving from the blood to the interstitial space exceeds the amount of fluid moving from this space back to the blood. Edema can be brought about in a number of different ways. Any change that increases capillary pressure will cause more fluid to move out of the capillaries.

This will produce edema. An increase in capillary pressure can be brought about by an increase in arterial pressure, vasodilation, or an increase in venous pressure. For example, if the left side of the heart cannot pump blood adequately, blood will build up in the pulmonary circulation behind the left side of the heart. This will lead to an increase in pulmonary capillary pressure and pulmonary edema.

Edema can also occur when there is a decrease in the concentration of plasma proteins. This occurs in some forms of kidney disease where plasma proteins are lost in the urine. Edema will also occur if protein accumulates in the interstitial fluid. This accumulation of protein can result from the capillaries becoming permeable to protein or through obstruction of the lymphatic channels draining a tissue.

## HEMORRHAGE

The various factors involved in the control of blood pressure, blood flow, and capillary exchange can be seen in the response to a hemorrhage. Figure 13-8 is a flow diagram that summarizes the various events that take place after a hemorrhage occurs.

As blood is lost, the amount of blood that flows back to the heart through the veins is reduced. Since stroke volume depends on the amount of blood in the ventricles at the time they contract, a reduction in venous return leads to a reduction in stroke volume. This, in turn, leads to a reduction in cardiac output. When cardiac output falls, arterial blood pressure also falls. The drop in blood pressure is detected by the baroreceptors in the carotid sinuses and aortic arch. The baroreceptors relay this information along the glossopharyngeal and vagus nerves to the vasomotor center in the medulla. Once the vasomotor center finds out that blood pressure has fallen, it activates efferent pathways to the effectors which can bring about an increase in blood pressure. The sympathetic nerves to the S-A node are activated and the vagus to the S-A node is inhibited. This brings about an increase in heart rate. Activation of the sympathetic nerves to the ventricular myocardium stimulates stronger contractions of the ventricles and an increase in stroke volume.

In addition to being directly stimulated by nerves, the heart is also stimulated by the hormones norepinephrine and epinephrine. These hormones are secreted as a result of stimulation of the adrenal medulla by the sympathetic nerves. The increase in heart rate and stroke volume aids in restoring the arterial blood pressure back to its original value.

At the same time that it activates pathways that increase car-

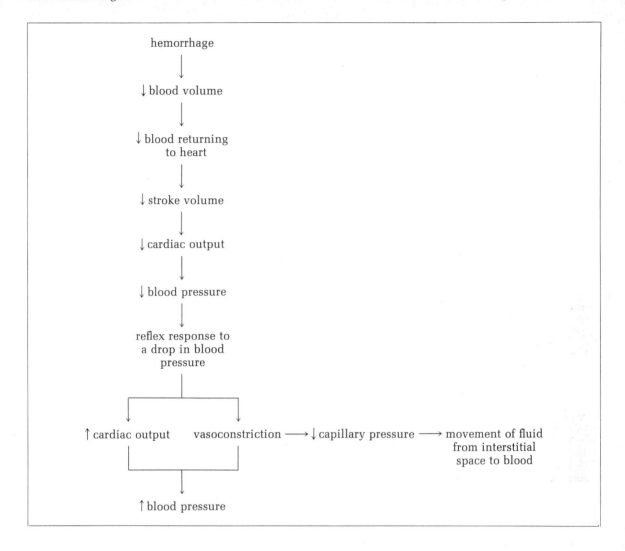

**Figure 13-8** Flow chart of the response to a hemorrhage.

diac output, the vasomotor center activates the sympathetic nerves to arterioles all over the body. These nerves bring about vasoconstriction which also helps restore blood pressure back to normal. The lungs, heart, and brain are the only places where vasoconstriction does not take place.

Figure 13-9 is a schematic illustration of how the body's response to a hemorrhage affects blood pressure and the blood flow to different tissues. In this illustration the arteries are represented as a large tank containing blood and the arterioles as the outlets through which blood flows to different tissues.

Figure 13-9a shows the situation prior to a hemorrhage. Arterial blood pressure is normal and blood is flowing to all of the tissues. The situation after a hemorrhage, but before the reflex

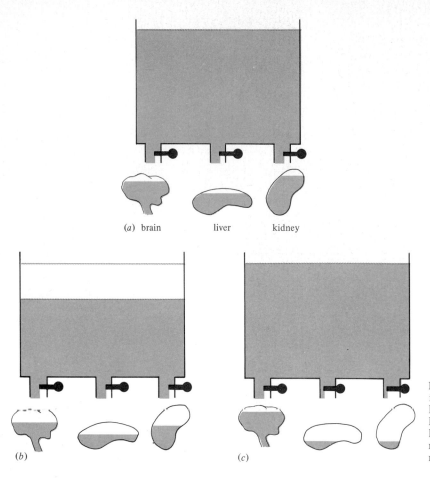

**Figure 13-9** Schematic illustration of the response to a hemorrhage. (a) Before hemorrhage. (b) After hemorrhage but before reflex response. (c) After reflex response to a hemorrhage.

response, is shown in Figure 13-9b. Because of the loss of blood, the amount of blood in the arteries is decreased and there is a concomitant drop in arterial blood pressure. Because of the drop in arterial pressure, the amount of blood flowing to the various tissues is reduced.

Figure 13-9c depicts the changes brought about as a result of the reflex response to hemorrhage. The increase in cardiac output helps fill the arteries and restore arterial pressure to its original value. The arterioles through which blood flows to most tissues are constricted so that blood leaves the arteries much less readily. This also increases arterial blood pressure. The arterioles to the lungs, heart, and brain do not constrict. Thus, as long as the arterial blood pressure can be kept close to its original value, the flow of blood to these vital organs will not be impaired.

Another way of saying this is that when blood volume drops, blood flow to most organs is reduced to ensure that an adequate amount of blood can flow to the most vital organs. If blood flow to these organs is significantly reduced, they can undergo irreversible damage and the person will die.

    The restoration of blood volume after a hemorrhage is brought about in three basic ways: fluid moves from the interstitial spaces into the blood; the kidney reduces the amount of urine formed; and as a result of activation of the thirst center, the person drinks liquids. A fairly large-scale movement of fluid from the interstitial spaces to the blood results from the fact that capillary blood pressure in most parts of the body falls after a hemorrhage. This is due directly to the drop in arterial blood pressure and the vasoconstriction which follows. When capillary blood pressure falls, the colloid osmotic pressure causing fluid to move into the blood becomes a stronger force than the capillary pressure pushing fluid out of the blood. Thus, there is an overall movement of fluid from the interstitial space into the blood. The role of the kidney and thirst center in maintaining blood volume will be discussed in Chapter 17.

## OBJECTIVES FOR THE STUDY OF BLOOD PRESSURE, BLOOD FLOW, AND CAPILLARY EXCHANGE

At the end of this unit you should be able to:

1. Define the term pressure.
2. Distinguish between systolic and diastolic pressure.
3. State the normal blood pressure for a young adult.
4. Define the term pulse pressure.
5. Explain the problems that arise if blood pressure is either too high or too low.
6. Explain the relationship between cardiac output and blood pressure.
7. Explain the relationship between vasoconstriction and blood pressure.
8. Describe the relationship between blood viscosity and blood pressure.
9. State the location of the baroreceptors.
10. Name the nerves that connect the baroreceptors to the brain.
11. State the location of the cardiovascular center in the brain.
12. Describe the various efferent pathways that can bring about a change in blood pressure.
13. Describe the steps involved in the reflex response to a drop in blood pressure.
14. Explain the relationship between blood pressure and capillary blood flow.
15. Explain the relationship between vasoconstriction, vasodilation, and capillary blood flow.
16. Explain the relationship between the level of metabolic activity in a particular area of the body and the amount of blood flowing to that area.
17. Describe the major force that moves fluid out of the capillaries.
18. Describe the major force that moves fluid into the capillaries.
19. Name two factors that can cause edema.
20. Describe the changes that take place as part of the body's response to a hemorrhage.

# 14

# The Respiratory System

RESPIRATORY TRACT
LUNGS
MECHANISM OF RESPIRATION
RESPIRATORY VOLUMES
COMPOSITION OF AIR
EXCHANGE AND TRANSPORT OF GASES
CONTROL OF RESPIRATION

The respiratory system has three major physiological functions. The first is to obtain oxygen from the outside air. This oxygen enters the blood from the respiratory tract and is transported to the cells where it is used in cellular respiration, the burning of food to obtain ATP. A second function of the respiratory system is to eliminate the carbon dioxide formed as a waste product in cellular respiration. The carbon dioxide formed in the cells is transported by the blood to the respiratory system which then eliminates it from the body. In addition to obtaining oxygen and eliminating carbon dioxide, the respiratory system plays an important role in regulating the pH (hydrogen concentration) of the blood. A final function of the respiratory system is the production of the sounds used in speech.

In looking at the respiratory system we will first discuss the anatomy of the respiratory tract, the passageway through which air travels as it enters and leaves the body. We will then discuss the following aspects of respiration: the mechanism of respiration, the forces that move air in and out of the body, the exchange of gases between the respiratory tract and the blood, the transport of gases within the blood, and the control of respiration.

## RESPIRATORY TRACT

Although air can enter the body through either the mouth or nose, it is the nose that is designed to process this entering air for use by the body. Figure 14-1 shows the structure of the nose. It consists of an external portion that protrudes from the face and an internal portion that is underneath the cranium and above the roof of the mouth. The external portion is formed at the top by the nasal bones and at the bottom by skin and cartilage. The nose is divided into right and left cavities by the **nasal septum.** This septum is formed anteriorly by skin and cartilage and posteriorly by the vomer bone and the perpendicular plate of the ethmoid bone.

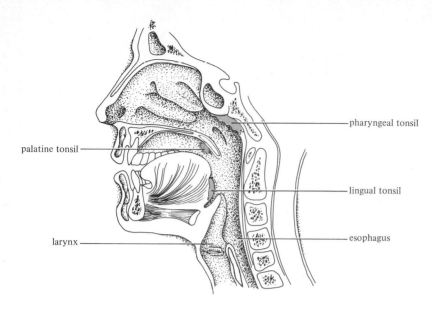

Figure 14-1 Nasal cavity and pharynx.

## Nasal Cavities

The nasal cavities open to the outside via the two external nares. Just inside these nares is a circle of coarse hairs that can filter large dust particles out of the entering air. The walls of the cavities at this point are formed by skin and cartilage. After a small distance from the external nares, the walls are composed of a ciliated mucous membrane. This membrane serves three important functions. Smaller dust particles that pass through the coarse hairs can be trapped by the mucus. The cilia move the mucus in a steady stream back toward the pharynx, at which point the mucus is swallowed and enters the digestive tract. Materials trapped in the mucus can be broken down by digestive enzymes or eliminated with the feces. The mucous membrane lining the nasal cavities also adds moisture to the entering air. It seems that very dry air can irritate the respiratory passageways. A final function of the mucous membrane is to adjust the temperature of the entering air to that of the body.

The surface of the mucous membrane lining the nasal cavities is enlarged by three projections, or conchae, which extend from the lateral walls of the cavities. These are the inferior, middle, and superior conchae.

In addition to lining the nasal cavities, the mucous membrane extends into the air sinuses, or spaces, of the bones which form the walls of the nose. These sinuses, as described in Chapter 5, are continuous with the nasal cavities. These sinuses serve to lighten the weight of the cranial bones, to increase the surface for processing entering air, and to resonate sound.

The nasolacrimal ducts, which drain tears from the eyes, open into the nasal cavities just below the inferior conchae.

Posteriorly, the nasal cavities open into the pharynx via the two posterior nares.

**Pharynx**

The pharynx, or throat, extends from the posterior nares to the top of the larynx and esophagus as illustrated in Figure 14-1. Its walls are formed of skeletal muscle with an internal mucous membrane lining. There are openings into the pharynx from the nose, mouth, eustachian tube, larynx, and esophagus. Because both air and food enter the body through the pharynx, the various openings must be controlled so that food moves down the esophagus and air moves into the larynx. Except during swallowing, the passage to the esophagus is kept closed and the passage to the larynx is kept open. In the next chapter we will discuss the changes that take place in the different passages to the pharynx during swallowing.

The pharynx is divided into three sections: the nasopharynx, the oropharynx, and the laryngopharynx.

The **nasopharynx** is that part of the pharynx below the posterior nares and above the soft palate. It is designed mainly for respiration. In the lateral walls are the openings of the eustachian tubes which connect the pharynx with the middle ear. These tubes are lined by a mucous membrane which is continuous with the mucous membrane of the nasopharynx and nasal cavities. Unfortunately the tubes can serve as a route by which infectious organisms that enter through the nose and can spread to the middle ear. Because these tubes are shorter and wider in children than they are in adults, children are particularly susceptible to having middle ear infections associated with respiratory infections.

Located on the posterior wall of the nasopharynx is a mass of lymphatic tissue known as the **pharyngeal tonsil.** As discussed in a previous chapter, one of the functions of lymphatic tissue is to destroy foreign organisms. Because the nasal passages are one of the principal routes by which infectious organisms can gain access to the body, the pharyngeal tonsil can help destroy these organisms as they enter. When this tonsil enlarges during the process of fighting infectious organisms, as can commonly occur in children, it is referred to as **adenoids.**

The **oropharynx** is located behind the mouth, extending from the soft palate to the level of the hyoid bone. Two arches, the **glossopalatine** and **pharyngopalatine** arches, form part of the side walls of the oropharynx. The glossopalatine arches extend from the soft palate to the base of the tongue. These arches form the entrance from the mouth to the oropharynx. Posterior to the glossopalatine arches are the pharyngopalatine arches which extend from the soft palate to the sides of the pharynx. Between these two sets of arches on each side there is a mass of lymphatic tissue known as the palatine tonsil. These are the tonsils that are swollen when one has tonsilitis and that are removed during a tonsilectomy. Because the tonsils are composed of protective lymphatic tissues, physicians are becoming more reluctant to remove this tissue than they were in the past.

In addition to the palatine tonsils, the oropharynx contains a second lymphatic mass, the **lingual tonsil.** The lingual tonsil is found behind the root of the tongue at the base of the oropharynx.

Taken together these tonsils—the pharyngeal tonsil, palatine tonsils, and lingual tonsil—form a circular protective ring of lymphatic tissue around the entrances to the larynx and esophagus. This ring is known as **Waldeyer's ring.**

The **laryngopharynx** is the lower part of the pharynx, between the hyoid bone and the openings to the larynx and esophagus.

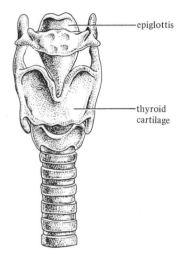

**Figure 14-2** External structure of the larynx (anterior view).

### Larynx

The larynx, or voicebox, is located in the neck at the top of the passageway from the pharynx to the lungs. It carries out two important functions: it protects the lungs against the entry of solid and liquid food and it contains the vocal cords used in speech. Figure 14-2 illustrates the structure of the larynx. It is formed by a number of cartilages which are held together by ligaments and controlled by skeletal muscles. The front of the larynx is formed by the thyroid cartilage, which consists of two plates that meet at an angle in the midline. These cartilages form the Adam's apple which can be felt at the front of the neck. In men the larynx is larger than it is in women and the thyroid cartilages meet at a sharper angle. For this reason men have a more prominent Adam's apple.

Extending upward from the thyroid cartilage is a second important cartilage, the **epiglottis.** The epiglottis closes off the entrance to the larynx when the larynx is lifted during swallowing.

The cavity of the larynx is lined with mucous membrane. As illustrated in Figure 14-3, there are two folds with anterior–posterior slits within this cavity. The top folds, called the **ventricular folds,** do not play a role in producing sound and are therefore referred to as false vocal cords. Sound is produced by vibrations of the lower folds, which are referred to as the true vocal folds or vocal cords. These folds are composed of a band of strong connective tissue surrounded by a mucous membrane. Forceful expulsion of air causes these folds to vibrate and produce sound. The nature of the sound produced depends on a number of factors: the force with which air is expelled; the tension of the folds; the shape of the larynx, which can be adjusted by skeletal muscles; the shape of the

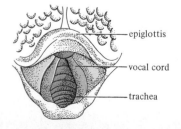

**Figure 14-3** Vocal cords.

**313** Respiratory tract

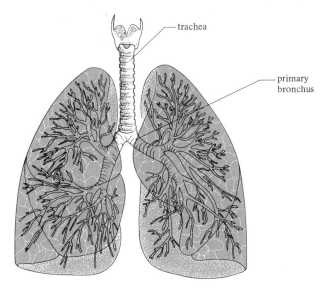

**Figure 14-4** Trachea and primary bronchi.

mouth; and the position of the tongue. If the larynx is removed for cancer, esophageal speech must be learned. This type of speech involves swallowing air into the esophagus and then forcing it out.

The vocal cords form the glottis which, in addition to the epiglottis, guards the passage to the lungs. During swallowing the slit between the folds, the **rima glottis,** is closed.

### Trachea

The trachea, or windpipe, extends from the larynx down through the neck and into the thorax. As seen in Figure 14-4, the trachea passes in front of the esophagus as it descends through the neck. It terminates by dividing into right and left bronchi which pass into the lungs.

The passageway of the trachea is kept open by the presence of a series of horseshoe-shaped cartilages within the tracheal wall. These cartilages do not join at the posterior wall of the trachea where it is adjacent to the esophagus. This permits the trachea to give somewhat as food passes down the esophagus.

The interior surface of the tracheal walls are lined with a ciliated mucous membrane. This membrane can trap dust or bacteria that have managed to pass through the nose, pharynx, and larynx. The cilia move the mucus up the trachea to the pharynx where it is swallowed.

When the respiratory passageway above the trachea is obstructed, a tube must be inserted into the trachea at the neck to permit the flow of air in and out of the lungs. Such a procedure is known as a **tracheostomy.** Care must be taken that the inserted tube does not become filled with mucus.

### Primary Bronchi

The two primary bronchi branch off the terminal end of the trachea and enter the two lungs. As can be seen in Figure 14-4, the right primary bronchus is wider and more vertical than the left primary bronchus. For this reason any particles passing down the trachea are more likely to enter the right lung than the left lung.

The structure of the primary bronchi is quite similar to that of the trachea. That is, the walls are reenforced by cartilage rings and the lining is composed of a ciliated mucous membrane.

Within the lungs the primary bronchi branch into smaller and smaller secondary bronchi. These secondary bronchi ultimately lead to the alveolar sacs where gas exchange takes place. We will discuss the respiratory passageway within the lungs, but first we must look at the location and external structure of the lungs.

## LUNGS

The lungs are located in the thorax on either side of the heart. They are cone-shaped with the base of the cone resting on the diaphragm and the apex of the cone just above the first rib.

A depression on the medial surface of each lung, the **hilus,** is the site at which the primary bronchus enters the lung. Blood vessels, lymphatics, and nerves also enter and leave the lungs at the hilus.

The space in the thorax between the lungs is called the mediastinum. It contains the esophagus, the heart, and the large blood vessel which carries blood to and from the heart. Figure 14-5 shows the position of the lungs within the thorax from a frontal and transverse view.

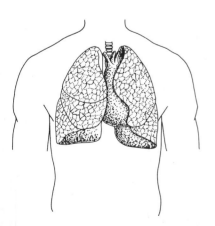
(a)

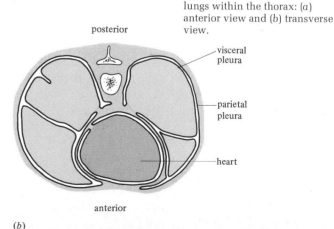
(b)

**Figure 14-5** Position of the lungs within the thorax: (a) anterior view and (b) transverse view.

The pleura, a serous membrane, covers the outer surface of each lung and folds back to line the inner wall of the thoracic cage as well as the top of the diaphragm. The portion of the pleura that covers the outer surface of the lungs is called the **visceral pleura,** whereas the portion lining the inner wall of the thoracic cage is called the **parietal pleura.** The space between the two layers of the pleura is called the **intrapleural space.** It is a very small space that is filled with fluid. Figure 14-5 shows the two layers of the pleura. It should be noted that the right and left pleural membranes are separate from each other.

Inflammation of the pleural membrane is known as **pleurisy.** In dry pleurisy there is not enough fluid between the two layers. causing them to rub painfully against each other during breathing. In pleurisy with effusion there is excess fluid in this space which can lead to edema of the lungs.

### Bronchial Tree

The interior of the lungs contains the branching bronchial tree, the alveoli, and the blood vessels which transport blood to and from the alveoli.

After they enter the lungs at the hilus, the primary bronchi branch into secondary bronchi. These secondary bronchi branch, in turn, into smaller and smaller bronchi which extend to all portions of the lungs. As the bronchi get smaller the structure of these tubes changes. The walls begin to contain less cartilage and more smooth muscle. **Bronchioles,** the smallest bronchi, are somewhat similar in structure and function to the arterioles in the vascular system. Their walls contain smooth muscle which can be used to constrict or dilate these tubes. Dilation of these tubes allows air to move more readily in and out of the lungs whereas constriction inhibits the flow of air. These bronchioles are regulated by the autonomic nervous system. Sympathetic stimulation causes bronchiolar dilation, whereas parasympathetic stimulation causes bronchiolar constriction.

The bronchioles enter the small lobules which form the smallest subdivisions of the lungs. Figure 14-6 illustrates the structure of a lobule. Within the lobule the bronchioles continue as alveolar ducts that lead to the alveoli.

### Alveoli

The alveoli are the tiny air sacs that form the terminal portion of the respiratory tract. There are approximately 300 million of these alveoli with a total surface area of about 70 square meters. This is about the size of a tennis court. It is at the alveoli that gas exchange between the respiratory tract and the adjacent capillaries takes place.

**316** THE RESPIRATORY SYSTEM

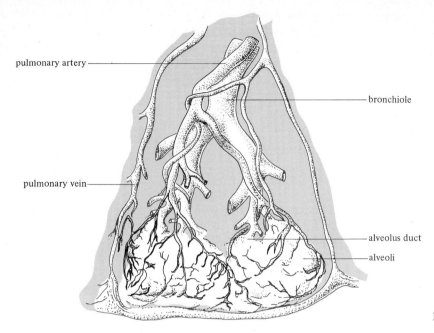

**Figure 14-6** Lobule of the lung.

The walls of the alveoli are composed of a single layer of endothelium. The alveoli are held open by the presence of a special secretion called **surfactant.** This secretion reduces the surface tension that would otherwise cause the alveoli to collapse. Occasionally children are born with hyaline membrane disease in which there is inadequate secretion of surfactant. In this situation many alveoli do not open and there is an insufficient surface for gas exchange.

## MECHANISM OF RESPIRATION

In order to explain adequately the mechanism by which air is moved in and out of the lungs we must first review a few basic physical principles about pressure. Pressure is defined as force per unit area. Hence for a liquid or gas it is the total force exerted against the walls of the container divided by the surface area of the container. The pressure exerted by a substance is equal in all directions. Essentially, the pressure of a gas is caused by the particles of gas striking the walls of the container.

As described in Chapter 13, the pressure of a substance in a container depends upon the amount of substance within the container and the volume of the container itself. If the volume of the container is kept constant and the amount of substance within it is

# 317 Mechanism of respiration

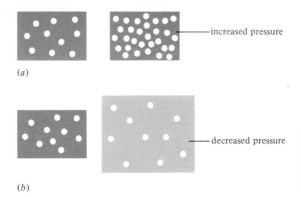

Figure 14-7 Effects of (a) the amount of substance present and (b) the volume of the container on the pressure in a container.

increased, as seen in Figure 14-7a, the pressure will increase. On the other hand, if the amount of substance is kept constant, but the volume of the container is increased, as in Figure 14-7b, the pressure will decrease. Thus, pressure increases when the amount of substance present increases and pressure decreases when the volume of the container increases. As already mentioned, any substance will flow from an area of higher pressure to an area of lower pressure.

In considering the flow of air during respiration, there are three relevant pressures. **Atmospheric pressure** is the pressure exerted by the air surrounding the body. At sea level it is equal to 760 mm Hg. It decreases with increasing altitude. **Intrapulmonary pressure** is the pressure within the respiratory space of the lungs (bronchi and alveoli). **Intrapleural pressure** is the pressure within the pleural space between the lungs and the walls of the thoracic cage.

Before looking at the events that take place during breathing, we must first look at the relationship between the various pressures between breaths, that is, when air is neither moving in nor out of the lungs. This is illustrated schematically in Figure 14-8. For illustrative purposes one can consider the respiratory tract an open tube connecting the interior of the lungs with the outside air.

Because air moves from an area of higher pressure to an area of lower pressure, intrapulmonary pressure must be equal to atmospheric pressure if there is no air movement. Intrapleural pressure, however, is less than atmospheric pressure — which means it is also less than intrapulmonary pressure. By definition a pressure less than atmospheric pressure is referred to as a **negative pressure.** It should be kept in mind that there is nothing "negative" about a negative pressure. It is merely less than atmospheric pressure. At rest the intrapleural pressure is approximately 756 mm Hg. This is equal to a pressure of $-4$ mm Hg, as atmospheric pressure is 760 mm Hg.

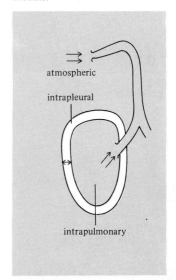

Figure 14-8 Schematic illustration of the relationship between atmospheric pressure, intrapulmonary pressure, and intrapleural pressure between breaths.

318   THE RESPIRATORY SYSTEM

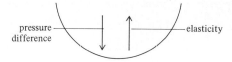

**Figure 14-9** Balance between pressure difference pushing lungs out and elasticity pulling lungs in.

We have said that intrapulmonary pressure is larger than intrapleural pressure. This pressure difference acts as a force which tends to expand the lungs, as illustrated in Figure 14-9. This force that tends to expand the lungs is equal and opposite to the elastic tendency of the lungs to contract. The walls of the lungs contain a large amount of elastic connective tissue which tends to contract the lungs. If this force were unopposed the lungs would collapse. The pressure difference between the intrapulmonary pressure and intrapleural pressure is a force that tends to expand the lungs. Between breaths this pressure difference pushing out on the lungs just balances the elastic force tending to collapse the lungs and they are held in a somewhat intermediate position.

Figure 14-10 illustrates what happens if the pleural membrane is broken from either the outside or the inside. In this situation air rushes into the intrapleural space and the intrapleural pressure becomes equal to atmospheric pressure. There is now no pressure difference between intrapulmonary pressure and intrapleural pressure, and therefore no force is pushing out on the lung walls. The

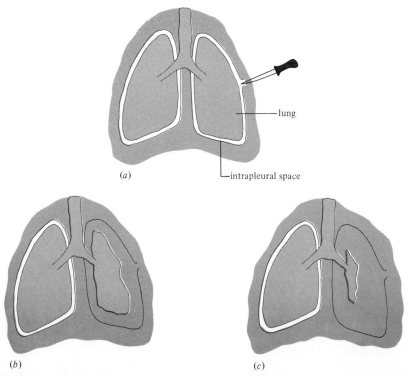

**Figure 14-10** Collapse of a lung after a wound. (a) Parietal pleura broken. (b) Air rushes into intrapleural space. (c) Lung collapses.

**319** Mechanism of respiration

elasticity of the lung is now unopposed and the lung collapses. Because the pleural membranes of the two lungs are separate, collapse of one lung does not cause collapse of the other lung.

**Inspiration**

Inspiration is the process by which air is moved into the lungs. This process is illustrated in Figure 14-11. In order for air to move into the lungs, the outside atmospheric pressure must be greater than the intrapulmonary pressure. Because the atmospheric pressure is fixed, the only way this pressure difference can be accomplished is by lowering intrapulmonary pressure below atmospheric pressure.

The initial event of inspiration is contraction of the inspiratory muscles, the diaphragm, and external intercostals. Contraction of the dome-shaped diaphragm causes it to flatten out, enlarging the thoracic cavity from top to bottom. Contraction of the external intercostals pulls the ribs upward and outward, enlarging the thoracic cavity from side to side and from front to back. As the boundaries of the thoracic cavity expand, the outer parietal layer of the pleura is pulled away from the inner visceral layer. This increases the volume of the intrapleural space. As mentioned pre-

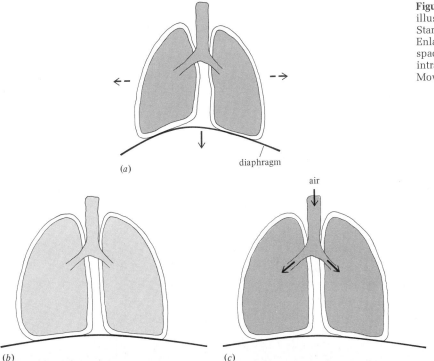

**Figure 14-11** Schematic illustration of inspiration. (a) Start of inspiration. (b) Enlargement of intrapulmonary space and lowering of intrapleural pressure. (c) Movement of air into lungs.

viously an increase in volume leads to a decrease in pressure. Thus as intrapleural volume is increased intrapleural pressure is decreased. This decrease in intrapleural pressure leads to a larger difference between intrapulmonary pressure and intrapleural pressure. Between breaths the difference between intrapulmonary and intrapleural pressure was an outward force that just balanced the inward elastic force of the lungs. When this pressure difference is increased by lowering intrapleural pressure, the outward force becomes greater than the inward elastic force. Once this happens the lungs begin to expand. Expansion of the lungs leads to an increase in intrapulmonary volume which is accompanied by a drop in intrapulmonary pressure. Prior to the start of inspiration, intrapulmonary pressure was equal to atmospheric pressure. Thus a fall in intrapulmonary pressure causes intrapulmonary pressure to become less than atmospheric pressure. Because the respiratory passage connecting the outside with the interior of the lungs is open, air will move in once intrapulmonary pressure is less than atmospheric pressure. As air moves into the lungs, intrapulmonary pressure begins to increase. Inspiration will cease when intrapulmonary pressure once again becomes equal to atmospheric pressure. At this point the lungs will be expanded and filled with an increased amount of air. Figure 14-12 sums up the major steps involved in inspiration.

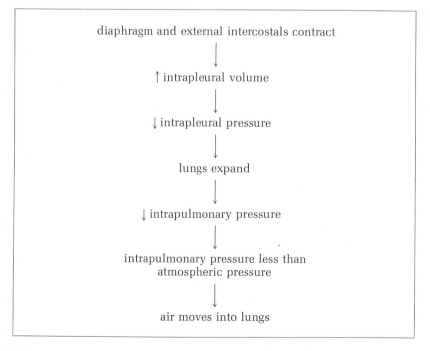

**Figure 14-12** Steps in inspiration.

## Expiration

For expiration to occur intrapulmonary pressure must become greater than atmospheric pressure. Normal expiration is a passive process; it does not involve contraction of muscles. The initial event of normal expiration is relaxation of the diaphragm and external intercostals. As these muscles relax there is no longer an adequate force to overcome the elastic recoil of the lungs. Thus, the lungs begin to contract. As the volume of the lungs decreases the intrapulmonary pressure increases. Remember that at the end of inspiration, intrapulmonary pressure was equal to atmospheric pressure. An increase in intrapulmonary pressure causes it to become larger than atmospheric pressure. Once intrapulmonary pressure exceeds atmospheric pressure, air moves out of the lungs and intrapulmonary pressure begins to fall. When intrapulmonary pressure falls to the same level as atmospheric pressure, expiration ceases. Figure 14-3 reviews the steps in expiration.

If there is a loss of lung elasticity, as in emphysema, or a narrowing of the respiratory passageway, as in asthma, expiration becomes an active process. Contraction of the abdominal muscles and the internal intercostal muscles forcefully reduces the size of the thoracic cavity and, thus, increases intrapulmonary pressure. This forces air out of the lungs.

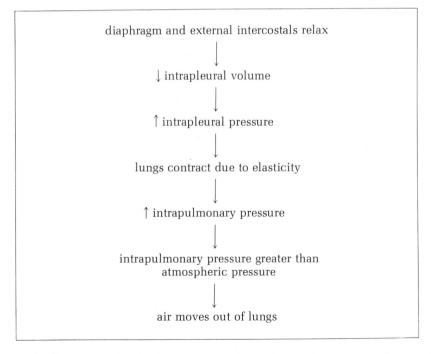

**Figure 14-13** Expiration.

## RESPIRATORY VOLUMES

In studying the capacity of a person's respiratory system to exchange air with the outside environment, a number of different types of measurements can be made. These are described below.

**Tidal volume** is the amount of air moved in and out of the lungs during a normal breath. For an average person the tidal volume is approximately 500 ml of air.

**Minute volume** is the amount of air moved in and out of the lungs during one minute. It is equal to the tidal volume times the number of breaths per minute. The normal respiratory rate is about 12 breaths per minute. Thus, normal minute volume is equal to 12 times 500, or 6000 ml per minute.

**Inspiratory reserve** is the amount of air that can be forcefully inspired above the tidal volume. For a normal person inspiratory reserve is equal to about 3000 ml.

**Expiratory reserve** is the amount of air above tidal volume that can be forcefully expired. Expiratory reserve averages about 1200 ml.

**Vital capacity** is the maximum amount of air a person can move in or out of the lungs. It can be measured as the amount of air expired after a maximal inspiration. The vital capacity is equal to the sum of the inspiratory reserve, tidal volume, and expiratory reserve. An average person has a vital capacity of about 4500–5500 ml.

**Residual air** is the amount of air left in the lungs after a maximum expiration. It is equal to about 1200 ml.

**Total lung capacity** is the total amount of air the lungs can hold. It is the sum of the vital capacity and the residual air, or approximately 5900 ml.

**Dead space** is air in the respiratory tract which is not available for gas exchange. The anatomical dead-space air is air that fills parts of the respiratory tract other than the alveoli. It is about 150 ml in men and 100 ml in women. Physiological dead-space air consists of the anatomical dead-space air plus air in any alveoli which are not exchanging gases with the blood. This can be caused by inadequate circulation to parts of the lungs.

## COMPOSITION OF AIR

**Atmospheric air.** The air in the atmosphere is approximately 79% nitrogen, 20% oxygen, and less than 1% carbon dioxide and other gases.

**Alveolar air.** As air enters the alveoli, oxygen is absorbed into the blood and carbon dioxide is released from the blood into the alveoli. Thus, alveolar air has less oxygen and more carbon dioxide than does atmospheric air. The oxygen composition is reduced to about 14% and the carbon dioxide composition increases to about 5.5%.

**Expired air.** During expiration, alveolar air is mixed with dead-space air in which no gas exchange takes place. Expired air contains about 15% oxygen and 4.5% carbon dioxide.

## EXCHANGE AND TRANSPORT OF GASES

One purpose of respiration is to obtain the oxygen the cells need for oxidizing food and to eliminate the carbon dioxide formed as a waste product of this process. Once oxygen is brought from the outside of the body to the alveoli, it diffuses from the alveoli into the adjacent capillaries. The blood then transports oxygen to the various cells of the body. Likewise, carbon dioxide formed in the various cells of the body is transported by the blood to the capillaries at the alveoli. It then diffuses into the alveoli and is eliminated from the body with the expired air. In this section we are going to look at the exchange of gases between the alveoli and the blood as well as the form in which the gases are transported within the blood.

The surface through which gas exchange occurs is illustrated in Figure 14-14. It consists of the alveolar wall, which is composed of a single layer of epithelial cells, a small interstitial space, the capillary basement membrane, and the single-layered epithelium of the capillary wall. This exchange surface is quite permeable to both oxygen and carbon dioxide and, thus, under normal conditions the gases pass freely between the alveoli and blood. The driving force that causes the gases to move is the difference between their alveolar and blood concentrations. Oxygen is at a higher concentration in the alveoli than in the blood entering the alveolar capillaries. Thus oxygen

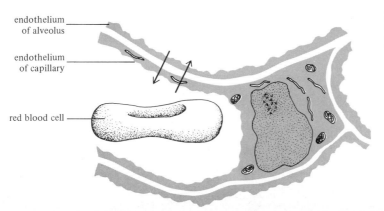

**Figure 14-14** Surface through which gas exchange takes place.

diffuses from the alveoli to the blood. On the other hand, the concentration of carbon dioxide is larger in the blood entering the alveolar capillaries than in the alveoli. Thus carbon dioxide diffuses into the alveoli. The concentrations of both oxygen and carbon dioxide in the blood leaving the alveolar capillaries are in equilibrium with the concentration of these gases in the alveoli.

**Transport of $O_2$**

Oxygen is relatively insoluble in water. However, the blood is able to transport a large amount of oxygen due to the presence of hemoglobin, a compound capable of combining with oxygen. When oxygen diffuses from the alveoli into the blood, only a small amount, approximately 3%, dissolves in the plasma. The remaining 97% enters the red blood cells and combines with hemoglobin. When oxygen-rich blood reaches the tissues, oxygen is released from hemoglobin and diffuses out of the blood and into the cells.

The affinity with which oxygen is bound by hemoglobin is measured by the **hemoglobin dissociation curve.** This curve shows the percentage of hemoglobin which contains oxygen at various oxygen concentrations. Figure 14-15 shows the hemoglobin dissociation curve. As can be seen, the percentage of hemoglobin-containing oxygen increases as oxygen concentration increases. The curve is very steep in the region between the venous and arterial oxygen concentration. This means that a small change in oxygen concentration over this region leads to a large change in the amount of oxygen bound to hemoglobin. The importance of this is that hemoglo-

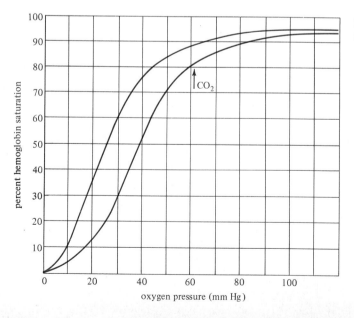

**Figure 14-15** Hemoglobin dissociation curves.

bin can quickly accept large amounts of oxygen as blood passes through the alveolar capillaries and unload large amounts of oxygen as blood passes through the tissue capillaries.

An increase in carbon dioxide concentration, or a decrease in pH, drives the curve to the right. This means that less oxygen is bound to hemoglobin at a particular oxygen concentration. The importance of this is that the rate at which hemoglobin unloads oxygen in the tissue capillaries is further increased as carbon dioxide enters the blood from the tissues.

An increase in temperature also drives the curve to the right. Thus in rapidly metabolizing tissue, the increase in temperature increases the amount of oxygen which leaves the blood and enters the cells.

Hypoxia refers to any situation in which an inadequate amount of oxygen is available for cellular use. There are four major types of hypoxia.

**Anemic hypoxia.** In anemic hypoxia the blood cannot carry adequate amounts of oxygen. This may be due to a loss of total blood volume, as after a hemorrhage, or an inadequate blood hemoglobin content, as in the various anemias.

**Stagnant hypoxia.** This is a hypoxia caused by sluggish circulation. It may be local, caused by an obstruction, or general, caused by cardiac problems.

**Hypoxic hypoxia.** In hypoxic hypoxia an insufficient amount of oxygen enters the alveoli. This can occur at high altitude or with respiratory disease.

**Histoxic hypoxia.** This occurs when the cells are not able to use oxygen, as in the presence of certain poisons such as cyanide.

### Transport of $CO_2$

Once carbon dioxide diffuses into the blood from the cells, it is transported in three different forms. A small amount of carbon dioxide dissolves in the plasma. The remaining carbon dioxide enters the red blood cells. Within the red blood cells a small amount of carbon dioxide combines with hemoglobin to form carbaminohemoglobin. However, most of the carbon dioxide within the red blood cells combines with water to form carbonic acid, as expressed in the equation

$$CO_2 + H_2O \longrightarrow H_2CO_3$$

This reaction takes place readily in red blood cells because of the presence of an enzyme called **carbonic anhydrase.** Once formed, carbonic acid dissociates into hydrogen ion and bicarbonate ion by the reaction

$$H_2CO_3 \longrightarrow H^+ + HCO_3^-$$

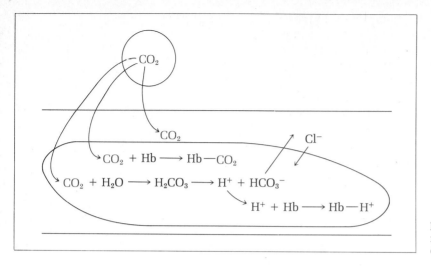

**Figure 14-16** Chemical reactions of carbon dioxide as it enters a tissue capillary.

The overall result of these reactions is that carbon dioxide that enters the red blood cells is converted to bicarbonate ion.

$$CO_2 + H_2O \longrightarrow H_2CO_3 \longrightarrow H^+ + HCO_3^-$$

Much of the bicarbonate formed in the red blood cells diffuses into the plasma. In order to preserve electrical balance, negative chloride ions diffuse from the plasma into the red blood cells as negative bicarbonate ions diffuse from the red blood cells into the plasma. This reaction is known as the chloride shift. The hydrogen ion formed as carbon dioxide is converted to bicarbonate is buffered by hemoglobin so that the blood does not become too acid.

In summary, carbon dioxide is transported by the blood in three forms. A small amount is dissolved in the plasma. An additional small amount is combined with hemoglobin. Most of the carbon dioxide is converted to bicarbonate ion in the red blood cells. Figure 14-16 illustrates the chemical reactions that take place as carbon dioxide enters the blood.

When venous blood reaches the alveoli, carbon dioxide diffuses out of the blood and into the alveoli. As this happens the chemical reactions that took place in the tissue capillaries are reversed. That is, carbon dioxide dissociates from hemoglobin and bicarbonate is converted back to carbon dioxide. Figure 14-17 illustrates the chemical reactions involving carbon dioxide that occur in the alveolar capillaries.

## CONTROL OF RESPIRATION

With regard to homeostasis, the function of the respiratory system is to keep the blood level of $O_2$, $CO_2$, and $H^+$ constant. This means

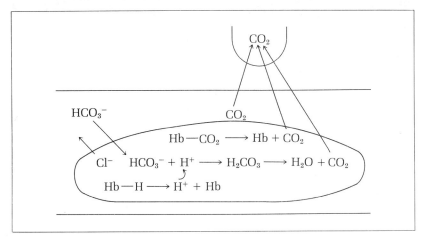

**Figure 14-17** Chemical reactions of carbon dioxide in a pulmonary capillary.

that if the blood level of any of these substances changes, respiration should be adjusted to bring the blood level back to normal. Therefore the rate and depth of respiration depends on the blood levels of $O_2$, $CO_2$, and $H^+$.

The basic reflexes which take place are summed up in Figure 14-18. A drop in blood $O_2$ stimulates respiration. The increase in respiration in turn serves to increase blood $O_2$. An increase in blood $CO_2$ or an increase in the amount of $H^+$ in the brain also leads to an increase in respiration. Increased respiration serves to eliminate the excess $CO_2$ or $H^+$ from the body.

The control of respiration, like other reflex activities, involves a number of elements. It involves stimuli, receptors, afferent pathways, an integrating center, efferent pathways, and effectors. As mentioned above, the stimuli for respiratory changes are changes in $O_2$, $CO_2$, and $H^+$.

**Respiratory receptors.** Blood levels of $O_2$ and $CO_2$ are measured by chemoreceptors in the carotid and aortic bodies. The carotid body is located at the junction of the internal and external carotids. The aortic body is located in the aortic arch. Receptors for changes in $H^+$ are located within the medulla of the brain. They seem to be sensitive to the $H^+$ content of the cerebral interstitial fluid. Under most circumstances, changes in the $H^+$ content of this fluid parallel changes in the $H^+$ content of the blood.

**Afferent pathways.** The carotid bodies are connected to the

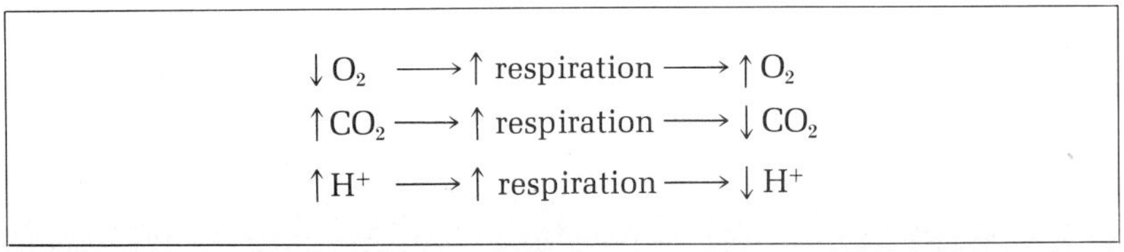

$$\downarrow O_2 \longrightarrow \uparrow \text{respiration} \longrightarrow \uparrow O_2$$
$$\uparrow CO_2 \longrightarrow \uparrow \text{respiration} \longrightarrow \downarrow CO_2$$
$$\uparrow H^+ \longrightarrow \uparrow \text{respiration} \longrightarrow \downarrow H^+$$

**Figure 14-18** Control of respiration.

respiratory centers in the pons and medulla by the glossopharyngeal, 9th cranial nerves. Information from the aortic body reaches the respiratory centers via the vagus, 10th cranial nerves.

**Respiratory control center.** Control over respiration is exerted by areas within the pons and medulla. It seems that a certain basic level of respiration is built into these areas. However, afferent information concerning changes in $O_2$, $CO_2$, and $H^+$ can alter this basic level. In addition, the respiratory center receives input from higher cerebral areas. Thus one can voluntarily alter respiration.

**Efferent pathways.** Respiration involves alternate contraction and relaxation of the diaphragm and external intercostal muscles. The phrenic nerve, which arises from the cervical plexus, controls the diaphragm. The intercostal nerves control the external intercostal muscles.

An additional factor that controls respiration is the stretch, or Hering-Breuer, reflex. Stretch receptors located in the walls of the lungs are stimulated when the lungs attain a certain size. Impulses from these receptors are conducted by the vagus nerve to the medulla where they inhibit further inspiration.

## OBJECTIVES FOR THE STUDY OF THE RESPIRATORY SYSTEM

At the end of this unit you should be able to:

1. State the major functions of the respiratory system.
2. Describe the internal structure of the nasal cavities.
3. Describe the functions of the mucous membrane lining the nasal cavities.
4. Name the structures that open into the pharynx.
5. Name the three divisions of the pharynx.
6. Name the different tonsils and describe the location of each.
7. State the functions of the tonsils.
8. State the two functions of the larynx.
9. State the location of the thyroid cartilage and the epiglottis.
10. Describe the structure of the vocal folds.
11. Describe the structure of the trachea.
12. Describe the internal structure of the lungs.
13. Distinguish between the locations of the visceral pleura and parietal pleura.
14. Name the portion of the respiratory tract in which gas is exchanged with the blood.
15. Explain the relationships between atmospheric pressure, intrapulmonary pressure, and intrapleural pressure when air is neither moving in nor moving out of the lungs.
16. Explain the sequence of events that takes place during inspiration and expiration.
17. Define the following terms: tidal volume, minute volume, inspiratory reserve, expiratory reserve, vital capacity, residual air, and dead space.
18. State the force that causes the exchange of gases between the alveoli and the blood.
19. Describe how oxygen is transported in the blood.
20. Name the three forms in which carbon dioxide is transported in the blood.
21. State the chemical reaction by which $CO_2$ is converted to bicarbonate.

## Objectives

22. Explain how changes in the blood levels of $O_2$, $CO_2$, or $H^+$ would change respiration.
23. State the location of the receptors sensitive to changes in respiration.
24. Explain how changes in respiration would change the blood levels of $O_2$, $CO_2$, and $H^+$.

# 15
# The Digestive System

DIGESTIVE TUBE STRUCTURE
PERISTALSIS
PERITONEUM
NUTRIENTS
THE DIGESTIVE TRACT
LIVER
PANCREAS
DEFECATION

The digestive system carries out three major functions. It breaks down the food we eat into small molecular and atomic components, it absorbs these components into the blood, and it eliminates undigested foods and certain metabolic wastes from the body. The term **digestion** refers to the breaking down of food. Mechanical digestion is the physical breakdown of food by such activities as chewing it or squeezing it with muscular contractions of the digestive tube. Chemical digestion is the breakdown of food by chemical reactions brought about by the action of enzymes. The term **absorption** refers to moving the digested food from the digestive tract to the blood and the term **elimination** refers to ridding the body of the undigested food along with certain metabolic wastes.

Figure 15-1 illustrates the structures which comprise the digestive system. The **digestive tract,** or **alimentary canal,** is a long continuous tube that extends through the body from the lips to the anus. It is composed of the mouth, pharynx, esophagus, stomach, small intestine, and large intestine. As the small and large intestines are coiled within the abdominal cavity, the total length of the system, about 30 feet, is greater than the length of the body. In addition to the digestive tract, the digestive systems contains three major accessory structures that secrete chemicals which aid in digestion. These are the salivary glands in the mouth, and the liver and pancreas in the abdominal cavity.

## DIGESTIVE TUBE STRUCTURE

Although there are variations among the different organs, the basic structure of the digestive tract wall is similar along the entire tract from the lower esophagus to the anus. As illustrated in Figure 15-2, the wall consists of four layers. The inner layer, the **mucosa,** is a mucous membrane with epithelium on the free surface and connective tissue beneath. Secretory glands are found in the mucosa throughout the length of the digestive tube. In addition, a small

**332** THE DIGESTIVE SYSTEM

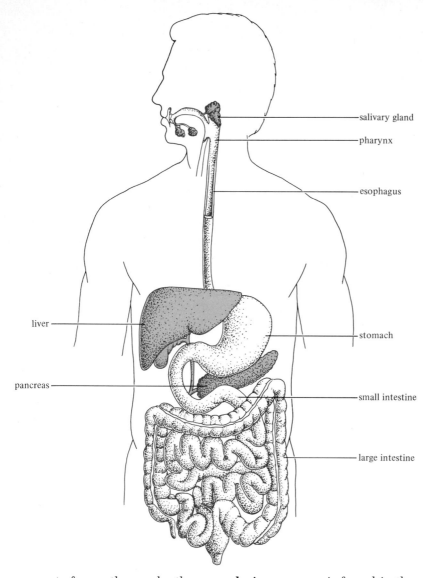

**Figure 15-1** The digestive system.

amount of smooth muscle, the **muscularis mucosae,** is found in the mucosa.

The **submucosal layer** adjacent to the mucosa is composed of connective tissue. It contains blood and lymphatic vessels. It also contains a network of interconnecting nerve fibers known as the **submucosal plexus,** or Meissner's plexus.

The main muscular layer of the digestive tube wall, the **muscularis externa,** is composed of two main bands of smooth muscle. The inner band is circular smooth muscle; the outer band is longitudinal smooth muscle. Between these bands is an extensive network of nerve fiber known as the **myenteric plexus** or Auer-

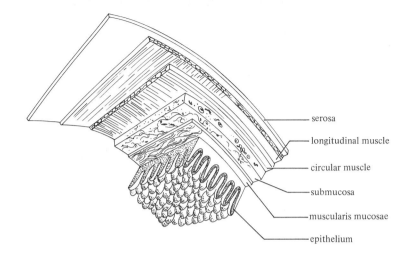

Figure 15-2 Layers of the digestive tract wall.

bach's plexus. This plexus extends throughout the tube and serves to help coordinate activities in one part of the tube with activities in another part.

The outer layer of the tube, the **serosa,** is a serosal membrane composed of connective tissue and epithelium.

### Sphincters

The various structures of the digestive tract are separated by constrictions known as sphincters. These sphincters are thick rings of circular smooth muscle that are normally kept in a state of tonic contraction. This prevents the backflow of substances from one portion of the tract to another. The sphincters are then relaxed and opened when substances are being pushed along the tract.

## PERISTALSIS

The basic means by which food is pushed along the digestive tract is **peristalsis.** Peristalsis consists of a wave of contraction which passes down along a portion of the digestive tube, pushing the contents of the tube ahead of it. It is often preceded by a wave of relaxation. Figure 15-3 illustrates the process of peristalsis. It should be kept in mind that peristalsis is a basic physiological process that occurs in tubes other than the digestive tube. For example, urine is moved along the ureters by peristalsis.

Figure 15-3 Peristalsis.

## PERITONEUM

The peritoneum is a serous membrane which lines the walls of the abdominal cavity and folds back to cover the outer surface of the

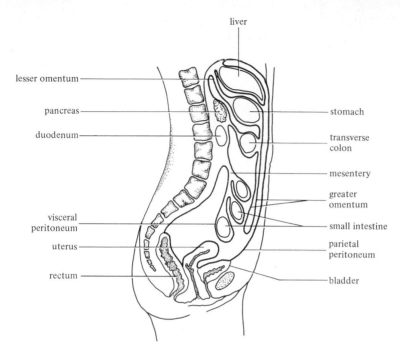

**Figure 15-4** Peritoneum.

abdominal organs. The portion of the peritoneum lining the abdominal walls is called the **parietal peritoneum,** and the portion of the peritoneum covering the outer surface of the abdominal organs is called the **visceral peritoneum.** Fluid between these two layers prevents friction when the internal organs move relative to the abdominal walls and each other. Figure 15-4 illustrates the peritoneum.

There are three major extensions of the visceral peritoneum which help anchor the abdominal organs. The **mesentery** is an extension from the back wall of the abdominal cavity which surrounds the intestine. The **greater omentum** extends downward from the stomach and lies like an apron over the front of the intestine. It folds backward and attaches to the transverse colon. Often the greater omentum contains a large amount of fat. The **lesser omentum** connects the stomach and duodenum to the liver at the back of the abdominal cavity.

## NUTRIENTS

The food we eat is used in the body for a number of different functions. The organic material is used as a source of energy within the cells, as a source for building new tissue, and as a source for synthesizing physiologically important chemicals such as hormones and enzymes. The inorganic salts and water serve as the medium in which the physiological activities of the body take place. At this point we will look at the major categories of materials that enter the digestive tract as food.

### Carbohydrates

Carbohydrates are ring structures composed of carbon, hydrogen, and oxygen atoms. The basic carbohydrate unit is a single ring called a monosaccharide, or simple sugar.

Most carbohydrate enters the body as polysaccharides, long chains of simple sugars. The three major carbohydrates entering the body are starch, cellulose, and glycogen, all of which are long chains of glucose that differ only in the way in which the glucose molecules are held together. **Starch** and **cellulose** are plant polysaccharides, while **glycogen** is an animal polysaccharide. In the digestive tract, starch and glycogen are broken down into individual glucose molecules which are then absorbed into the blood. The human digestive tract is not capable of digesting cellulose.

Some carbohydrate enters the body as disaccharide, two simple sugars bound together. The principal disaccharides are **sucrose** (cane sugar), composed of the simple sugars glucose and fructose; **lactose** (milk sugar), composed of glucose and galactose; and **maltose,** composed of two glucose molecules. Disaccharides are digested to their component simple sugars which are then absorbed.

### Proteins

Proteins are composed of chains of amino acids linked together by peptide bonds. All amino acids are composed of carbon, hydrogen, oxygen, and nitrogen atoms. Some amino acids also contain sulfur. There are approximately 20 different amino acids, some of which can be synthesized in the body, and others of which must be supplied by dietary proteins. Those synthesized in the body are called **nonessential amino acids.** Those which must be supplied by the diet are called **essential amino acids.**

Most animal proteins we eat are **complete** proteins which contain all the essential amino acids. Plant proteins are generally **incomplete** proteins lacking one or more essential amino acids.

### Lipids

Lipids are chemicals that are not soluble in water. The major types of lipids which are found in food are the fats, or triglycerides, the compound lipids, and the steroids.

**Triglycerides** are composed of three fatty acids bound to a glycerol molecule. In the digestive tract the fatty acids are split off the glycerol and the individual components are then absorbed.

Compound lipids are combinations of lipids with other substances such as carbohydrates or phosphates. The compound lipids are broken down into their individual components in the digestive tract.

**Steroids** are composed of four interconnected carbon rings. A number of hormones, vitamin D, and cholesterol are all steroids. These compounds are absorbed intact or in slightly modified form.

### Vitamins

Vitamins are chemicals that are needed for certain metabolic reactions to take place. Some vitamins act as enzymes. Others act as coenzymes, chemicals that aid enzymes. Still others serve as the precursors from which enzymes are made.

Although the body is capable of synthesizing small amounts of certain vitamins, the bulk of needed vitamins must be obtained in the diet. A deficiency in any of the vitamins may lead to disease. On the other hand, any vitamin excess will not strengthen the body and, in fact, can lead to illness. In general, the body will excrete any excess vitamin. However, the body has difficulty excreting certain vitamins such as vitamin A. Excesses of these vitamins deposit in the tissues and can cause tissue damage.

Vitamins are usually divided into two large categories: **fat-soluble vitamins** and **water-soluble vitamins.** Absorption of fat-soluble vitamins depends on the processes necessary for fat digestion and absorption.

#### Fat-Soluble Vitamins

**Vitamin A.** This vitamin is formed in the liver from the carotenoid pigments found in leafy green vegetables, yellow vegetables, and yellow fruits. We can obtain the vitamin either by taking in the pigment and forming the vitamin in our liver or by eating other animals' livers which contain relatively large amounts of stored vitamin A.

Vitamin A is necessary for the synthesis of rhodopsin, the light-sensitive pigment of the rod cells in the retina. Lack of vitamin A can lead to an inability to see at low light levels (night blindness). Vitamin A is also necessary for normal growth and development, particularly for epithelial tissue. Vitamin A deficiency can lead to excessive keratinization of the epithelium.

**Vitamin D.** Vitamin D is formed in the skin in response to ultraviolet light. Because most of us do not get enough exposure to ultraviolet light, it is necessary to add synthetic vitamin D to foods.

Vitamin D is necessary for the absorption of calcium from the digestive tract. It is also necessary for the proper calcification of bones and teeth. A deficiency of vitamin D in children, called rickets, is characterized by soft and fragile bones. Children with rickets often are bowlegged. In infants with rickets the fontanels do not close. Vitamin D deficiency in adults, called osteomalicia, is also characterized by soft and fragile bones.

**Vitamin E.** Vitamin E is found in leafy green vegetables, vegetable oils, unmilled cereals, and eggs. Although the exact function of vitamin E is unknown, it seems necessary for normal growth and development. There is no disease that can clearly be attributed to a lack of vitamin E. However, some people feel a vitamin E deficiency is at least in part responsible for cystic fibrosis. In general, there is more than an adequate amount of vitamin E in the diet.

**Vitamin K.** Vitamin K is found in liver, spinach, and cabbage. It is also synthesized to a small degree by bacteria in the large intestine. Vitamin K is necessary for prothrombin synthesis in the liver. A deficiency in vitamin K is sometimes seen in newborn infants because the bacterial flora of the large intestine has not had time to develop. These infants have difficulty forming blood clots.

**Water-Soluble Vitamins**

**B-complex.** A number of vitamins are classified on the basis of their water solubility and nitrogen content as B vitamins. Although these vitamins are found in similar sources, their relative proportions vary from source to source. Good sources of B vitamins are liver, kidney, lean meat, eggs, yeast, whole grain cereals, and milk products. Some of the more important B vitamins are described below.

**Thiamine.** This vitamin is necessary for the conversion of glucose to pyruvic acid in the breakdown of carbohydrate. Deficiency of thiamine leads to central nervous system disorders and heart failure. This disease is called **beriberi.**

**Riboflavin.** Riboflavin is used to make FAD, one of the compounds that can transport hydrogen from one chemical to another. It is rare to find riboflavin deficiency among humans.

**Niacin (nicotinic acid).** This is used to manufacture NAD, the major hydrogen-transporting compound in cells. A lack of NAD interferes with a large number of chemical reactions. **Pellegra,** the disease caused by a lack of niacin, is not very common in America, but is common in countries in which the typical diet is lacking in meat and milk products.

**Vitamin $B_{12}$.** Vitamin $B_{12}$ is essential for the synthesis of DNA. It is especially important in rapidly dividing tissue such as the hemopoietic tissue that manufactures red blood cells. Only a small amount of vitamin $B_{12}$ is required in the diet and its absence is rarely seen. However, some people lack the ability to manufacture intrinsic factor, the stomach substance necessary for $B_{12}$ absorption.

The absence of intrinsic factor and the resulting inability to absorb vitamin $B_{12}$ lead to pernicious anemia. It can be corrected by giving injections of vitamin $B_{12}$.

**Vitamin C (ascorbic acid).** Vitamin C differs from the water-soluble B vitamins in that it lacks nitrogen. It is found in citrus fruits and certain vegetables such as broccoli, potatoes, and cabbage. It is necessary for maintaining intercellular protein and for the production of hemoglobin. Vitamin C deficiency, known as **scurvy,** is characterized by easily damaged tissues which hemorrhage and by a failure of wound healing. Because of its role in building and strengthening tissue, some people feel that vitamin C helps protect against colds.

## THE DIGESTIVE TRACT

### Mouth

The mouth, or buccal cavity, extends from the lips in front to the oropharynx at the back. The top of the mouth is formed anteriorly by the hard palate and posteriorly by the soft palate. As we shall see later, the soft palate is lifted to close off the posterior nares during swallowing. The sides of the mouth are formed by the cheeks and the floor of the mouth is formed by the tongue. A mucous membrane covers the entire interior surface of the buccal cavity.

The tongue is formed of striated muscle covered with a mucous membrane. This membrane forms a fold, the **frenulum linguae,** that attaches the tongue to the bottom of the mouth. In certain individuals the frenulum linguae is abnormally short and prevents free movement of the tongue; this is called tonguetie.

The dorsal surface of the tongue is covered with projections called papillae. There are three types of papillae. The **filiform papillae** are tall, narrow projections arranged in parallel rows across the tongue. These projections are slightly sticky and aid in licking. Nerve endings for touch are found within the filiform papillae. The **fungiform papillae,** scattered among the filiform papillae, contain taste buds. Taste buds are also found in the **vallate papillae,** arranged in a V-shaped row at the back of the tongue.

The tongue carries out a number of important functions. It pushes food from one side of the mouth during chewing, it pushes food into the pharynx as the first step in swallowing, and it helps shape the resonating cavity of the mouth during speech.

### Teeth

A child's first set of teeth begins to appear at about six months of age. These teeth, called the **deciduous teeth,** grow in at the rate

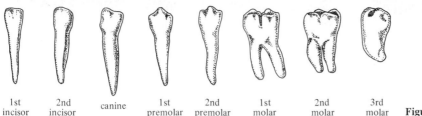

1st incisor    2nd incisor    canine    1st premolar    2nd premolar    1st molar    2nd molar    3rd molar

**Figure 15-5** Teeth.

of about one a month until all 20 appear.

From the ages of 6 to 13, the deciduous teeth are lost and the permanent teeth appear. Figure 15-5 shows the permanent teeth of an adult. Starting at the midline and moving in a lateral direction these teeth are: two **incisors** which are used to cut food, a **canine** which is used to cut and shred food, two **premolars** (bicuspids) which cut and tear food, and three **molars** which crush and grind food. The last molar tooth is called a wisdom tooth. In some individuals the wisdom teeth never break through the gums.

The structure of a tooth is shown in Figure 15-6. The portion above the gum is known as the **crown.** It is covered by an outer layer of **enamel.** Below the enamel is a thick layer of calcified connective tissue called **dentin.** The dentin makes up the bulk of the crown. At the core of the crown there is soft connective tissue known as the **pulp.** Blood vessels and nerves are found within the pulp.

The **roots** of the teeth are embedded in the alveolar processes of the jaw bones. The root is the portion of the tooth below the gums. It is covered by a thin layer of cementum and is bound to

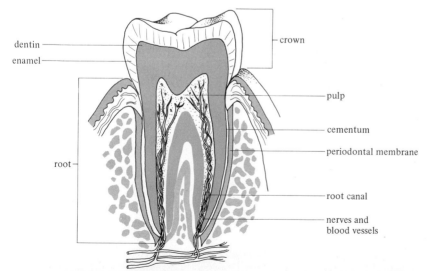

**Figure 15-6** Internal structure of a tooth.

bone by the **periodontal membrane.** There is dentin underneath the cementum. The **root canal** lies at the center of the root. It contains pulp and is continuous with the pulp of the crown. Nerves and blood vessels enter the tooth through the root canal and then pass on to the pulp of the crown.

**Salivary Glands**

There are three pairs of salivary glands in the mouth; these are illustrated in Figure 15-7. The **parotid glands** lie below and in front of the ears. Their secretions pass into the mouth through a duct known as Stenson's duct. Inflammation of the parotid glands is known as mumps. The **submandibular glands** are located on the inner surface of the lateral portion of the mandible. Their ducts, Wharton's ducts, open into the floor of the mouth just behind the lower incisors. The **sublingual glands** lie underneath the membrane on the floor of the mouth. Many small ducts conduct the secretions of the sublingual glands into the mouth.

The salivary glands secrete about 1000–1500 ml of saliva each day. In addition to water and electrolytes, the saliva contains two important substances: ptyalin and mucus. **Ptyalin,** or salivary amylase, is an enzyme which begins the breakdown of carbohydrates. However, it only functions at a pH of 6–7 and thus becomes inactivated once food is swallowed and exposed to the acid contents of the stomach. The **mucus** in the saliva serves to lubricate the food so it can be more readily swallowed.

Saliva contains a high concentration of calcium which prevents the removal of calcium from the enamel of the teeth. In addition to the functions mentioned above, saliva acts to dissolve food, protects the mouth against dryness, and aids in the movements of the tongue in speech.

Salivary secretion can be initiated by a number of different stimuli. The presence of food in the mouth, the taste or smell of

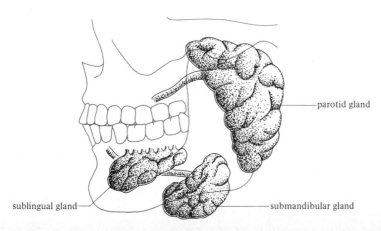

**Figure 15-7** Salivary glands.

food, or even the thought of food can lead to salivary secretion. These stimuli activate the salivary center located at the boundary of the medulla and pons in the brainstem. The salivary center then activates the salivary glands through efferent neurons of the parasympathetic system. These neurons pass through the facial and glossopharyngeal nerves to the salivary glands.

### Pharynx

The structure of the pharynx is discussed in detail in Chapter 14 and is illustrated in Figure 14-1. It contains openings to the internal nares, eustachian tubes, mouth, larynx, and esophagus. Three sets of tonsils, the pharyngeal, palatine, and lingual tonsils, aid in the destruction of foreign organisms that enter the pharynx through the nose and mouth. Below the internal mucous membrane of the pharynx there are three skeletal muscles which help move food from the pharynx to the esophagus during swallowing: these are the **superior, middle,** and **inferior** constrictor muscles.

In order to understand the changes that take place in the pharynx during swallowing, we must first look at the anatomical situation of the pharynx when one is not swallowing, that is, during normal breathing. This is illustrated in Figure 15-8a. The soft palate is hanging down so that the internal nares are open. The opening to the larynx is also open. This means that there is an open passageway through the pharynx from the nose to the trachea and air can move freely in and out of the lungs. The esophagus is closed off from the pharynx by constriction of the **hypopharyngeal sphincter** at the top of the esophagus.

### Swallowing (Deglutition)

Swallowing, or deglutition, involves the coordinated activities of a large number of different muscles. After chewing, a small amount of food called a **bolus** is pushed to the back of the mouth by

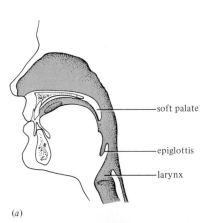
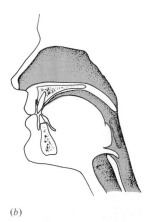

(a)  (b)

**Figure 15-8** Swallowing.

the tongue. This triggers the swallowing reflex, illustrated in Figure 15-8b. Once started the swallowing reflex goes to completion over a period of about one second. The soft palate is pulled upward to close off the internal nares. This prevents food from entering the nostrils. Food is prevented from entering the trachea in two ways. The vocal cords are pulled together to close the rima glottis. In addition, the entire larynx is raised to pull the epiglottis over the opening to the larynx. As food enters the pharynx it is pushed toward the esophagus by the coordinated contraction of the constrictor muscles in the pharyngeal wall. The hypopharyngeal sphincter relaxes so that the bolus can pass from the pharynx into the esophagus.

### Esophagus

The esophagus extends downward from the laryngopharynx behind the trachea. It passes through the mediastinum and joins the stomach just below the diaphragm. The total length of the esophagus is approximately 10 inches.

At the top of the esophagus there is a sphincter, the **hypopharyngeal sphincter,** which opens and closes the esophagus during swallowing. The muscular layer of the upper one-third of the esophagus is skeletal muscle. However, this muscle is under the control of the vagus nerve, part of the parasympathetic division of the autonomic nervous system. The lower two-thirds of the esophagus has a structure similar to that described at the beginning of this chapter for the digestive tube in general. Two layers of smooth muscle are found in the wall. Although there is no distinct anatomical sphincter at the bottom of the esophagus, the circular smooth muscle is hypertrophied for a short distance above and below the diaphragm. This muscle is usually kept tonically constricted; thus it effectively serves as a sphincter which prevents the movement of stomach contents back into the esophagus. The constricted smooth muscle is considered a physiological sphincter. It is known as the **gastroesophageal** or **cardiac sphincter.**

Movement of food down the esophagus is brought about by peristalsis. As a bolus of food enters the esophagus through the hypopharyngeal sphincter, the peristalsis which began in the pharynx continues down the esophagus. This wave of peristalsis pushes the food ahead of it. If the bolus is quite watery, it can move by gravity down the esophagus at a faster rate than the peristaltic wave. It normally takes the peristaltic wave about 5–10 seconds to pass down the esophagus.

Peristalsis in the esophagus is controlled by the coordinated firing of vagal efferents to the esophagus. If any food gets stuck in the esophagus, afferent neurons are activated. This leads to a forceful wave of peristalsis known as secondary peristalsis.

When the bolus reaches the lower end of the esophagus, the gastroesophageal sphincter relaxes and the bolus moves on into the stomach. The sphincter then closes again and prevents the reflux of stomach contents into the esophagus. **Achlasia** refers to the condition in which the gastroesophageal sphincter does not relax. Food gets stuck behind the sphincter and only dribbles into the stomach slowly. In this situation surgical intervention may be necessary.

**Stomach**

The stomach is located in the upper left-hand portion of the abdomen just below the diaphragm. It occupies a portion of the epigastric and left hypochondriac region.

**Structure.** The major parts of the stomach are illustrated in Figure 15-9. The esophagus opens into the stomach at the **cardiac orifice** and the region around the orifice is known as the **cardiac region.** That portion of the stomach which extends above the cardiac region is known as the **fundus.** The central, major portion of the stomach is known as the **body** of the stomach. The narrow portion of the stomach leading into the small intestine is known as the **pylorus.** A sphincter, the **pyloric sphincter,** separates the stomach from the small intestine. The larger, lateral curve of the stomach is the **greater curvature,** and the smaller, medial curve is the **lesser curvature.** In a fasting man the stomach has a volume of about 50 ml. It can expand to hold 1000–2000 ml.

In the empty stomach the mucous membrane forms folds known as rugae. These folds are stretched out as the stomach fills. The outer epithelial cells are tall columnar cells. Depressions between these cells, known as the **gastric pits,** contain the glandular cells that form the gastric secretions. As can be seen in Figure

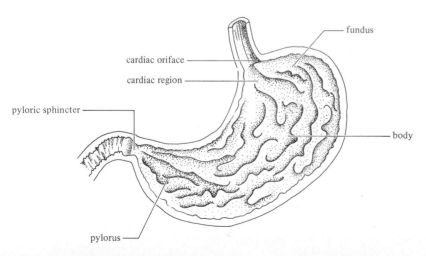

**Figure 15-9** Regions of the stomach.

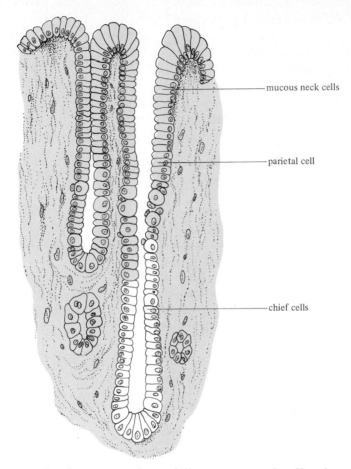

**Figure 15-10** Gastric gland.

15-10, a gastric gland contains three different types of cells: the **mucous neck cells** secrete an alkaline mucus which coats and protects the stomach; the **chief cells,** or **zygomatic cells,** secrete **pepsinogen,** an inactive form of the protein-digesting enzyme pepsin; the **parietal cells,** or **oxyntic cells,** secrete a concentrated solution of hydrochloric acid.

**Function.** The stomach carries out three major functions: it stores food, digests food, and delivers food to the small intestine at a rate that the small intestine can handle it. Digestion in the stomach is both mechanical and chemical. Mechanical digestion involves contractions of the stomach wall which break up food and mix it with the digestive secretions.

Chemical digestion in the stomach involves primarily the partial breakdown of protein by pepsin and hydrochloric acid. Pepsinogen from the chief cells and hydrochloric acid from the parietal cells enter the lumen of the stomach through the tubes within the gastric pits. In the stomach, hydrochloric acid converts pepsinogen to the active proteolytic enzyme pepsin. Pepsin and hydrochloric acid then begin the breakdown of protein—a process which is later completed in the small intestine. These secretions also help destroy bacteria which enter the stomach with the food.

The protein-digesting enzymes present in the small intestine are capable of the complete breakdown of protein. Thus protein digestion can take place even in the absence of gastric secretion such as occurs if the stomach is removed because of ulceration.

The presence of pepsin and hydrochloric acid in the stomach presents a potential danger to the stomach itself. If these substances come in contact with the stomach wall, they can start to digest the stomach wall tissues. The stomach is protected against pepsin and hydrochloric acid in two basic ways: epithelial cells on the surface of the stomach and in the neck of the gastric pits secrete a thick alkaline mucus which neutralizes the acid and prevents acid and pepsin from coming in contact with the cells themselves. In addition, there is very rapid mitosis in the gastric epithelium so that any cells that are destroyed are immediately replaced. Ulceration is probably more often a result of a breakdown in the protective lining than it is a result of excessive acid secretion.

**Control of Acid and Pepsinogen Secretion**

Control of hydrochloric acid and pepsinogen secretion can be divided into three phases. The **cephalic phase** refers to the influence of the brain on secretion. When there is food present in the mouth or when one smells or even thinks of food, increased gastric secretion is brought about through stimulation of the gastric glands by efferents in the vagus nerve. This leads to the presence of acid and pepsin in the stomach even before food enters the stomach.

The **gastric phase** of secretion is brought about by the presence of food in the stomach. It is controlled by the hormone **gastrin** which is produced in the mucosa of the pyloric region of the stomach. Gastrin is released in response to stretching of the antrum caused by the presence of food in this region, or in response to specific substances in the food, particularly protein. Coffee and alcohol are also potent stimulants of gastrin releases. Once released, gastrin is transported by the blood to the body of the stomach where it stimulates the secretion of hydrochloric acid and pepsinogen.

The **intestinal phase** of acid secretion refers to the influence of the small intestine on gastric secretion. If the material present in the duodenum of the small intestine is too acid, a hormone is released by the intestinal mucosa. This hormone is carried by the blood to the body of the stomach where it inhibits further acid secretion. This serves as a protective device for the small intestine which is not as well protected against acid as the stomach. The total volume of gastric secretions in response to all the stimuli mentioned above is approximately 2000–3000 ml per day. Figure

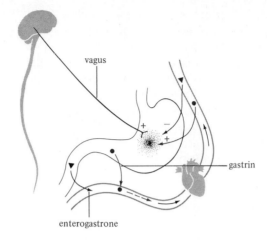

**Figure 15-11** Control of gastric secretion.

15-11 is a schematic illustration of the control of acid and pepsinogen secretion.

### Emptying of Stomach

Once food is mixed with the gastric secretions, it is known as **chyme.** The rate at which chyme leaves the stomach and enters the small intestine primarily depends on the status of the chyme already present in the small intestine. Basal peristaltic contractions of the smooth muscle in the stomach drive food toward the duodenum. However, the volume and composition of the chyme already present in the duodenum can influence the force of these contractions. If the volume of the intestinal chyme is too large, if it is too acid, too concentrated, or if it contains too much fat, reflex inhibition of gastric emptying is brought about. This reflex is the result of both hormonal and nervous activity. Figure 15-12 illustrates the manner in which the duodenum can inhibit gastric emptying.

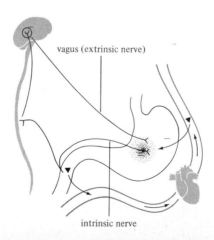

**Figure 15-12** Inhibition of gastric emptying by the duodenum.

An increase in the volume, acidity, concentration, or fat content of the chyme in the duodenum leads to the release of the hormone **enterogastrone** from the duodenal mucosa. This hormone then acts on the stomach to reduce the force of gastric contractions. Nervous inhibition of the stomach involves an intrinsic neuronal pathway in the myenteric plexus of the digestive tube wall and an extrinsic neuronal pathway involving afferent and efferent neurons in the vagus nerve. In both cases stimuli from the duodenum lead to nervous inhibition of gastric contractions.

## LIVER

The liver is located just below the diaphragm with the bulk of the liver lying to the left of the midline. It is in the epigastric and right hypochondriac regions. Weighing 3–3½ pounds, the liver is the largest organ in the body.

Figure 15-13 illustrates the gross structure of the liver. It is divided by the **falciform ligament** into right and left lobes. The undersurface of the liver is concave and contains the impressions of the underlying organs.

The visceral peritoneum, which covers the liver, forms folds that serve as ligaments attaching the liver to the diaphragm and the anterior abdominal wall. Underneath the peritoneum there is a capsule of connective tissue known as Glisson's capsule. This connective tissue enters the liver at the porta, a fissure which passes in a transverse direction across the inferior portion of the right lobe of the liver. Within the liver the connective tissue divides and forms a branching network. The portal vein and hepatic artery enter the liver in the connective tissue at the porta. They then divide into smaller branches which follow the connective tissue network of the

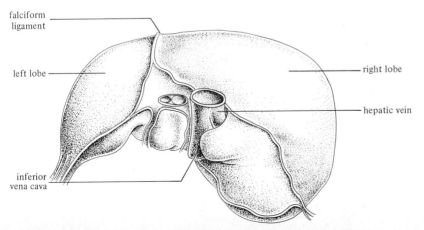

**Figure 15-13** Gross structure of the liver (posterior view).

liver. Lymphatic vessels are also found in this network and pass out at the porta. The hepatic veins which transport venous blood to the interior vena cava leave the liver by a different route, through a groove in the posterior portion of the liver.

The basic structural and functional unit of the liver is the liver lobule. These lobules are separated by the connective tissue network described previously. Figure 15-14 illustrates the structure of a liver lobule. The connective tissue which surrounds the lobule contains branches of the hepatic portal vein which brings venous blood from the digestive tract to the liver and branches of the hepatic artery which brings arterial blood to the liver. It also contains the beginning of bile ducts which ultimately transport bile from the liver to the small intestine. The lobule itself is formed by plates of hepatic cells which extend from the center of the lobule to the periphery. Between these hepatic cells there are very permeable capillary-type vessels, known as the **hepatic sinuses.** These sinuses are lined with phagocytic cells known as **Kupffer's cells.** Blood flows into the hepatic sinuses from the branches of the portal vein and hepatic artery at the periphery of the lobule. As the blood flows through the sinuses there is an exchange of material between the blood and the hepatic cells. In addition, the phagocytic cells can remove any foreign organisms, such as bacteria, which gain access to the blood. From the hepatic sinuses blood flows into the central vein at the center of the lobule. The central veins empty into the

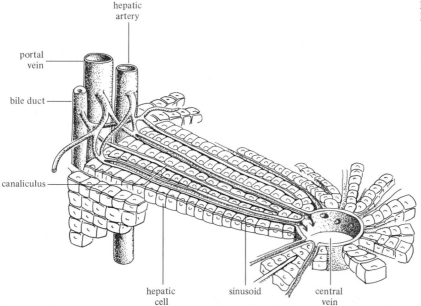

**Figure 15-14** Structure of a liver lobule.

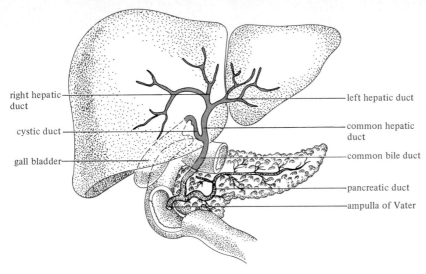

**Figure 15-15** Ducts draining the liver, gallbladder, and pancreas.

hepatic veins which in turn empty into the inferior vena cava.

Bile formed in the hepatic cells enters the bile canaliculi which are terminal tubes between some of the plates of cells. The canaliculi drain into the bile ductules located in the connective tissue at the periphery of the lobule. These ductules come together to form the right and left hepatic ducts which drain bile from the two lobes of the liver. The right and left hepatic ducts join to form the **common hepatic duct,** as illustrated in Figure 15-15. This duct descends about 1½ inches before joining the **cystic duct** from the gallbladder to form the **common bile duct.** The common bile duct descends behind the head of the pancreas, where it is usually joined by the **pancreatic duct,** to form the **ampulla of Vater** which opens into the small intestine. Two sphincters, the **sphincters of Boyden and Odi,** regulate the flow of bile from the ducts into the small intestine. The sphincter of Odi is at the ampulla of Vater and the sphincter of Boyden is in the common bile duct just before it joins the pancreatic duct.

### Gallbladder

The gallbladder is a muscular sac located in a depression in the bottom surface of the right lobe of the liver. When the sphincters of Boyden and Odi are constricted, bile formed in the liver flows into the gallbladder where it is stored and concentrated. As food enters the small intestine, the gallbladder contracts and empties its stored bile through the cystic duct and common bile duct into the duodenum.

### Liver Function

The liver carries out an enormous number of different functions. These functions are described in the sections of the text in which they are relevant. For example, the production of blood proteins is discussed in the section on the blood. In this section we are going to consider the production of bile, the digestive secretion formed in the liver.

The liver produces approximately 600–800 ml of bile a day. In addition to water and electrolytes, the bile contains three important constituents: bile salts, bile pigments, and cholesterol. The **bile salts** are important for both the digestion and the absorption of fat. They aid in the digestion of fat by breaking up, or emulsifying, large fat droplets. Emulsification of fat increases the surface area which can be attacked by the fat-digesting lipases produced in the pancreas. Emulsification of fat by bile salts is similar to the process by which soap emulsifies grease. In addition to their role in fat digestion, the bile salts play an essential role in fat absorption by making small fat particles water soluble. Bile salts which enter the duodenum are reabsorbed in the ileum of the small intestine. They are then transported back to the liver and reused in the formation of bile.

The **bile pigments,** bilirubin and biliverdin, are formed in the liver and spleen from the breakdown of hemoglobin. They are excretory products and play no role in digestion and absorption. The color of bile and stool is due to the presence of these pigments. If the bile ducts are occluded for any reason, the bile pigments accumulate in the blood and lead to jaundice. A very large concentration of these substances can cause them to deposit in and damage tissues.

Cholesterol is found in the bile as a by-product of bile salt formation and secretion. It plays no known role in digestion and absorption. Because cholesterol is quite insoluble in water, an alteration in the composition of bile can cause cholesterol to come out of solution as the bile is concentrated in the gallbladder. This leads to the formation of gallstones. If these stones leave the gallbladder, they can occlude the ducts leading to the small intestine thereby blocking the flow of bile.

### Control of Bile Secretion

The liver is constantly forming and secreting bile. If the digestive tract is empty, the sphincters of Boyden and Odi are constricted and the bile flows into the gallbladder where it is stored and concentrated. As one begins to eat, the vagus nerve stimulates the liver to increase its rate of bile production. When food, particularly if it contains fat, enters the small intestine, the gallbladder contracts and the sphincters relax so that bile flows into the small intestine. This is brought about by a hormone **cholecytokinin** which is released by the duodenal mucosa in response to fat. Cholecystokinin stimulates gallbladder contraction and relaxation

# PANCREAS

of the sphincters. Contraction of the gallbladder is also brought about by vagal stimulation. This, however, is not as potent a stimulus as cholecystokinin.

## PANCREAS

The pancreas extends toward the left from the concave curve of the duodenum to the spleen. It lies below and behind the stomach.

The exocrine portion of the pancreas consists of groups of epithelial cells, known as acini, clustered around the lumen of a duct, as illustrated in Figure 15-16. Ducts from the various acini empty into the pancreatic duct which in turn joins the common bile duct to form the ampulla of Vater. In some individuals the pancreatic duct empties directly into the small intestine.

The two most important components of the 2000 ml of fluid the pancreas secretes each day are sodium bicarbonate and the various digestive enzymes. Sodium bicarbonate from the pancreas serves to neutralize the hydrochloric acid which enters the duodenum from the stomach. This serves to protect the duodenal lining against the effects of acid and to provide a pH of 6–7 in which the pancreatic enzymes can function optimally.

The pancreas produces enzymes which are capable of digesting protein, carbohydrate, and fat. In order to prevent digestion of the pancreas itself, the protein-digesting, proteolytic enzymes of the pancreas are produced and secreted in an inactive form. These enzymes are **trypsinogen, chymotrypsinogen,** and **procarboxypolypeptidase.** When these enzymes enter the small intestine, they are converted to their active form. **Enterokinase,** a substance produced by the intestinal mucosa, converts trypsinogen to

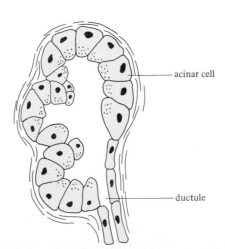

**Figure 15-16** Acinus of the pancreas.

**trypsin.** Trypsin in turn causes the conversion of chymotrypsinogen to **chymotrypsin,** procarboxypolypeptidase to **carboxypolypeptidase,** and further conversion of trypsinogen to trypsin. Once formed in the small intestine, trypsin, chymotrypsin, and carboxypolypeptidase digest proteins to small peptides.

In order to prevent the formation of trypsin and the subsequent activation of the pancreatic proteolytic enzymes in the cells and ducts of the pancreas, the pancreas produces a substance known as **trypsin inhibitor.** If one of the ducts of the pancreas is blocked, trypsin inhibitor may not be able to prevent the formation of trypsin in the large amount of backed-up pancreatic secretion. In this case the enzymes become activated and start to destroy the pancreas, a condition known as **acute pancreatitis.** If blood vessels in the pancreas are destroyed, this can lead to hemorrhage and shock.

**Ribonuclease** and **deoxyribonuclease** produced by the pancreas digest nucleic acids in the small intestine.

**Pancreatic amylase** digests polysaccharides to disaccharides in the small intestine.

**Pancreatic lipase** digests simple fats to their component fatty acids and glycerol in the small intestine.

### Control of Pancreatic Secretion

Pancreatic secretion is controlled primarily by hormones. When food enters the small intestine, two hormones which influence the pancreas are released by the duodenal mucosa. One of these, **secretin,** stimulates the pancreas to secrete large amounts of sodium bicarbonate, whereas the other, **pancreozymin,** stimulates the pancreas to secrete digestive enzymes.

### Small Intestine

The small intestine is a coiled tube, approximately 23 feet long, located in the center of the abdomen. It extends from the stomach to the large intestine. The small intestine is anchored to the posterior wall of the abdominal cavity by the mesentery and is covered in front by the greater omentum.

There are three divisions of the small intestine: the duodenum, the jejunum, and the ileum. The **duodenum** is the first portion of the small intestine. It extends for approximately 10 inches from the stomach, forming a concave curve to the right which contains the head of the pancreas. Secretions from the gallbladder, liver, and pancreas enter the duodenum through the ampulla of Vater. The next two-fifths of the small intestine is the **jejunum** and the remainder is the **ileum.** The ileum joins the cecum of the large intestine at a right angle in the lower right-hand corner of the abdomen.

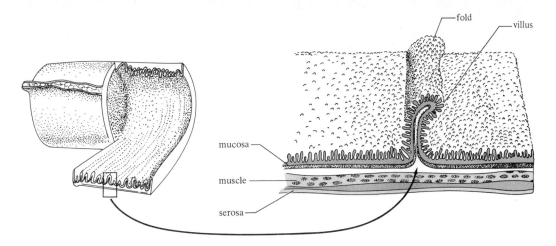

Figure 15-17 illustrates the internal structure of the small intestine. The mucosa forms a series of circular folds—which increase the total surface area. This surface area is further increased by the presence of **villi,** tiny fingerlike projections of mucosa. Figure 15-18 illustrates the structure of a villus. Each villus contains a small artery, a network of capillaries, a vein, and lymphatic vessels. The surface of the villus is covered by columnar epithelial cells which have miniscule projections called microvilli on their free surface. Microvilli increase the surface area of the small intestine even further. Because the small intestine is the primary site at which substances are absorbed from the digestive tract into the

**Figure 15-17** Internal structure of the small intestine.

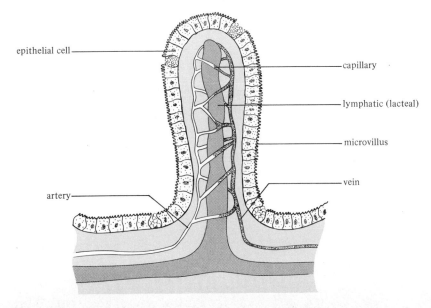

**Figure 15-18** Structure of a villus.

blood, the microvilli, villi, and mucosal folds provide a large surface for absorption.

Intestinal secretions are formed in the crypts of Lieberkuhn, mucosal depressions located between the villi. These secretions contain mucus, electrolytes, and water. They do not contain digestive enzymes. The total volume of intestinal secretions is approximately 3000 ml of fluid a day.

**Digestion and Absorption in the Small Intestine**

The small intestine is the primary site of both digestion and absorption. Mechanical digestion in the small intestine is brought about by segmentation contractions. These contractions are random constrictions of the intestinal circular smooth muscle which break up and mix the chyme which is present in the lumen.

Chemical digestion in the small intestine is brought about by enzymes which enter the small intestine as part of the bile and pancreatic juice and by enzymes present in the epithelial cells of

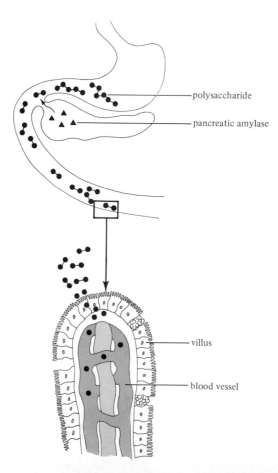

**Figure 15-19** Schematic illustration of the digestion and absorption of carbohydrate in the small intestine.

the intestinal villi. We will consider the digestion and absorption of the major components of the food separately.

**Carbohydrates.** The digestion of carbohydrate in the small intestine is illustrated schematically in Figure 15-19. Glycogen and starch are digested by pancreatic amylase to the disaccharide maltose. The enzymes **maltase, lactase,** and **sucrase** which digest the disaccharides maltose, lactose, and sucrose are located in the borders of the epithelial cells of the villi.

As the disaccharides come into contact with the border cells, they are digested by these enzymes to the monosaccharides glucose, fructose, and galactose. The monosaccharides are then absorbed by active transport into the portal blood flowing through the capillaries of the villi.

**Protein.** As illustrated in Figure 15-20, the digestion of protein is begun by hydrochloric acid and pepsin in the stomach. In the small intestine, protein digestion is continued by the pancreatic enzymes trypsin, chymotrypsin and carboxypolypeptidase. The end

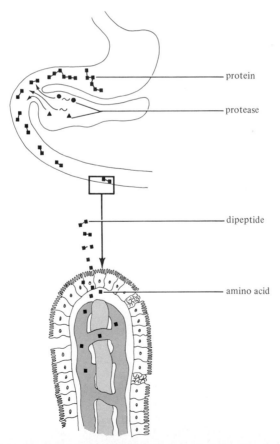

**Figure 15-20** Schematic illustration of the digestion and absorption of protein in the small intestine.

result of digestion by these enzymes is the breakup of protein into small peptides and a few individual amino acids. Enzymes, such as aminopeptidase and dipeptidases, present in the mucosal border digest the small peptides into individual amino acids. The individual amino acids are then absorbed by active transport into the portal blood. There seem to be a number of different types of transport systems for different types of amino acids.

**Fat.** Fat digestion and absorption is illustrated in Figure 15-21. Large fat droplets are broken up (emulsified) by the action of the bile salts. This provides smaller particles with a much larger total surface area for enzymatic action. Pancreatic lipase then attacks the simple fats or triglycerides and digests them to their component

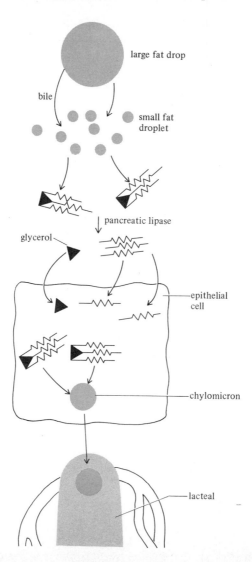

**Figure 15-21** Schematic illustration of the digestion and absorption of fat in the small intestine.

fatty acids and glycerol. Some fat molecules are only digested to monoglycerides.

Fatty acids and monoglycerides are absorbed into the epithelial cells of the intestinal mucosa. Within these cells, fatty acids and glycerol are resynthesized into fats. The fats are then combined with cholesterol and phospholipid into tiny droplets called chylomicrons. These chylomicrons are then coated by a layer of protein which makes them water soluble.

Chylomicrons pass from the mucosal cells into the central **lacteals,** or lymph vessels, of the villi. They are carried through the lymphatic system and eventually enter the blood via the thoracic duct.

**Electrolytes.** Sodium is absorbed by active transport from the intestine. Other electrolytes are absorbed, either actively or passively, as a secondary result of sodium absorption. Certain electrolytes, such as sodium and potassium, are absorbed independently of the amount already in the body. That is, the amount of sodium absorption depends on the amount of sodium in the food, not on the amount already in the body. Other electrolytes such as calcium and iron are absorbed to the degree they are needed. As described in Chapter 9, calcium absorption is controlled by the hormone parathormone. When blood calcium falls, the parathyroids release parathormone. Among other things parathormone stimulates the rate at which calcium is absorbed through the intestinal mucosa. The rate at which iron is absorbed also depends on the amount of iron already present in the body, although the mechanism is not understood. In general, only a small portion of ingested iron is absorbed.

**Water.** Water moves freely in both directions across the intestinal mucosa. Absorption of water depends on the active transport of glucose, amino acids, and sodium into the blood. As these substances are actively absorbed, water follows by osmosis.

### Large Intestine

The large intestine extends for a length of $4\frac{1}{2}$–5 feet from the ileum to the anus. It is divided into four major proportions: the cecum, colon, rectum, and anal canal. Figure 15-22 illustrates the various parts of the large intestine.

The **cecum** is a blind pouch which extends downward 2–3 inches from the border of the ileum and large intestine. It is separated from the ileum by a sphincter, the **ileocecal valve.** A coiled thin tube, the **appendix,** extends downward from the cecum. As the contents of the appendix are not moved along with the flow of material through the large intestine, a large colony of bacteria tends to develop. Occasionally this leads to **appendicitis,** an inflammation of the appendix.

The **colon** is subdivided into four parts: the ascending colon,

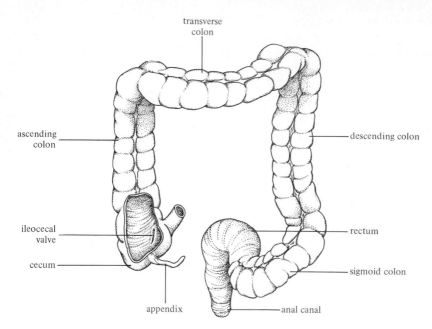

**Figure 15-22** Regions of the large intestine.

transverse colon, descending colon, and sigmoid colon. The **ascending colon** passes up the right side of the abdominal cavity to just under the liver. At the liver the colon makes a right angle, the **hepatic flexure,** and continues as the **transverse colon** from the right to the left side of the abdominal cavity. Just under the spleen the colon forms a second right angle, the **splenic flexure,** and the **descending colon** passes down the left side of the abdominal cavity to the top of the pelvis. The **sigmoid colon,** shaped like an "s," passes through the pelvis from the descending colon to the rectum.

The **rectum** extends a distance of about 5 or 6 inches from the sigmoid colon to the anal canal. Although cancer of the intestine occurs in any of its parts, it is most commonly found in the rectum.

The anal canal forms the final 1–1½ inches of the large intestine, extending from the rectum to the anal opening as illustrated in Figure 15-23. It contains two sphincters, the internal and external anal sphincters, which close off the anal opening. The **internal anal sphincter** is a thickening of the circular, smooth muscle of the intestinal wall. It is not under voluntary control. The **external anal sphincter** is a band of skeletal muscle which surrounds the base of the anal canal. This sphincter is under voluntary control and can prevent defecation until the opportunity is available.

The structure of the large intestine is illustrated in Figure 15-24. There are no villi in the mucosa. The only secretion formed in the large intestine is a mucus secretion formed in goblet cells

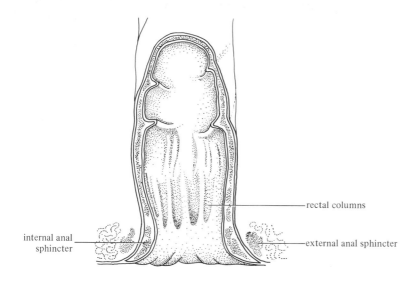

**Figure 15-23** Anal canal.

located in the mucosal lining. The longitudinal muscle in the muscular layer of the intestinal wall is arranged in three flat bands called the **teniae coli.** These bands are not as long as the other tissues of the intestinal wall and, thus, cause the other tissue to be gathered into pouches called **haustra.**

In the anal canal the mucosa of the large intestine is gathered into vertical folds known as the rectal columns (Figure 15-24). The hemorrhoidal (rectal) veins are located underneath these folds. Occasionally the hemorrhoidal veins become dilated and irritated, a condition known as **hemorrhoids.**

The primary functions of the large intestine are the absorption of salt and water and the storage of feces prior to defecation. Approximately 300–500 ml of isotonic chyme enter the large intestine from the ileum each day. As the chyme moves through the

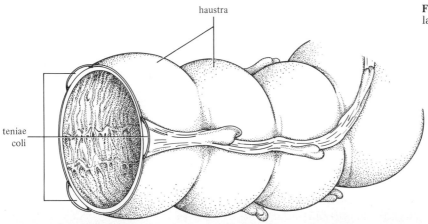

**Figure 15-24** Structure of the large intestine.

large intestine, salt and water are absorbed. Approximately 150 grams of feces are formed each day as salt and water are removed from the chyme. The absorptive capacity of the large intestine is put to use when drugs are administered in the form of enemas and suppositories.

### Movement Through the Large Intestine

The ileocecal sphincter is normally closed. This prevents the free flow of material between the small and large intestine. As waves of peristalsis pass down the small intestine, the ileocecal valve opens briefly and allows a small amount of chyme to enter the large intestine. When food moves from the stomach into the small intestine, a reflex, the **gastroileac reflex,** leads to intensified peristalsis in the ileum and relaxation of the ileocecal sphincter. Thus, the rate at which chyme moves from the small intestine to the large intestine is increased when food moves from the stomach to the duodenum.

Weak waves of peristalsis and segmentation contractions normally occur in the large intestine. They serve to mix the chyme and expose it to the mucosa for absorption. The primary force which moves material through the large intestine are very strong waves of peristalsis known as **mass movements.** These mass movements move a large amount of material for long distances through the large intestine. They usually occur two or three times a day and are initiated when food enters the stomach and small intestine by reflexes called the **gastrocolic** and **duodenocolic** reflexes. Mass movements seem to be particularly strong after breakfast.

The relatively slow movement of substances through the large intestine leads to the development of a large bacterial colony. These bacteria are harmless in that they cannot cross the mucosal lining and gain access to the blood. The bacteria in the small intestine are capable of synthesizing vitamin K and some of the B vitamins. These vitamins can be absorbed and used by the body. The bacteria present a potential danger in that any breakage of the intestinal wall, such as after a wound, allows the bacteria to enter the abdominal cavity. Bacteria in the abdomen can cause **peritonitis** which, if not controlled, can be fatal.

# DEFECATION

Defecation is a reflex that is initiated by the filling of the rectum with fecal matter. As the rectum fills, afferent neuronal endings are stimulated which in turn stimulate peristaltic contraction of the descending and sigmoid colon. This stimulation is brought about by way of intrinsic nerves in the myenteric plexus and by way of spinal pathways involving the parasympathetic nervous system. As the peristaltic waves pass down the descending colon, sigmoid colon, rectum, and anal canal the internal anal sphincter relaxes. If the external anal sphincter is relaxed, feces can then leave the body through the anus. Defecation is facilitated by forcefully contracting the muscles of the abdominal wall.

The frequency of defecation depends on the individual. Some people defecate two or three times a day, and others defecate once every two or three days. The popular concern over constipation and the widespread use of laxatives is basically unfounded. There are no "poisons" that build up in the body if one does not defecate frequently. Diarrhea, on the other hand, can present very serious problems, particularly the loss of fluid and electrolytes. The total volume of digestive secretions is approximately 8000 ml of fluid a day. If these secretions are not reabsorbed along the digestive tract, they will leave the body with the feces. When substances are moved through the digestive tract very quickly, as they are during diarrhea, there is not adequate time for reabsorption.

### Transit Time

If one administers a standard test meal, it first reaches the cecum in about 4 hours. All of the meal is in the colon after 8 to 9 hours. Within the colon, the meal begins to reach the hepatic flexure about 6 hours after eating, the splenic flexure after 9 hours, and the sigmoid colon after 12 hours. The meal tends to remain a fairly long time in the rectum. Seventy-two hours after eating, about 25% of the meal still remains in the rectum.

## OBJECTIVES FOR THE STUDY OF THE DIGESTIVE SYSTEM

At the end of this unit you should be able to:

1. State the three major functions of the digestive system.
2. Name the four layers of the digestive tube wall and describe the structure of each.
3. Describe the process of peristalsis.
4. Distinguish between the location of the visceral and parietal peritoneum.
5. Describe the location of the mesentery, greater omentum, and lesser omentum.
6. Distinguish between polysaccharides, disaccharides, and monosaccharides.
7. State whether each of the following is a polysaccharide, disaccharide, or monosaccharide: glycogen, starch, cellulose, sucrose, maltose, lactose, glucose, fructose, and galactose.
8. Distinguish between essential and nonessential amino acids.
9. State the general role of vitamins in metabolism.
10. State the main function of each of the following vitamins: A, D, E, K, thiamine, riboflavin, niacin, $B_{12}$, and C.
11. Describe the structure of the tongue.
12. State the major function of the tongue.
13. Name all the teeth.

14. Describe the structure of a tooth.
15. State the location of each of the three types of salivary glands.
16. Name the enzyme secreted by the salivary glands and state its function.
17. State the function of the mucus secreted by the salivary glands.
18. Describe the process of swallowing.
19. Name the sphincters located at the top and bottom of the esophagus.
20. Describe how food is moved along the esophagus.
21. State the location of the following regions of the stomach: cardiac, fundus, body, and pylorus.
22. State the three major functions of the stomach.
23. State the function of the HCl and pepsinogen secreted by the stomach.
24. Describe how the stomach is protected against the effects of HCl and pepsinogen.
25. Name the valve that separates the stomach and small intestine.
26. Describe the various mechanisms by which HCl and pepsinogen secretion are controlled.
27. Describe the mechanisms by which gastric emptying is controlled.
28. Describe the structure of the liver.
29. Describe the structure of a lobule of the liver.
30. Name and identify from a drawing the various ducts that connect the liver, gallbladder, and pancreas with the small intestine.
31. State the location and function of the gallbladder.
32. Name the three major components of the bile.
33. Describe the digestive action of the bile salts.
34. Describe the mechanisms by which the secretion of bile into the small intestine is controlled.
35. State the location of the pancreas.
36. State the function of the sodium bicarbonate secreted by the pancreas.
37. Name the active forms of the proteolytic enzymes secreted by the pancreas.
38. Describe the functions of pancreatic amylase and pancreatic lipase.
39. Describe the mechanisms by which pancreatic secretion is controlled.
40. Name the three major regions of the small intestine.
41. Describe the internal structure of the small intestine and the purpose of this type of structure.
42. Describe the structure of a villus.
43. Describe the processes by which carbohydrate, protein, and fat are digested and absorbed in the small intestine.
44. Describe the basic processes by which the absorption of salt and water take place.
45. Name the four parts of the large intestine.
46. Name the valve that separates the small and large intestines.
47. Distinguish between the internal and external anal sphincters in terms of location and the manner in which they are controlled.
48. State the functions of the large intestine.
49. Describe the mechanism of defecation.

# 16
# Energy Metabolism

CONVERSION OF FOOD ENERGY TO ATP
STORAGE OF ENERGY
USAGE OF STORED ENERGY
CONTROL OF METABOLISM
DIABETES MELLITUS
ENERGY BALANCE
REGULATION OF BODY TEMPERATURE

Energy is defined as the ability to do work, that is, to bring about some type of change. In Chapter 3 we described how cells transfer the chemical energy of glucose, fat, and amino acids to the high-energy compound adenosine triphosphate (ATP). The ATP is then used as a source of energy for cellular work. In this chapter we are going to look at patterns by which the body stores the energy of the food absorbed after a meal and then uses this stored energy at a later time. We are also going to look at energy balance, the relationship between the amount of energy taken in as food, and the amount of energy the body uses to do work. Finally we will discuss the control of body temperature, the relationship between the heat energy produced in the body and the heat energy exchanged with the environment.

## CONVERSION OF FOOD ENERGY TO ATP

In order to understand the processes by which food energy is stored and then later used it is necessary to have a clear understanding of the basic chemical processes by which glucose, fats, and amino acids are broken down in order to form ATP. These processes are reviewed here.

Figure 16-1 reviews the stages involved in the breakdown of glucose. During glycolysis, one six-carbon glucose molecule is broken down to two three-carbon pyruvic acid molecules. Glycolysis does not require oxygen; on the other hand, it only leads to the formation of two ATP molecules per glucose molecule. The complete breakdown of glucose involves the Krebs cycle, the series of reactions in which pyruvic acid is converted to acetyl-CoA which is in turn converted to carbon dioxide and water. ATP is formed during oxidative phosphorylation, the series of reactions in which hydrogen removed during glycolysis and the Krebs cycle is combined with molecular oxygen. Thirty-four molecules of ATP are formed by oxidative phosphorylation during the complete breakdown of one glucose molecule. These 34 ATPs plus the

# 365 Conversion of food energy to ATP

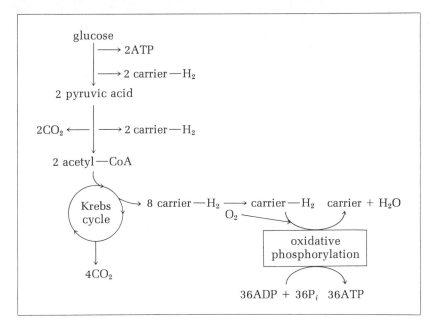

**Figure 16-1** Formation of ATP from glucose.

two formed directly during glycolysis and two more formed directly during the Krebs cycle combine to make a total of 38 molecules of ATP per glucose molecule.

The basic steps in the breakdown of fat are shown in Figure 16-2. Fat molecules are first split into their component fatty acids

**Figure 16-2** Formation of ATP from fat.

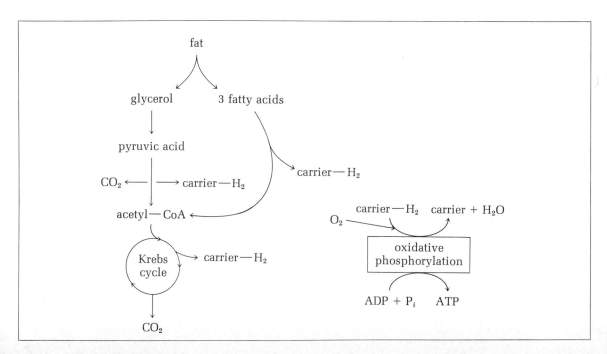

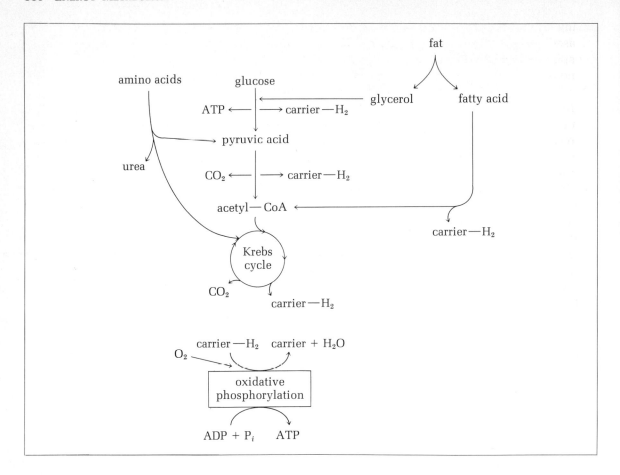

and glycerol. Glycerol is converted to one of the glycolytic intermediates which in turn is converted to pyruvic acid. The pyruvic acid can then enter the Krebs cycle. Fatty acids are converted by a series of reactions to acetyl-CoA which can enter the Krebs cycle.

**Figure 16-3** Formation of ATP from glucose, fat, and amino acids.

**Figure 16-4** Absorptive state.

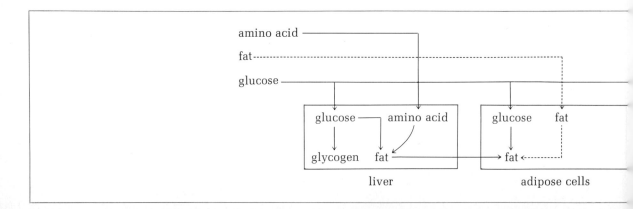

**367** Storage of energy

During the conversion of fatty acids to acetyl-CoA, hydrogens are released. These hydrogens are combined with oxygen by oxidative phosphorylation. Fat is a higher-energy compound than glucose or amino acids in that it forms twice as much ATP per unit weight.

The breakdown of amino acids is more complicated than that of glucose and fat in that each amino acid is broken down by its own particular series of reactions. All of these reactions involve the removal of nitrogen and the formation of a compound which can enter the Krebs cycle and produce ATP. The enzymes for removing nitrogen from amino acids are found primarily in the liver. Hence the liver is the primary site of amino acid breakdown. The nitrogen that is removed is used either to form other nitrogen-containing compounds needed by the body or to form the nitrogen waste product urea. Urea formed in the liver is excreted from the body by the kidney. Figure 16-3 sums up the basic steps involved in the conversion of the energy found in glucose, fats, and amino acids to the energy of ATP. It should be recalled that the conversion is not 100%. Over 50% of the chemical energy stored in the food appears as heat during the formation of ATP.

## STORAGE OF ENERGY

Our eating patterns are such that we take in the food necessary to supply the day's energy needs in two or three major meals. During the time that these meals are being absorbed from the digestive tract, glucose, fats, and amino acids are entering the blood at a rate faster than they are needed to meet the immediate energy needs of the cells. The body deals with this situation by storing the excess food for use at a later time. The pattern of energy metabolism that takes place during the absorption of food from the digestive tract is called the **absorptive state.**

Figure 16-4 illustrates the primary events that occur during

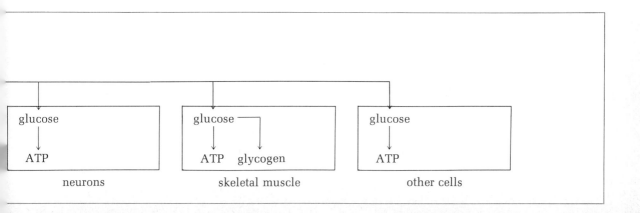

the absorptive state. Most cells of the body use glucose as their source of ATP. However, the amount of glucose entering the blood after a typical meal is greater than the amount necessary to meet cellular needs. The excess glucose is stored as either glycogen or fat. Glycogen, a polysaccharide chain of glucose molecules, is formed from glucose primarily in the liver and skeletal muscles. Glucose is converted to fat in the liver and in the adipose cells. In this conversion glucose is used to make both glycerol and acetyl-CoA. The acetyl-CoA is then used to synthesize fatty acids. Once formed the fatty acids are combined with glycerol to form fat. Although glucose can be converted to fat in both liver and adipose cells, all of the fat is normally stored in the adipose cells.

During the absorptive state very little fat is burned by the cells to form ATP. The fat that is absorbed into the blood from the digestive tract is transported to the adipose cells where it is stored for later use. Only if there is very little carbohydrate in the meal will fat be used as a source of ATP during the absorptive state.

Amino acids that enter the blood during the absorptive state are transported to cells all over the body where they are used in the synthesis of cellular proteins. However, the cells synthesize protein only to the extent that it is needed for the structure and function of the cell. That is, amino acids are *not* converted to protein for storage. Any excess amino acid not needed for protein synthesis is either converted to fat in the liver or burned to provide ATP for the liver.

During the absorptive state most of the excess food energy is stored as fat. The reason for this is that fat stores energy much more efficiently than either glycogen or protein. One gram of fat contains twice as much stored chemical energy as one gram of glycogen or one gram of protein; therefore it takes only half as much weight to store excess energy as fat as it would to store it as glycogen or protein.

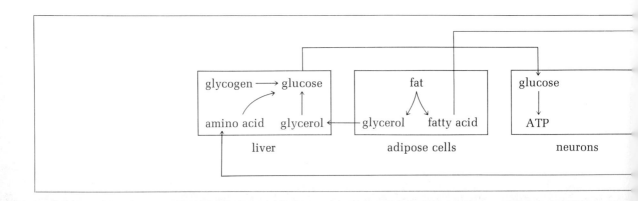

# USAGE OF STORED ENERGY

The pattern of energy metabolism that occurs when food is no longer entering the blood from the digestive tract is called the **postabsorptive state**. In the postabsorptive state, the body must draw on the energy stored during the absorptive state.

Figure 16-5 illustrates the pattern of events that occurs during the postabsorptive state. This pattern is best understood in light of the fact that a stable level of blood glucose must be maintained. This is because cells of the nervous system can only use glucose as a source for producing ATP. Because glucose is not entering the blood from the digestive tract during the postabsorptive state, the glucose used by the nervous system must be provided by energy stores already present in the body.

Blood glucose is maintained during the postabsorptive state by a number of different mechanisms. Glycogen stored in the liver can be broken down to glucose molecules which are then released into the blood. The glycogen stored in skeletal muscle does not serve directly as a source of blood glucose. This is because glycogen is broken down in skeletal muscle in such a way that glucose is not available to enter the blood. However, the glycogen does serve as a source of energy for the skeletal muscles themselves, thereby saving the glucose formed in the liver for the use of the nervous system. In addition, some of the glycogen breakdown in the skeletal muscle involves glycolysis with the formation of pyruvic acid and lactic acid. The pyruvic acid and lactic acid can enter the blood. Once in the blood, the pyruvic acid and lactic acid are transported to the liver where they can be used to synthesize glucose.

So far we have said that in the postabsorptive state the liver produces glucose from glycogen, pyruvic acid, and lactic acid. In addition to these sources, the liver also makes glucose from glycerol and from amino acids. During the postabsorptive state, fat is

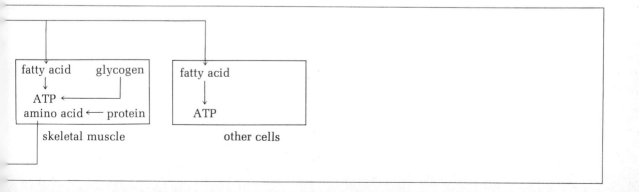

**Figure 16-5** Postabsorptive state.

broken down in the adipose cells to glycerol and fatty acids. The glycerol is then transported to the liver where it is used in the synthesis of glucose. If the postabsorptive state is prolonged, for example, during fasting, protein begins to break down into amino acids. The amino acids are then used by the liver to form glucose.

The glucose produced by the liver during the postabsorptive state is used almost exclusively by the nervous system. Other types of cells rely on fatty acids as a source of producing ATP. As mentioned above, fat is broken down to glycerol and fatty acids in the adipose cells during the postabsorptive state. The glycerol is converted to glucose by the liver and the fatty acids are burned to form ATP in all cells except those of the nervous system.

The use of fatty acids as a source of ATP by most cells during the postabsorptive state is not always direct. This is illustrated in Figure 16-6. To a large degree the initial breakdown of fatty acids takes place in the liver. That is, fatty acids are transported from the adipose cells to the liver where they are broken down to acetyl-CoA. Two acetyl groups are then combined in the liver to form a **ketone.** Once formed, the ketones enter the blood and are transported to cells all over the body. In the cells the ketones are converted back to acetyl groups which can combine with Coenzyme A and enter the Krebs cycle leading to the formation of ATP. One danger of ketone formation is that the ketones are quite acidic. Any situation in which ketones are formed in large numbers can lead to metabolic acidosis.

In summary, during the postabsorptive state, glucose, which is necessary for use by the cells of the nervous system, is formed in the liver. The liver forms glucose from glycogen, pyruvic acid, lactic acid, glycerol, and amino acids. In order to save the glucose formed in the liver for the nervous system, other cells use fatty

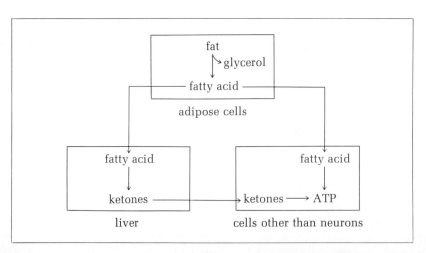

**Figure 16-6** Ketone formation.

acids as a source of ATP. The fatty acids are either used directly in the cells or are first converted by the liver to ketones which are then used by cells.

## CONTROL OF METABOLISM

In our description of the events of the absorptive and postabsorptive state, we did not describe what determines which events take place. That is, during the absorptive state most cells use glucose as a source of energy, whereas during the postabsorptive state most cells utilize fatty acids. What tells the cells which one to use at a particular time? During the absorptive state fat is deposited in the adipose cells, and during the postabsorptive state fat is released from adipose cells. What tells the adipose cells when to deposit fat and when to release it?

Control over the events of the absorptive and postabsorptive states is exerted by both the endocrine system and the nervous system. The hormones involved in the control of metabolism are insulin, glucagon, epinephrine, norepinephrine, growth hormone, cortisol, and thyroxine. Each of these hormones is described in Chapter 9. At this point we are only going to look at their role in regulating metabolism.

### Hormonal Control of Metabolism

**Insulin.** The primary metabolic effect of insulin, which is released by the beta cell of the islets of Langerhans in the pancreas, is to allow glucose to enter cells. All cells of the body, except those of the nervous system and possibly the liver, depend on the presence of insulin for glucose entry. Once glucose enters a cell, its fate depends upon the particular type of cell it enters. In most cells glucose either is burned to form ATP or is converted to glycogen. In liver and adipose cells, glucose is also converted to fat. In addition to allowing glucose into cells, insulin has a number of other effects. It promotes the entry of amino acids into cells along with protein synthesis, and it blocks the breakdown of fat.

The overall result of an increase in the blood level of insulin is that glucose moves into cells and the blood level of glucose drops. This is what happens during the absorptive state.

The rate at which the pancreas secretes insulin into the blood depends primarily on the existing blood level of glucose. As shown in Figure 16-7, when the blood level of glucose goes up, the pancreas increases its rate of insulin secretion. An increase in the blood level of insulin in turn allows more glucose to enter cells. Within the cells, glucose either is burned to form ATP or is stored. The entry of glucose into cells lowers the blood level of glucose.

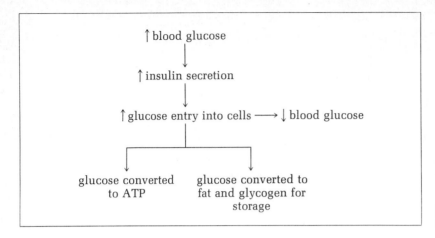

Figure 16-7 Role of insulin in the regulation of blood glucose.

If the blood level of glucose begins to drop below normal, the pancreas stops secreting insulin. In the absence of insulin glucose no longer enters cells other than those of the nervous system. This helps prevent any further decline in the blood level of glucose.

From the point of view of homeostasis, insulin acts to maintain a constant blood level of glucose. When the concentration of glucose goes up, for example, after a meal, insulin is released. The insulin increases the rate at which glucose enters cells and blood glucose starts to return to normal. If the blood glucose begins to fall too much, the pancreas stops secreting insulin. This helps to prevent any further drop in blood glucose. As a result of the action of insulin, and other hormones to be discussed below, blood glucose is maintained in the range of 80–120 mg per 100 ml.

**Glucagon.** Glucagon released by the alpha cells of the islets of Langerhans in the pancreas has effects that are in many ways exactly the opposite of those of insulin. It stimulates the conversion of glycogen and amino acids to glucose in the liver and it stimulates fat breakdown in adipose cells. Both of these effects lead to an increase in blood glucose.

The primary stimulus for glucagon release is a decrease in blood glucose. Figure 16-8 illustrates the role of glucagon in the regulation of metabolism. When blood glucose drops, glucagon is released. Glucagon stimulates the liver to form glucose from glycogen and amino acids, and it stimulates fat breakdown in the adipose cells. These activities serve to raise blood glucose back to normal.

**Epinephrine and norepinephrine.** These hormones, released by the adrenal medulla, act to raise blood glucose by stimulating the breakdown of glycogen and fat. They are released by the adrenal medulla in response to stimulation by sympathetic nerves. Figure 16-9 illustrates the role of epinephrine and norepinephrine in

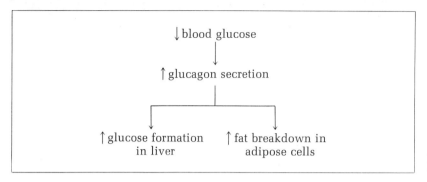

Figure 16-8 Role of glucagon in the regulation of metabolism.

the regulation of metabolism. A drop in blood glucose can be detected by receptors in the hypothalamus. This leads to activation of the sympathetic nerves to the adrenal medulla and the release of epinephrine and norepinephrine. These hormones then stimulate the breakdown of glycogen and fat. The breakdown of glycogen in the liver allows glucose to be released into the blood thereby causing an increase in blood glucose.

When these hormones are released in response to a drop in blood glucose, one gets not only an increase in blood glucose but also all of the other effects of the hormones. The effects include increased irritability and increased heart rate. Thus 4 or 5 hours after a meal, when blood glucose begins to fall and epinephrine is secreted, a person often becomes irritable and nervous. This phenomenon is particularly acute in children.

**Growth hormone.** In addition to growth-promoting effects, growth hormone plays a role in the regulation of energy metabolism. It is an insulin antagonist in that it blocks glucose entry into all cells except those of the nervous system, and it stimulates the breakdown of fat in the adipose cells.

The release of growth hormone by the pituitary is controlled by growth hormone releasing factor from the hypothalamus. This

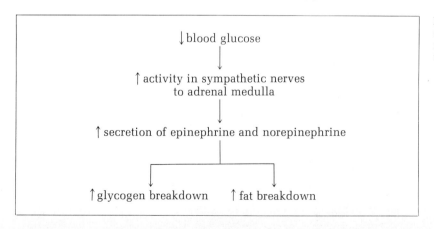

Figure 16-9 Role of epinephrine and norepinephrine in the regulation of metabolism.

releasing factor is secreted by the hypothalamus in response to a drop in blood glucose as detected by the hypothalamic glucose receptors. Thus a drop in the blood glucose level leads to an increase in growth hormone which in turn leads to an increase in blood glucose because of the blockage of glucose entry into cells and the stimulation of fat breakdown.

**Glucocorticoids.** The glucocorticoids (primarily cortisol) from the adrenal cortex are essential for the synthesis of glucose by the liver. The glucocorticoids seem to act by causing protein breakdown in most tissues, particularly in skeletal muscle. This makes amino acids available for glucose formation in the liver. The glucocorticoids may also directly stimulate the enzymes used to make glucose in the liver.

### Nervous System Control of Metabolism

Nervous system control of metabolism is exerted both directly and indirectly. There are sympathetic nerves which innervate the adipose cells. Stimulation by these nerves causes fat breakdown and the release of glycerol and fatty acids into the blood. This stimulation occurs when blood glucose drops or during any stressful situation such as exercise.

Indirectly the nervous system affects metabolism through the sympathetic nerves, which regulate the adrenal medulla. Stimulation of the adrenal medulla by these nerves leads to the release of epinephrine and norepinephrine into the blood. These hormones then stimulate glycogen and fat breakdown.

### Summary

Figure 16-10 sums up the events of the absorptive state. As a meal is absorbed from the digestive tract, the blood level of glucose begins to increase. This increase stimulates the release of insulin

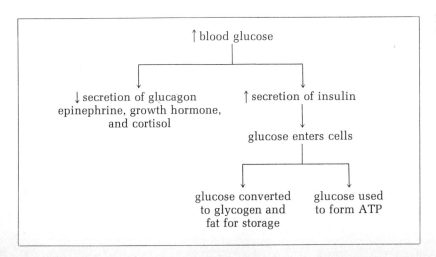

**Figure 16-10** Hormonal control of the absorptive state.

# Control of metabolism

and inhibits the release of glucagon, epinephrine, growth hormone, and cortisol. The high level of insulin allows glucose to enter cells readily. In most of the cells, glucose is burned to form ATP. Excess glucose is converted to glycogen and fat. Fat absorbed during the absorptive state is stored as fat. Amino acids are used to synthesize necessary proteins. Excess amino acids are converted to fat for storage. The low levels of glucagon, epinephrine, growth hormone, and cortisol guarantee that the glycogen and fat which are being formed during the absorptive state are not broken down during this time, but saved for later use.

Control over the postabsorptive state is summed up in Figure 16-11. Once food is no longer being absorbed from the digestive tract, the blood glucose level begins to fall. This inhibits the secretion of insulin and stimulates the secretion of glucagon, epinephrine, growth hormone, and cortisol. The drop in blood glucose also activates the sympathetic nerves to the adipose cells.

As insulin secretion is inhibited and the blood level of insulin falls, glucose can no longer enter cells other than those of the nervous system. Blockage of glucose entry into cells is brought about not only by the fall in insulin, but also by the presence of growth hormone. The secretion of glucagon, epinephrine, and growth hormone during the postabsorptive state leads to glycogen and fat breakdown. Glycogen breakdown in the liver provides a

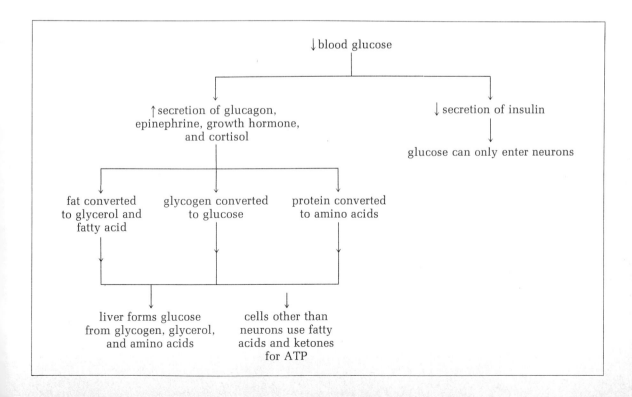

**Figure 16-11** Hormonal control of the postabsorptive state.

source of glucose for the blood, whereas glycogen breakdown in other cells, particularly skeletal muscle cells, provides energy for those cells. The breakdown of fat in the adipose cells provides glycerol which is used by the liver to synthesize glucose and provides fatty acids which are used as a source of energy by most cells. Fat breakdown in the adipose cells is brought about by sympathetic nervous stimulation as well as by hormonal stimulation. The secretion of glucocorticoids during the postabsorptive state stimulates protein breakdown, thus providing a source of amino acids for glucose synthesis in the liver. The net result of all of the events in the postabsorptive state is that blood glucose is maintained at a relatively constant level. This guarantees a continuous, adequate supply of glucose for use by the cells of the nervous system.

## DIABETES MELLITUS

Diabetes mellitus is a disease in which there is insufficient secretion of insulin relative to the level of blood glucose. There are two basic types of diabetes: **juvenile diabetes** which occurs in children, and **late-onset diabetes** which begins to manifest itself later in life. Juvenile diabetes seems to be a genetic defect in which the pancreas is not able to produce insulin. In late-onset diabetes the blood level of insulin may be low, normal, or even high. However, the blood level of glucose is always high and the pancreas is not able to secrete enough insulin to lower blood glucose to normal. Late-onset diabetes is usually associated with obesity and often will disappear if the patient loses weight. In diabetes, blood glucose can go as high as 1200 mg per 100 ml.

The primary physiological problems of a diabetic are the loss of salt and water, the development of acidosis, and blood vessel damage. Salt and water loss and its consequences are illustrated schematically in Figure 16-12. Normally all of the glucose filtered by the kidney is reabsorbed into the blood. However, in diabetes the blood concentration of glucose is so high that all of the filtered glucose cannot be reabsorbed and thus is excreted in the urine. For reasons that will become clear in the next chapter when we discuss the way the kidney works the loss of glucose in the urine leads to the loss of excess salt and water in the urine. In other words, the high blood level of glucose acts as a diuretic. The loss of salt and water in diabetes leads to a drop in blood volume. In order to maintain blood pressure in the face of a drop in blood volume, there is reflex vasoconstriction, particularly in the limbs. This causes a decreased blood flow to the limbs which can lead to ulceration, as seen in Figure 16-13.

Acidosis occurs in a diabetic because he is constantly in a

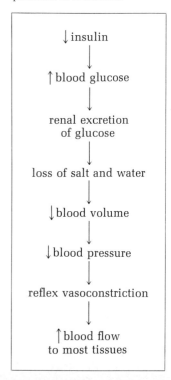

**Figure 16-12** Salt and water problems of a diabetic.

postabsorptive state of metabolism. Because the blood level of insulin is insufficient to allow glucose to enter cells other than those of the nervous system, fatty acids and ketones must be used by most cells as a source of energy. In diabetes there is an increase in the rate of glucagon, epinephrine, growth hormone, and cortisol secretion. These hormones cause fat breakdown in the adipose cells and thus make fatty acids available for cellular use. An unfortunate consequence of these hormones is that they stimulate the liver to synthesize glucose. This is totally inappropriate, considering that blood glucose is already too high.

Both fatty acids and ketones are quite acidic. As the blood level of these substances increases, metabolic acidosis develops. The body responds to this with an increase in respiration and the formation of an acid urine. However, if diabetes goes untreated, the acidosis is not adequately compensated for and the pH of the blood falls. Once the pH of the blood falls to about 7.1, a person will go into a coma and die.

In addition to the salt and water and the acid-base problems, a diabetic also has problems with blood vessels and is usually very susceptible to infections. The blood vessel problems include both a tendency to arteriosclerosis and degenerative damage to small vessels. To a large extent the vascular problems of a diabetic continue even if he is treated with insulin or other hypoglycemic agents.

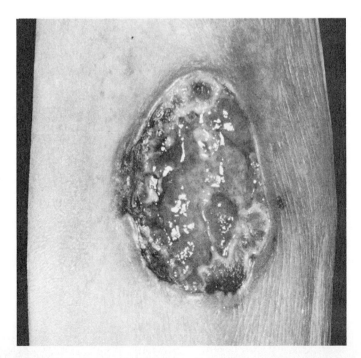

**Figure 16-13** Ulceration in the skin of a diabetic. (Photo by Martin M. Rotker, Taurus Photo.)

## ENERGY BALANCE

Energy balance refers to the relationship between the amount of energy taken into the body and the amount of energy used by the body. Physiologically, energy is measured in Calories. One Calorie is the amount of heat energy required to raise the temperature of 1 liter of water 1°C.

Energy is taken into the body in the form of food. The amount of food eaten is controlled primarily by the appetite centers in the hypothalamus. In the lateral region of the hypothalamus, there is an eating center. When this center is active a person desires to eat. The medial portion of the hypothalamus contains a satiety center. This satiety center can inhibit the eating center. At this point no one is quite sure as to what stimuli serve to inform the appetite centers of the energy state of the body. One theory is that the medial satiety center responds to the blood level of glucose. When blood glucose is high the neurons of the satiety center are activated. These neurons in turn inhibit the eating center and thus there is no desire to eat. On the other hand, when blood glucose falls, the neurons of the satiety center stop firing. This ends the inhibition of the lateral eating center. Once this center is active a person desires to eat. The blood glucose theory of appetite control implies that changes in blood glucose are indicative of the energy needs of the body. A high blood level of glucose indicates that there is enough food energy in the body, and a low level of blood glucose indicates that food energy is needed.

It seems that appetite and the need for food energy depends on more than the blood level of glucose. Other theories have postulated that the blood level of fatty acids and body temperature play an important role in regulating appetite. Higher centers in the cerebral cortex also play an important role in regulating appetite and eating. Clearly, all kinds of emotional problems can lead to either an excessive or a deficient amount of eating.

The energy output of the body is referred to as the **metabolic rate,** the number of Calories burned per unit of time. A person's metabolic rate depends on how much energy is required for the basic processes of living and how much additional energy is being used for specialized activities. The basic energy of living is referred to as the **basal metabolic rate.** It is measured under standard conditions so that it can be compared to that of other people of the same age, sex, and body size. These conditions are: a comfortable room temperature, no food for 12 hours before the test, a morning hour after a good night's sleep, and lying down with minimal muscle tone.

A number of factors can increase the metabolic rate above the basal level. The most significant of these is muscular activity. Eating also increases the metabolic rate. This is known as the specific dynamic action (SDA) of food. The increase in metabolic rate that ensues after a meal is caused in part by the digestive process and in part by the processing of food by the liver. Protein produces a higher SDA than fat or carbohydrate. After a heavy meal the SDA can increase the metabolic rate by as much as 30% or 40%.

The metabolic rate depends on the blood levels of hormones, particularly epinephrine and thyroxine. Both of these hormones significantly increase the metabolic rate.

Finally, metabolic rate depends on the external temperature. If the external temperature is either too high or too low, the metabolic rate will be increased. A low external temperature leads to the muscular activity of shivering, and a high external temperature speeds up all metabolic activity.

A person's body weight ultimately depends upon the balance between the amount of food eaten and the amount of energy expended. The only way to lose weight is to burn more calories than are taken in. There is good evidence that a large number of people are overweight as a result of an insufficient level of activity rather than from overeating. This is particularly true for people who work at sedentary jobs.

## REGULATION OF BODY TEMPERATURE

The various physiological processes of the body can only proceed normally within a very narrow temperature range. Oral temperature is usually in a range between 97°F and 99°F with rectal temperature approximately one degree higher. By various mechanisms, the body is able to maintain a temperature in this range in the face of rather large fluctuations in the external temperature. A nude person can maintain body temperature with an external temperature as low as 55°F or as high as 150°F in dry air. With proper clothing an even greater variety of temperatures can be tolerated.

Body temperature obviously depends upon the relationship between the amount of heat gained by the body and the amount of heat lost. Within the body there is a constant level of heat production owing to metabolic activity. The higher the metabolic rate the more heat the body produces. In addition to internal heat production, the body can also gain heat if the external temperature is higher than the body temperature.

The body loses heat by four basic processes: radiation, conduction, convection, and evaporation. **Radiation** is the emission of heat through infrared heat waves. As long as the body temperature is above the external temperature, heat will be lost by radiation. However, if the external temperature is higher than the body temperature, heat will be gained by radiation. **Conduction** is the transfer of heat from one substance to another by direct contact. When the external temperature is below body temperature, heat is lost to the air by conduction. However, just as for radiation, heat is gained by conduction when the external temperature is higher than the body temperature. **Convection** refers to the movement of air surrounding the body. If the air next to the body is heated by conduction, it will move away and be replaced by cooler air. This is because hot air expands and rises. Any other factor, such as wind, that moves the air will increase the amount of heat that can be lost by conduction. The **evaporation** of water from the surface of the body uses up heat. Thus, evaporation is a means of heat loss. Water is constantly evaporating insensibly, that is, without our being aware of it. When there is a need to lose excessive heat, the amount of evaporative heat loss can be increased greatly by sweating. If the external temperature is above body temperature, evaporation becomes the only means for losing heat. The amount of evaporation depends upon the humidity (water content) of the external air. This is why one can tolerate high temperatures in dry weather, but not in humid weather.

Control over body temperature involves adjusting heat production and heat loss in response to changes in the external and internal temperatures, as illustrated in Figure 16-14. The external temperature is measured by hot and cold temperature receptors in the skin. This information is then relayed along afferent pathways to the temperature control center in the hypothalamus. The internal, or core, temperature is measured by neurons which are part of the hypothalamic temperature control center.

If the temperature falls, efferents to the skeletal muscles stimulate an increase in muscle tone and shivering. This increases body temperature. In addition, sympathetic efferents to the arterioles of the blood vessels in the skin cause vasoconstriction. This decreases

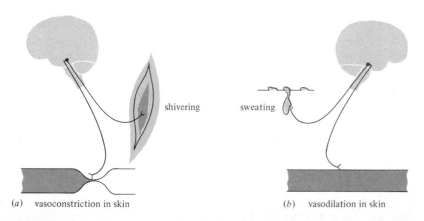

**Figure 16-14** Control of body temperature. (a) Response to cold. (b) Response to heat.

blood flow to the skin and thus reduces the amount of heat lost to the environment.

An increase in temperature leads to a response which is the opposite of that described above. Muscle tone is somewhat decreased, which helps reduce heat production. Blood flow to the skin is increased, thus facilitating heat loss to the environment. Sympathetic nerves stimulate the sweat glands to secrete more sweat. This increases the amount of heat lost by evaporation.

### Fever

Substances released by pathogenic bacteria or by the tissues they destroy or by both act on the hypothalamus to reset the level at which it maintains body temperature. These substances which are proteins and lipopolysaccharides are called **pyrogens.**

The body responds to the pyrogens by elevating its temperature to a new level. This is done by increasing heat production through shivering and decreasing heat loss by reducing the blood flow to the skin. As the body temperature is rising the person feels cold until it reaches the new level.

When the fever starts to break, either by itself or in response to drugs such as aspirin, the hypothalamus is reset back to normal. The person then feels warm and invokes the mechanisms necessary for heat loss. Muscle tone is decreased and blood flow to the skin is increased. Profuse sweating leads to heat loss by evaporation. The protective significance of increasing body temperature in response to bacterial and viral invasion is not entirely clear. It is possible that bacteria and viruses do not function as well at the higher temperatures.

## OBJECTIVES FOR THE STUDY OF ENERGY METABOLISM

At the end of this unit you should be able to:

1. Define the term absorptive state.
2. Name the major source for ATP production during the absorptive state.
3. Explain how excess glucose is stored during the absorptive state.
4. Describe what happens to fat and to amino acids during the absorptive state.
5. Define the term postabsorptive state.
6. Name the various substances from which the liver can make glucose during the postabsorptive state.
7. Name the sources of ATP production in cells other than those of the nervous system during the postabsorptive state.
8. Name the source of ATP production in nervous system cells during the absorptive state.
9. Describe the major physiological action of insulin.
10. State the stimulus for insulin secretion by the pancreas.

11. State whether there is a high or low concentration of each of the following hormones during the absorptive state: insulin, glucagon, epinephrine, growth hormone, and cortisol.
12. State whether there is a high or low concentration of each of the following hormones during the postabsorptive state: insulin, epinephrine, glucagon, growth hormone, and cortisol.
13. Explain why a diabetic loses salt and water and develops acidosis.
14. Define the term metabolic rate.
15. Describe the conditions under which basal metabolic rate is measured.
16. Describe at least three factors that can bring about an increase in metabolic rate.
17. Describe how heat is gained and lost by the body.
18. Describe the mechanisms by which heat production and heat loss are adjusted to maintain a constant body temperature.

# 17

# The Urinary System

FUNCTIONS OF THE URINARY SYSTEM
ANATOMY OF THE URINARY SYSTEM
BASIC PROCESSES OF URINE FORMATION
CONTROL OF EXCRETION
FLUID AND ELECTROLYTE BALANCE
ACID-BASE BALANCE
MICTURITION
DISEASES OF THE KIDNEY
ARTIFICIAL KIDNEY

The urinary system consists of the two kidneys, the two ureters, the bladder, and the urethra. Urine is completely formed within the kidneys and then flows through the ureters into the bladder for storage. During urination the bladder contracts and urine flows out of the body through the urethra.

## FUNCTIONS OF THE URINARY SYSTEM

Two major physiological functions are carried out by the urinary system: the elimination of metabolic wastes and the regulation of fluid and electrolyte balance. Various metabolic waste products are formed as a result of the numerous chemical reactions that take place in the cells of the body. Many of these wastes can become toxic if they accumulate in large amounts and therefore must be eliminated.

The major route for the removal of metabolic wastes is through the urinary system. As blood flows through the kidneys, metabolic wastes are removed and then eliminated from the body as part of the urine. It should be kept in mind, however, that certain wastes are eliminated by other routes. For example, carbon dioxide formed during the Krebs cycle is eliminated by the respiratory system, and bilirubin formed during the breakdown of hemoglobin is eliminated by the digestive system.

**Urea** is the principal nitrogenous waste product. It is formed in the liver during the breakdown of amino acids. Any time large amounts of body protein are being broken down, for example, during starvation, a large amount of urea will appear in the urine. **Uric acid** is formed from nucleic acids, the large molecules which make up DNA and RNA. A third nitrogen waste, **creatinine,** is formed from creatine. In Chapter 6 we discussed how creatine phosphate is a storage form of energy in muscles. The **steroids** in the urine are formed from the breakdown of the various steroid hormones such as aldosterone, estrogen, progesterone, and testosterone. In the past, analysis of the amount of the various steroids in the urine has been used to evaluate the blood level of the steroid hormones. Whenever the body is using fatty acids as the primary source of cellular energy, ketones begin to appear in the urine. It should be recalled from Chapter 16 that ketones are formed during fatty acid breakdown in the liver.

In addition to removing waste products from the blood, the kidneys play an essential role in regulating the electrolyte composition, osmolarity, and volume of the extracellular fluid. This fluid provides the internal environment in which the cells function. Throughout this book we have discussed the fact that the internal environment is kept stable by various homeostatic processes so that the cells can function properly. The kidneys play a major role in maintaining the stability of the internal environment.

Table 17-1 shows the composition of the intracellular fluid, the interstitial fluid, and plasma. Examination of this table will show that the fluid within the cells, the intracellular fluid, has quite a different composition than the fluid outside the cells, the interstitial fluid and plasma. Within the cells, potassium is the major positive ion or electrolyte. Most of the negative charge within cells is found on protein and hydrogen phosphate. Approximately two-thirds of the body fluid is found within cells.

In the extracellular fluid, sodium is the major positive ion and chloride and bicarbonate are the major negative ions. The electrolyte composition of plasma and interstitial fluid is essentially identical. Plasma differs from interstitial fluid in that it contains a certain amount of protein. Approximately one-third of the body fluid is extracellular. One-quarter of the extracellular fluid is plasma, and the other three-quarters is interstitial fluid.

Because water travels freely in and out of cells, the water concentration of the extracellular fluid must be kept exceedingly constant to prevent either swelling or shrinking of the cells. The water concentration is usually expressed in terms of **osmolarity,** the total

**Table 17-1** Composition of Body Fluids

| | Positive Ions (meq/liter) | | | | | | |
|---|---|---|---|---|---|---|---|
| | $Na^+$ | $K^+$ | $Ca^{2+}$ | $Mg^{2+}$ | | | |
| Intracellular fluid | 11 | 162 | 2 | 28 | | | |
| Extracellular fluid | | | | | | | |
|   Interstitial fluid | 145 | 4 | 5 | 2 | | | |
|   Plasma | 142 | 4 | 5 | 2 | | | |
| | Negative Ions (meq/liter) | | | | | | |
| | $Cl^-$ | $HCO_3^-$ | $HPO_4^{2-}$ | $SO_4^{2-}$ | Protein | Others | |
| Intracellular fluid | 3 | 10 | 102 | 20 | 65 | 3 | |
| Extracellular fluid | | | | | | | |
|   Interstitial fluid | 114 | 31 | 2 | 1 | 1 | 7 | |
|   Plasma | 101 | 27 | 2 | 1 | 16 | 6 | |

solute concentration. As the osmolarity increases, the water concentration decreases, and as osmolarity decreases, water concentration increases. One of the most important functions of the kidneys is to regulate the osmolarity of the extracellular fluid.

In this chapter we are going to look at the structure of the various organs of the urinary system and the mechanism by which the kidneys form urine. We shall pay particular attention to the way in which the kidneys regulate the fluid and electrolyte composition of the extracellular fluid in order to maintain a stable internal environment.

## ANATOMY OF THE URINARY SYSTEM

### Kidneys

The urinary system is illustrated in Figure 17-1. The two kidneys are located on the posterior wall of the abdominal cavity on either side of the vertebral column. They are just below the diaphragm and are loosely attached to the abdominal wall so that they move up and down with the diaphragm during breathing.

Both the inner and outer surfaces of the kidney are formed by concave surfaces facing in a medial direction. The inner, or medial, concave surface contains a notch called the **hilus**. At the hilus, blood vessels, nerves, and lymphatics enter and leave the kidneys. In addition the ureters which transport urine from the kidneys to the bladder leave the kidneys at the hilus.

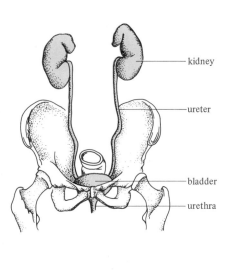

**Figure 17-1** Urinary system.

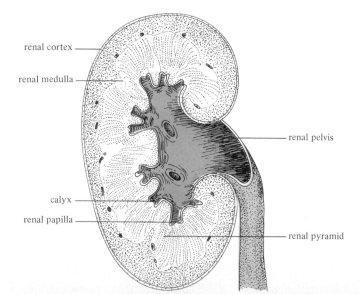

**Figure 17-2** Internal structure of the kidney.

**387** Anatomy of the urinary system

The entire kidney is covered by a fibrous capsule and is surrounded by adipose tissue. The outer portion of the kidney is called the **renal cortex,** and the inner portion is called the **renal medulla.**

Figure 17-2 shows the gross internal structure of the kidney. At the core of the medial surface the expanded upper end of the ureter forms the **renal pelvis.** The **calyces** are extensions of the pelvis into the renal medulla. Approximately 12 **renal pyramids** open through **renal papillae** into the calyces. The renal pyramids contain the collecting ducts of the renal tubules which will be discussed subsequently. Thus urine formed in the kidney tubules flows through the papillae into the calyces and then into the pelvis. From the pelvis the urine flows through the ureter to the bladder.

Each kidney is composed of approximately 1 million tiny **renal tubules** and their associated blood vessels. Figure 17-3 illustrates the structure of a renal tubule. The beginning of the tubule in the renal cortex is formed by a cup-shaped structure known as Bowman's capsule. This capsule leads into the convoluted **proximal tubule** which dips toward the medulla to form the **loop of**

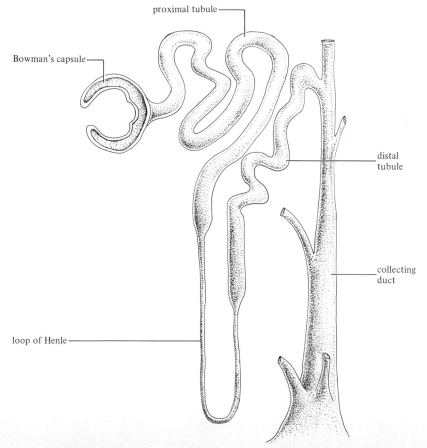

**Figure 17-3** Renal tubule.

**388** THE URINARY SYSTEM

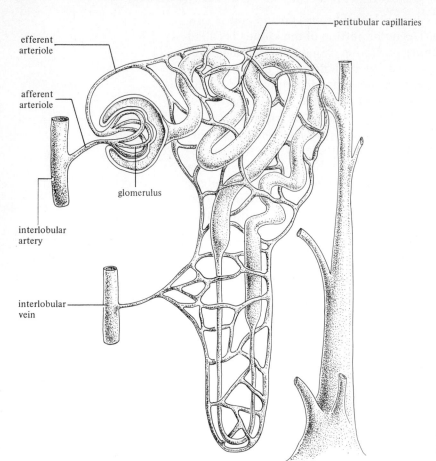

**Figure 17-4** Blood supply of a renal tubule.

Henle. In some tubules the loop of Henle extends deep into the medulla, whereas in others it only extends to the edge of the medulla. The loop of Henle joins with the **distal tubule** which, in turn, empties into a **collecting duct.** Urine is deposited into each collecting duct from a number of distal tubules. The collecting duct passes through the renal medulla as part of a renal pyramid. Eventually urine flows from the collecting ducts into the renal pelvis. In summary, each tubule is composed of a Bowman's capsule, proximal tubule, loop of Henle, distal tubule, and collecting duct.

Figure 17-4 illustrates the blood vessels associated with the renal tubules. Within the kidney the renal artery branches into smaller and smaller arteries. The smallest of these, the interlobular arteries, transport blood to the renal tubules. There are two sets of capillaries associated with the renal tubules. The **glomerulus** is a ball of capillaries embedded within Bowman's capsule. A second set of capillaries, the peritubular capillaries, surround the remainder of the tubule. Blood flow through the glomerulus and peri-

# Anatomy of the urinary system

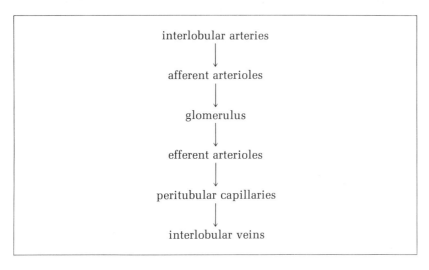

Figure 17-5 Flow of blood through the kidney.

tubular capillaries is controlled by two arterioles: the **afferent arteriole** and the **efferent arteriole.** The afferent arteriole is between the interlobular arteries and the glomerulus, whereas the efferent arteriole is between the glomerulus and the peritubular capillaries. Blood flows from the interlobular artery through the afferent arteriole and into the glomerulus. From the glomerulus, blood flows through the efferent arteriole into the peritubular capillaries. Blood from the peritubular capillaries drains into small veins which eventually converge to form the renal vein. Figure 17-5 sums up the flow of blood through the kidney. Each tubule and its associated glomerulus is referred to as a nephron. The nephrons are the functional units of the kidney; that is, they form the urine.

### Ureter

The ureters extend a distance of about 10–12 inches from the hilus of the kidneys to the base of the bladder. They pass along the posterior abdominal wall to the brim of the pelvis and then in a downward and medial direction along the pelvic floor to the posterior base of the bladder (see Figure 17-1). The ureters are lined by a mucous membrane that is continuous with the membrane lining the bladder and urethra. Three layers of smooth muscle, an inner longitudinal, middle circular, and outer longitudinal layer, form the bulk of the walls of the ureter. The outer surface is formed of fibrous connective tissue.

Waves of peristalsis pass along the ureter from the kidney to the bladder. These waves of muscular contraction are responsible for moving urine through the ureters. Although there are afferent and efferent neurons connected to the ureters, the peristaltic waves are not under nervous control. If a "stone" (a crystal of hardened

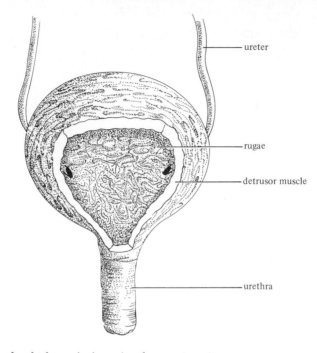

**Figure 17-6** Urinary bladder.

salts) formed in the kidney lodges in the ureter, it can cause extreme pain when it activates the afferent neurons.

### Bladder

The bladder is located in the pelvic cavity just behind the pubic bones (Figure 17-1). Figure 17-6 illustrates the structure of the bladder. Its mucous lining is folded into rugae when the bladder is empty and stretched into a smooth surface as the bladder fills with urine. The wall of the bladder contains a thick layer of smooth muscle known as the **detrusor muscle.** Contractions of the detrusor muscle expel urine during urination. Fibers of the detrusor muscle pass around the urethra to form the internal sphincter of the bladder. These fibers are arranged in such a way that contraction of the detrusor muscle opens the internal sphincter. The external spincter of the bladder is formed by a band of skeletal muscle which encircles the urethra. This sphincter is under voluntary control, except in the infant, and thus assures that urination occurs at the proper time. The **trigone** is a triangular area at the bottom of the bladder. It is bounded by the openings from the ureters at the back and the opening into the urethra in front.

### Urethra

The urethra is the tube through which urine flows out of the body from the bladder. In males the urethra is part of both the urinary system and the reproductive system. That is, both urine and semen flow through the male urethra. The structure of the male urethra is discussed in the next chapter and is illustrated in Figure 18-1.

The female urethra (Figure 18-10) is only excretory in function. It is approximately 1½ inches long and is located just anterior to the vagina. The posterior wall of the urethra is united with the anterior wall of the vagina.

The urethra in both males and females is lined with a mucous membrane and is composed of smooth muscle.

## BASIC PROCESSES OF URINE FORMATION

The formation of urine involves three basic processes: glomerular filtration, tubular reabsorption, and tubular secretion. Each of these processes is illustrated in Figure 17-7.

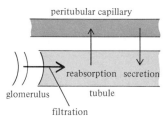

Figure 17-7 Basic processes of urine formation.

### Glomerular Filtration

Glomerular filtration is the movement of substances from the blood into the renal tubule at the glomerulus. As the name implies, this is a filtration process; substances below a certain size move into the tubules readily whereas those that are too large are held in the blood. Essentially all substances smaller than the plasma proteins enter the renal tubules. The plasma proteins and blood cells are not able to enter the tubules. Thus, if one examines the fluid at the very beginning of the proximal tubule, it will be identical to plasma except it will not contain the plasma proteins.

Glomerular filtration is brought about by a pressure difference between the fluid in the glomerular capillaries and the fluid in Bowman's capsule. The capillary pressure in the glomerulus is about 90 mm Hg. This is considerably higher than the pressure in most capillaries because the afferent arterioles are normally quite dilated. The capillary pressure pushing fluid into the tubule is opposed by the hydrostatic pressure of the fluid in the tubule and by an osmotic force caused by the presence of protein in the blood, but not in the tubule. These two forces pushing fluid from the tubule to the glomerulus are equal to a pressure of 45 mm Hg. Thus, the overall pressure pushing fluid into the renal tubule from the glomerulus is equal to 90 mm Hg − 45 mm Hg, or 45 mm Hg. Various degrees of constricting and dilating the afferent and efferent arterioles can change the glomerular capillary pressure and thus change the amount of glomerular filtration. Changes in arterial blood pressure will also change the amount of filtration.

Normally about 1 liter of blood a minute flows through the kidneys. This is one-fifth of the entire cardiac output of 5 liters per minute. As the blood flows through the kidney, about 125 ml of

**392** THE URINARY SYSTEM

fluid enter the tubules each minute. This is known as the glomerular filtration rate (GFR). A GFR of 125 ml per minute means that during one day, 180 liters of fluid enter the renal tubules. This is four times the total body water, 15 times the extracellular volume, and 60 times the plasma volume. On the other hand, a person normally loses about 1½ liters of fluid a day in the urine. In other words, 180 liters of fluid enter the renal tubules each day and 1½ liters leave the tubules as urine. The other 178½ liters of fluid are reabsorbed back into the blood.

### Tubular Reabsorption

Tubular reabsorption is the movement of substances from the renal tubules into the peritubular capillaries. It is a process that involves both active and passive transport. Certain substances such as sodium, potassium, glucose, and amino acids are actively transported out of the renal tubules and into the peritubular capillaries. That is, energy is expended by the cells of the tubules to move these substances from an area of lower concentration to an area of higher concentration. As illustrated in Figure 17-8, the active transport of these substances is followed by the passive reabsorption of the remaining filtered substances. When positive sodium and potassium ions are actively transported into the peritubular capillaries, an electrical potential is set up so that the capillaries are slightly

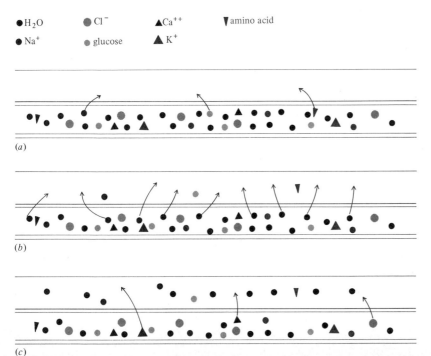

**Figure 17-8** Schematic illustration of tubular reabsorption. (a) Active transport of sodium, glucose, and amino acids. (b) Movement of water by osmosis. (c) Diffusion of remaining solute.

positive with respect to the tubules. Negative chloride and bicarbonate ions are then attracted to the positive charge and move into the capillaries. The active transport of glucose, sodium, and potassium raises the solute concentration of the peritubular capillaries and lowers the solute concentration of the tubules. As this happens the water concentration of the tubules becomes larger than the water concentration of the capillaries. Water then moves by osmosis from the renal tubules into the peritubular capillaries. This means that the reabsorption of water depends upon the reabsorption of solute. As sodium is the most abundant solute filtered into the tubules, the reabsorption of water depends primarily on the reabsorption of sodium. Anything that interferes with sodium reabsorption will interfere with water reabsorption. Likewise excessive sodium reabsorption leads to excessive water reabsorption. The exact amount of each substance reabsorbed depends on its level in the body.

### Tubular Secretion

Tubular secretion is the movement of substances from the peritubular capillaries into the renal tubules. It is brought about by both active and passive transport. Only certain substances are secreted into the tubules. The most important of these are hydrogen, potassium, and foreign chemicals such as drugs.

## CONTROL OF EXCRETION

The amount of any substance excreted depends upon the amount of that substance filtered, reabsorbed, and secreted. Figure 17-9 gives the relationship between these factors. The amount excreted is equal to the amount filtered plus the amount secreted minus the amount reabsorbed. In other words, an increase in the filtration or secretion of a substance leads to the appearance of more of that substance in the urine. An increase in the reabsorption of a substance leads to less of that substance appearing in the urine. For each constituent of the blood filtration, reabsorption, and secretion are controlled such that the proper blood level of that substance is maintained.

**Figure 17-9** Relationship between the amount of any substance excreted and the amount of that substance filtered, reabsorbed, and secreted.

### Control of Filtration

The degree to which any substance is filtered depends on two factors: the total amount of plasma filtered and the concentration

amount excreted = amount filtered − amount reabsorbed + amount secreted

**Table 17-2** Control of Filtration

|  | Situation | | |
| --- | --- | --- | --- |
|  | 1 | 2 | 3 |
| Concentration in blood | 10 mg/liter | 12 mg/liter | 10 mg/liter |
| Glomerular filtration rate | 100 ml/min | 100 ml/min | 120 ml/min |
| Amount of substance filtered | 1 mg/min | 1.2 mg/min | 1.2 mg/min |

of that substance in the plasma. Table 17-2 shows how these factors affect filtration. As more plasma is filtered there will be an increase in the amount of each plasma constituent (except protein) that enters the renal tubules. Likewise as the concentration of any substance in the plasma increases, so will the amount of that substance that is filtered into the renal tubules. This means that if the plasma concentration of a substance increases above normal, more of that substance will be filtered and thus more will be excreted. This helps return the plasma level back to normal.

The amount of plasma filtered depends on the glomerular capillary pressure. This pressure can be adjusted by constriction or dilation of the afferent and efferent arterioles. Dilation of the afferent arterioles or constriction of the efferent arterioles increases glomerular capillary pressure and therefore increases the glomerular filtration rate. Constriction of the afferent arterioles lowers glomerular capillary pressure and decreases the filtration rate. The afferent and efferent arterioles are controlled by sympathetic efferents of the autonomic nervous system. Impulses in these neurons lead to constriction of the arterioles.

### Control of Reabsorption

The amount that any substance will be reabsorbed depends on the amount filtered and in some cases on specific hormones. To a certain degree, reabsorption follows filtration so that if more of a substance is filtered, more will be reabsorbed. For many substances there is what is called a transport maximum. That is, reabsorption follows filtration up to a point. However, there is a maximum capacity for reabsorption. Beyond this point, any excess filtered substance cannot be reabsorbed and appears in the urine. For example, under normal circumstances the kidney reabsorbs all of the filtered glucose so that no glucose appears in the urine. If the plasma glucose goes up, more glucose is filtered but the excess is also reabsorbed. However, if plasma glucose exceeds 300 mg/ml, the reabsorbing ability of the kidney is exceeded and all of the excess glucose appears in the urine.

The rate of reabsorption of sodium, water, and calcium is under hormonal control. We will discuss the hormones that control the reabsorption of these substances in later sections of this chapter.

### Control of Secretion

Only certain substances are secreted into the renal tubules. The rate of secretion for these substances depends primarily on how concentrated they are in the blood. As the blood concentration increases, more of the substance is secreted and, therefore, more is excreted. To a certain degree, there is competition for secretion between different substances. For example, hydrogen and potassium seem to compete for secretion in the distal tubule. This means that if potassium secretion is increased, hydrogen secretion is decreased and vice versa. This will be described in more detail when we discuss renal handling of hydrogen and potassium.

## FLUID AND ELECTROLYTE BALANCE

One of the primary functions of the kidney is to regulate the fluid and electrolyte balance of the body. In this section we are going to look at the factors that affect the amount of the important electrolytes and water that exist in the body and how the kidneys act to maintain these substances at a stable level.

### Sodium Balance

Under normal conditions the extracellular sodium concentration is maintained at about 142 milliequivalents (meq) per liter. Because the intake of sodium is quite variable, the output of sodium must be regulated quite carefully. A small amount of sodium is lost through the sweat and feces. However, this loss is uncontrolled; it is not adjusted to the sodium needs of the body. Ultimately, the sodium concentration of the body is controlled by the kidney. The amount of sodium in the urine can be varied from 1 meq/day on a low-sodium diet to 400 meq/day on a high-sodium diet.

Because sodium is not secreted by the kidney, the amount of sodium excreted depends upon the amount filtered and the amount reabsorbed. The amount of sodium filtered depends on the glomerular filtration rate and the concentration of sodium in the plasma. Any situation that leads to an increase in sodium filtration will also lead to an increase in sodium excretion. Likewise a decrease in sodium filtration will lead to a decrease in sodium excretion.

Sodium reabsorption depends on sodium filtration and on the hormone aldosterone. To a certain extent a fixed proportion of the filtered sodium is reabsorbed. However, this proportion can be varied by aldosterone which promotes sodium reabsorption. In any situation in which sodium must be retained, aldosterone is secreted by the adrenal cortex and acts on the kidney to increase sodium

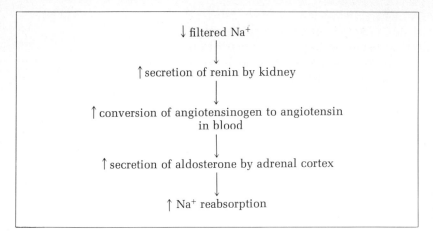

**Figure 17-10** Control of aldosterone secretion.

reabsorption. An increase in sodium reabsorption leads to a decrease in the amount of sodium excreted.

The rate of aldosterone secretion by the adrenal cortex is controlled in a somewhat complicated way. If there is any decrease in the amount of sodium filtered into the kidney, the kidney responds by secreting the hormone **renin.** Renin is released by the juxtaglomerular apparatus, a group of cells at the point at which the distal tubule lies adjacent to the afferent arteriole. The juxtaglomerular apparatus is composed of tubular epithelial cells and cells of the afferent arteriole. When renin enters the blood, it converts a plasma protein, angiotensinogen, into angiotensin. The angiotensin then stimulates the adrenal cortex to secrete aldosterone. Figure 17-10 sums up the control of aldosterone secretion.

Let us try to put all of this together by examining how the kidney responds to a drop in the sodium concentration of the extracellular fluid. The events of this response are shown in the flow chart of Figure 17-11. A decrease in the sodium concentration of the blood leads to a decrease in the amount of sodium filtered. If less sodium enters the renal tubules, less sodium will be excreted. Moreover, the decrease in filtered sodium leads to the release of renin by the kidney. Renin converts angiotensinogen to angiotensin, which in turn stimulates the adrenal cortex to increase its secretion of aldosterone. The aldosterone acts on the kidney to promote sodium reabsorption and thus decrease the amount of

**Figure 17-11** Renal response to a drop in extracellular sodium concentration.

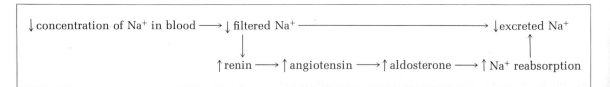

sodium excreted. Thus when the level of sodium in the blood falls there is a decrease in sodium filtration and a proportional increase in sodium reabsorption. Both of these factors act to reduce the amount of sodium excreted in the urine. This conserves sodium and helps elevate extracellular sodium back to normal.

**Water Balance**

The total amount of water in the body must be controlled in such a way as to meet two requirements: the volume of water and the osmolarity or water concentration should both be kept constant. Table 17-3 shows the various routes by which water enters and leaves the body. Almost equal amounts of water enter the body as part of liquid and of solid food. It should be kept in mind that so-called solid food contains a great deal of water. A cucumber is approximately 90% water and a rare steak is approximately 70% water. A smaller amount of water appears as the end result of metabolic reactions in the cells.

The major routes by which water leaves the body are through evaporation from the skin and lungs (insensible loss) and through the urine. Additional smaller amounts of water leave through the sweat and feces. Under normal conditions water loss exactly balances water gain.

Of the various routes by which water enters and leaves the body, only two are regulated to keep the body in water balance. Water intake is regulated by the thirst center in the hypothalamus and water output is regulated by the kidney.

Activation of the thirst center in the hypothalamus leads to a sense of dryness and a desire to drink. There are two types of stimuli that can activate the thirst center: a decrease in blood volume, and an increase in extracellular fluid osmolarity. A drop in blood volume is detected by volume receptors in the veins and atria. This information is relayed to the thirst center in the hypothalamus and there is a conscious desire to drink. Obviously, fluid intake helps restore blood volume to normal.

Changes in extracellular osmolarity are detected by **osmoreceptors** located in the hypothalamus itself. An increase in osmolarity activates the thirst center and leads to drinking. As additional water is taken into the body, fluids are diluted back to normal osmolarity.

Water output is controlled by the kidney. Because water is not secreted into the renal tubules, the total amount of water excreted by the kidney depends upon the amount of water filtered and the amount of water reabsorbed. As mentioned previously, the glomerular filtration rate, and thus the amount of water filtered, depends upon the glomerular capillary pressure. In any situation in which blood volume drops, there is reflex vasoconstriction as the

**Table 17-3** Water Intake and Output

| INTAKE | ml |
|---|---|
| Liquid | 1000 |
| Solid food | 1200 |
| Metabolism | 300 |
| TOTAL | 2500 |
| OUTPUT | ml |
| Lungs | 350 |
| Feces | 150 |
| Skin | 500 |
| Urine | 1500 |
| TOTAL | 2500 |

body attempts to maintain blood pressure. Vasoconstriction at the kidney leads to a reduction in the glomerular capillary pressure and a lower filtration rate. If less water is filtered, less water will be excreted. This helps restore the blood volume to normal.

Water reabsorption depends on two factors: sodium reabsorption and the amount of antidiuretic hormone in the blood. Because water reabsorption is dependent upon sodium reabsorption, the total volume of the extracellular fluid is regulated to a large extent through the regulation of sodium reabsorption. In the previous section we described how sodium reabsorption is regulated by aldosterone. Thus in situations in which the extracellular fluid volume drops, aldosterone is secreted and most of the filtered sodium and water are reabsorbed. Likewise if there is an expansion of the extracellular fluid volume, aldosterone secretion is inhibited and a larger percentage of the filtered sodium and water are excreted.

A second factor that affects water reabsorption is the blood level of antidiuretic hormone (ADH). ADH, a hormone secreted by the posterior pituitary gland, allows a certain amount of water to be reabsorbed independent of sodium reabsorption. This is illustrated in Figure 17-12. For reasons that are beyond the scope of this book to explain, there is a gradient of increasing solute concentration from the cortex to the medulla of the kidney. In addition, the fluid leaving the loop of Henle and entering the distal tubule is hypotonic to the blood. The reabsorption of water throughout the distal tubule and collecting duct depends on the presence of ADH. If there is little or no ADH present, water cannot pass through the tubular wall and consequently is not reabsorbed. This means that the hypotonic fluid which leaves the loop of Henle passes through the distal tubule and collecting duct without further water reabsorption. In this situation a large amount of dilute urine is formed. On

**Figure 17-12** Action of ADH on the kidney. (a) Low blood level of ADH. (b) High blood level of ADH.

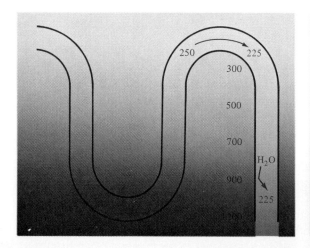
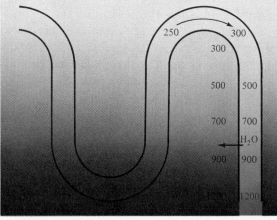

the other hand, in the presence of ADH water can pass through the walls of the distal tubule and collecting duct quite readily. Thus when the blood level of ADH is high, water is reabsorbed as fluid flows through the distal tubule and collecting duct. As the collecting duct descends toward the medulla of the kidney, the tubular fluid is exposed to an increasingly concentrated interstitial fluid. When ADH is present, water will move out of the collecting duct by osmosis and the urine will become quite concentrated. Thus, a high blood level of ADH leads to the formation of a small amount of concentrated urine.

The two stimuli for ADH secretion are a drop in blood volume or an increase in extracellular osmolarity. Blood volume is detected by volume receptors in the atria and veins. Osmolarity is detected by the hypothalamic osmoreceptors.

Renal handling of water is summed up in Figures 17-13 and 17-14. Figure 17-13 is a flow chart showing how the kidney responds to a drop in blood volume. When blood volume drops, blood pressure also begins to drop. The cardiovascular center responds by stimulating vasoconstriction. At the kidney, vasoconstriction leads to a decrease in glomerular capillary pressure and thus a decrease in the glomerular filtration rate. As less water is filtered into the renal tubules, less water will be excreted by the kidney. In addition, the reduction in glomerular filtration rate leads to a decrease in the amount of sodium filtered. The kidney responds to the decrease in filtered sodium by secreting renin. Renin causes the formation of angiotensin in the blood which in turn stimulates the adrenal cortex to secrete aldosterone. The aldosterone then promotes sodium and water reabsorption at the kidney. An increase in the amount of water reabsorbed decreases the amount

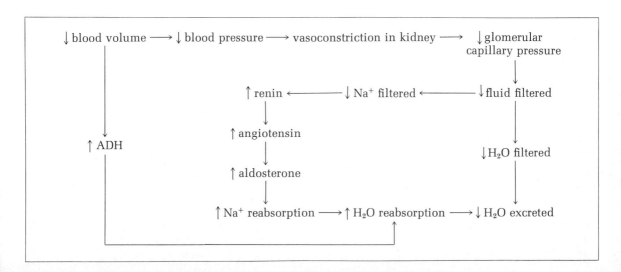

**Figure 17-13** Renal response to a drop in blood volume.

```
↓ osmolarity ⟶ ↓ ADH ⟶ ↓ H₂O reabsorption ⟶ dilute urine
```

**Figure 17-14** Renal response to a drop in blood osmolarity.

of water excreted. Water reabsorption is further increased by the release of ADH by the posterior pituitary in response to stimulation by the atrial and venous volume receptors.

Figure 17-14 is a flow chart of how the kidney responds to a drop in the osmolarity of the blood, as might occur after drinking a large amount of pure water. As blood osmolarity drops, the posterior pituitary greatly reduces its secretion of ADH. In the absence of ADH, water is not reabsorbed in the distal tubule and collecting duct. Thus a large amount of dilute urine is formed. This enables the body to rid itself of water without losing salt and the concentration of the body fluids is returned to normal.

### Potassium Balance

Most of the body potassium is located within the cells. Under normal conditions, extracellular potassium is maintained at a level of 4.5–5.0 meq/liters.

Potassium intake is less variable than sodium intake, usually being somewhere between 50 and 150 mg/day. Control of potassium balance is exerted by the kidney.

In general, almost all of the potassium which is filtered into the renal tubules is reabsorbed into the peritubular capillaries. This means that the amount of potassium excreted in the urine depends on the amount of potassium secreted from the peritubular capillaries into the renal tubules. As the blood level of potassium increases above normal, the rate at which potassium is secreted is also increased. This leads to an increase in potassium excretion and the return of blood potassium to normal. Aldosterone plays a role in the regulation of potassium balance. The adrenal cortex responds to an increase in blood potassium by secreting aldosterone. Aldosterone in turn promotes potassium secretion and thus potassium excretion. Figure 17-15 describes the renal response to an increase in plasma potassium.

Increased extracellular potassium, hyperkalemia, can result from cellular damage with the release of potassium from the damaged cells into the extracellular fluid. This is commonly seen after

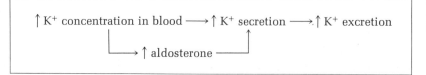

**Figure 17-15** Renal response to an increase in extracellular potassium.

burns. This increase leads to hyperexcitability of nerve and muscle cells with spastic paralysis. If the level of potassium gets high enough, it can cause cardiac arrest.

Any decrease in the extracellular fluid concentration of potassium interferes with the ability of nerve cells and muscle cells to conduct impulses. Decreased extracellular potassium (hypokalemia) can be brought about by a shift of potassium from the extracellular fluid to the intracellular fluid or by a loss of potassium from the body as occurs with certain diuretics. The primary symptoms of hypokalemia are muscular weakness and cardiac arrhythmias.

### Calcium Balance

Normally about 99% of the calcium in the body is contained in the matrix of bones. Extracellular calcium concentration is maintained between 4.5 and 5.2 meq/liter.

The body regulates calcium differently from the way it regulates sodium and potassium in that calcium absorption from the digestive tract into the blood depends on how much of it is needed. When the extracellular calcium concentration is low, parathormone is released from the parathyroid glands. Parathormone promotes calcium absorption. Likewise, when extracellular calcium is high, the parathyroids secrete very little parathormone and only a small amount of calcium is absorbed into the blood.

Under normal circumstances the bulk of the calcium contained in the food is not absorbed and leaves the body with the feces. Renal output of calcium is controlled primarily by the effects of parathormone on calcium reabsorption. When parathormone is released in response to a drop in extracellular calcium, it stimulates the kidney to increase its rate of calcium reabsorption.

Calcium competes with sodium for entry into nerve and muscle cells. Thus when there is too much calcium, sodium does not enter these cells readily and it is harder for them to generate impulses. This leads to muscular weakness and depressed reflexes. Calcium excess can lead to the formation of stones in the kidney as it tries to eliminate the calcium from the body.

Calcium deficiency makes nerves and muscles hyperexcitable and thus leads to twitching and, occasionally, convulsions. As calcium moves out of the bones in an attempt to raise the extracellular calcium concentration, the bones become weak.

## ACID-BASE BALANCE

One of the most closely regulated properties of the extracellular fluid is its hydrogen ion concentration. Before discussing how the

$$HCl \longrightarrow H^+ + Cl^-$$
hydrochloric acid

$$H_2CO_3 \longrightarrow H^+ + HCO_3^-$$
carbonic acid

**Figure 17-16** Examples of acids.

hydrogen ion concentration is regulated, we must first review a few basic definitions which were given in Chapter 2.

An acid is any chemical compound which can release hydrogen ion. In the equations given in Figure 17-16, HCl, hydrochloric acid, and $H_2CO_3$, carbonic acid, are both acids since they release hydrogen ion. It should be stressed that an acid is the chemical which releases hydrogen ion, not the hydrogen ion itself.

A base is any chemical compound which can combine with hydrogen ion. In the equations given in Figure 17-17, $OH^-$, hydroxyl ion, and $HCO_3^-$, bicarbonate ion, are the bases since they combine with hydrogen ion.

Hydrogen ion concentration is normally measured in terms of pH units. These units are sometimes confusing because the pH of a solution decreases as its hydrogen ion concentration increases. A solution with a pH of 7 is a neutral solution; its concentration of hydrogen ions is equal to its concentration of hydroxyl ions. A pH of less than 7 indicates an acidic solution; there are more hydrogen ions than hydroxyl ions. Conversely, a pH greater than 7 indicates a basic solution; there are more hydroxyl ions than hydrogen ions. Because of the way in which pH units are determined, a change of one pH unit indicates a 10-fold change in the hydrogen ion concentration.

Under normal circumstances the pH of the extracellular fluid is in the range 7.35–7.45. This is a hydrogen ion concentration of approximately 0.00004 mmoles/liter, a very small amount. Any change in pH, either above or below this level, leads to serious disturbances. The maximum pH range that can be tolerated for even a few minutes is 7.0–7.8.

$$OH^- + H^+ \longrightarrow H_2O$$
hydroxyl ion

$$HCO_3^- + H^+ \longrightarrow H_2CO_3$$
bicarbonate ion

**Figure 17-17** Examples of bases.

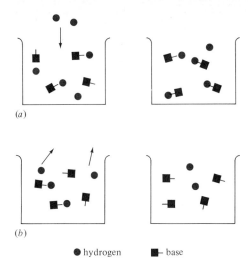

**Figure 17-18** Schematic illustration of the response of a buffer system to the addition or loss of hydrogen ion. (a) Addition of hydrogen ion. (b) Loss of hydrogen ion.

Maintenance of the proper pH involves three basic mechanisms: chemical buffers, respiratory adjustments, and kidney adjustments. **Chemical buffers** consist of acid-base pairs that prevent major changes in pH if hydrogen ion is added to or removed from a solution. Figure 17-18 indicates the mechanism by which an acid-base buffer functions. If extra hydrogen ion is added to the solution, the base pair of the buffer combines with it and removes it from the solution. In contrast if hydrogen ion is for some reason removed from the solution, the acid releases hydrogen ion to replace a portion of the amount of hydrogen ion lost.

The three important buffer systems in the body are carbonic acid-bicarbonate ($H_2CO_3$–$HCO_3^-$), dihydrogen phosphate-hydrogen phosphate ($H_2PO_4^-$–$HPO_4^{2-}$), and intracellular protein. Figure 17-19 indicates how each of these systems responds to the addition and removal of hydrogen ion from the body fluids. Extra hydrogen ion added to the body fluids is removed as it combines with bicarbonate, hydrogen phosphate, and protein bases. Hydrogen ion lost from the body fluids is replaced by the release of hydrogen ion from carbonic acid, dihydrogen phosphate, and acid proteins within cells. The carbonic acid-bicarbonate system is particularly important because the concentration of both elements can be ad-

$H^+ + HCO_3^- \longrightarrow H_2CO_3$   $H_2CO_3 \longrightarrow H^+ + HCO_3^-$

$H^+ + HPO_4^{2-} \longrightarrow H_2PO_4^-$   $H_2PO_4^- \longrightarrow H^+ + HPO_4^{2-}$

$H^+ + protein^- \longrightarrow H - protein$   $H - protein \longrightarrow H^+ + protein^-$

(a)   (b)

**Figure 17-19** Response of important body buffers to the addition and loss of hydrogen ion. (a) Addition of hydrogen ion. (b) Loss of hydrogen ion.

$$\uparrow \text{respiration} \longrightarrow \downarrow CO_2 \longrightarrow \downarrow H_2CO_3 \longrightarrow \downarrow H^+ (\uparrow pH)$$
$$\downarrow \text{respiration} \longrightarrow \uparrow CO_2 \longrightarrow \uparrow H_2CO_3 \longrightarrow \uparrow H^+ (\downarrow pH)$$

**Figure 17-20** Effects of changes in respiration on the pH of the body fluids.

justed to the body's needs. Carbonic acid concentration is regulated by respiration, and bicarbonate concentration is regulated by the kidney.

### Respiratory Control of pH

Respiration affects pH by determining the amount of carbon dioxide in the extracellular fluid. Because carbon dioxide combines with water to form carbonic acid, the level of carbonic acid is affected by the level of carbon dioxide. The level of carbonic acid in the extracellular fluid in turn affects the hydrogen ion level. Figure 17-20 describes the effects of increased and decreased respiration on pH.

An increase in respiration causes the removal of carbon dioxide from the extracellular fluid at an accelerated rate. This decreases the extracellular concentration of carbon dioxide. As the carbon dioxide level falls, the carbonic acid level also falls. This in turn causes a drop in hydrogen ion concentration (increased pH) because there is less acid present in the body.

The reverse situation takes place if respiration is depressed. Carbon dioxide is retained and thus the carbonic acid and hydrogen ion concentration become greater.

The hydrogen ion concentration of the extracellular fluid is one of the prime factors that determine the degree of respiration; therefore adjustments in respiration can be used to adjust the pH. If the hydrogen ion level increases (decreased pH), increased respiration in turn lowers the hydrogen ion level back toward normal. Likewise a drop in hydrogen ion concentration (increased pH) inhibits respiration. As the degree of respiration is lowered the hydrogen ion concentration begins to increase back to normal. Figure 17-21 sums up the role of respiration in the control of pH.

### Renal Control of pH

Figure 17-22 illustrates how hydrogen ion is handled by the kidney. The renal tubular cells secrete hydrogen ion into the tu-

$$\uparrow H^+ \longrightarrow \uparrow \text{respiration} \longrightarrow \downarrow H^+$$
$$\downarrow H^+ \longrightarrow \downarrow \text{respiration} \longrightarrow \uparrow H^+$$

**Figure 17-21** Role of respiration in the control of pH.

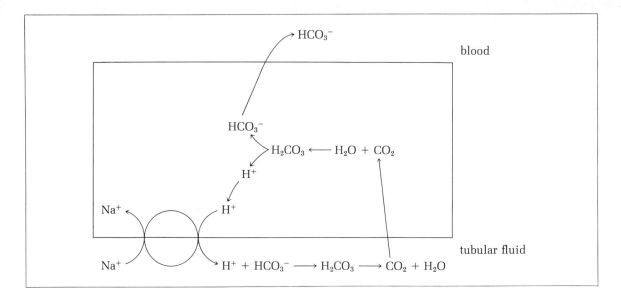

Figure 17-22 Relationship between hydrogen secretion and bicarbonate reabsorption.

bular fluid in exchange for sodium ions. Within the fluid the hydrogen ion combines with bicarbonate ion to form carbonic acid. The carbonic acid then dissociates into carbon dioxide and water. Carbon dioxide formed in the tubular fluid diffuses into the tubular cells and recombines with water to form carbonic acid. This carbonic acid then dissociates to bicarbonate ion and hydrogen ion. The bicarbonate ion is absorbed into the peritubular capillaries and the hydrogen ion is again secreted into the tubular fluid. In examining Figure 17-22, it can be seen that the overall effect of hydrogen ion secretion is the reabsorption of bicarbonate ion.

The rate of hydrogen ion secretion depends on the hydrogen ion concentration of the body fluids. If the concentration is low, the amount of bicarbonate flowing through the tubule will exceed the amount of hydrogen ion being secreted into the tubule. In this situation there will not be enough hydrogen to combine with all of the bicarbonate and the excess bicarbonate will appear in the urine. This is a protective mechanism since the loss of base from the body will help restore the hydrogen level back to normal.

Under most circumstances the amount of hydrogen ion secreted into the tubule exceeds the amount of bicarbonate filtered into the tubule. All of the hydrogen ion in excess of the amount needed to reabsorb bicarbonate remains in the tubular fluid and is excreted in the urine. The pH of the tubular fluid cannot fall below 4.5. This means that in situations in which a large amount of hydrogen ion must be excreted, some of the hydrogen ion must be combined with basic buffers in the tubular fluid. The two basic buffers found in the urine are ammonia and hydrogen phosphate.

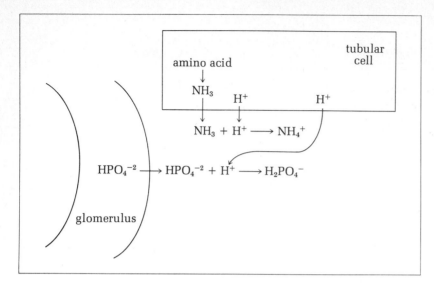

**Figure 17-23** Buffering of hydrogen ions in the urine.

Figure 17-23 illustrates the tubular buffering of hydrogen ion. The amount of ammonia produced by the kidney increases whenever large amounts of hydrogen must be excreted.

### Disturbances in Acid-Base Balance

The presence of excess hydrogen ion in the body fluids is known as **acidosis** and a lowering of hydrogen ion concentration is known as **alkalosis**. Acidosis and alkalosis can be brought about by respiratory and metabolic disturbances.

**Respiratory acidosis.** Any situation which interferes with respiration will lead to the retention of carbon dioxide and a buildup in the level of carbonic acid in the blood. This increases the hydrogen ion concentration and produces acidosis. The body compensates for the acidosis through the combination of hydrogen ion with basic buffers and the production of an acid urine. Respiratory acidosis is brought about by respiratory diseases such as emphysema and asthma.

**Metabolic acidosis.** Metabolic acidosis refers to any situation in which the hydrogen ion level increases for other than respiratory reasons. In diabetes mellitus and in starvation the fatty acid and ketone levels of the blood rise as these substances become the primary source of cellular energy. Both fatty acids and ketones are quite acidic and produce a metabolic acidosis. Diarrhea can produce an acidosis when the bicarbonate secreted by the pancreas is lost in the feces and not reabsorbed into the blood, thus leading to the loss of base. Kidney disease can produce acidosis if the kidneys lose their ability to excrete the hydrogen ion normally generated by the body.

Metabolic acidosis is compensated, in part, by the combination of hydrogen ion with basic buffers. In addition, the increase in hydrogen ion stimulates respiration which lowers the blood level of carbon dioxide. This helps reduce the hydrogen ion concentration. In metabolic acidosis, of the type not caused by kidney disease, the kidneys play the major role in excreting the excess hydrogen ion from the body.

**Respiratory alkalosis.** If respiration is increased, such as when one ascends to a high altitude or when one is anxious, the carbon dioxide level of the blood is decreased below normal and respiratory alkalosis develops. Compensation involves the release of hydrogen ion by basic buffers and the excretion of bicarbonate in the urine.

**Metabolic alkalosis.** Metabolic alkalosis is much less common than metabolic acidosis. It can be caused by the ingestion of sodium bicarbonate or by vomiting out the acid contents of the stomach. Respiration is somewhat depressed in compensation; however, there is a limit to this because respiration must be adequate to maintain the blood level of oxygen. The kidney responds to metabolic alkalosis by excreting bicarbonate and conserving hydrogen.

## MICTURITION

Micturition, or urination, is the process by which the bladder contracts and expels urine from the body through the urethra. When the bladder fills to a volume of 300–400 ml, stretch receptors are stimulated and impulses are conducted along afferent neurons to the sacral region of the spinal cord. These in turn activate parasympathetic efferents which stimulate contraction of the bladder and a desire to urinate. Higher centers in the brain can prevent urination at an inappropriate time by partially inhibiting the reflex and by maintaining closure of the extrinsic urinary sphincter. When there is an opportunity to urinate, the higher centers facilitate the urinary reflex and inhibit the external sphincter. Urination can be aided by voluntary contraction of the abdominal muscles.

## DISEASES OF THE KIDNEY

Kidney disease involves the loss of functional nephrons and often changes in those nephrons that remain functioning. In **glomerulonephritis,** many glomeruli are inflamed and nonfunctional. **Pyelonephritis** involves inflammation and obstruction of the renal tubules. Kidney disease can also be brought about by disease of the blood vessels leading to the nephrons.

Whatever the cause, there are certain general phenomena common to most forms of kidney disease. When some of the nephrons become nonfunctional, the glomeruli of the remaining nephrons become more permeable to protein. This leads to proteinurea, the presence of protein in the urine. This loss of protein in the urine can lead to a depletion of plasma protein and a drop in blood volume as the osmotic pressure pulling water from the interstitial fluid to the blood is decreased. Edema occurs as fluid accumulates in the interstitial space. The loss of functional nephrons also leads to a loss of the ability of the kidney to concentrate the urine. This leads to polyuria, the loss of a large amount of dilute urine. In the advanced stages of kidney disease, anuria ensues as the kidneys stop making urine. Poorly functioning kidneys are unable to excrete urea and other nitrogenous wastes and thereby cause a buildup of these substances in the blood. **Uremia,** the buildup of these wastes in the blood, can lead to lethargy and confusion. The functioning of the nervous system is further impaired by the acidosis which develops as the kidney is not able to excrete hydrogen ion. The uremia and acidosis caused by severe kidney disease eventually lead to coma and death if proper treatment is not obtained.

## ARTIFICIAL KIDNEY

The artificial kidney is designed to perform the essential functions carried out by the normal kidney. As illustrated in Figure 17-24, blood is pumped out of one of the patient's arteries through a coiled cellophane tube immersed in a bath and then returned to the body

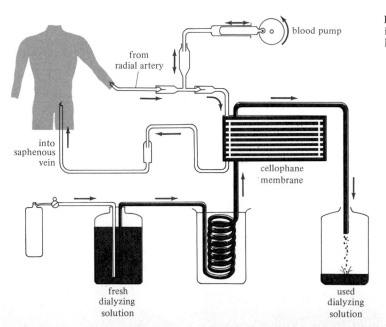

**Figure 17-24** Schematic illustration of the artificial kidney.

through a vein. The bath through which the tube passes is similar in composition to extracellular fluid except that it does not contain nitrogenous wastes. As blood flows through the cellophane tubing, any substances which are present in the blood in excess will diffuse out of the tubing into the bath; likewise any substances which are insufficiently concentrated in the blood will diffuse into the blood from the bath. Nitrogenous wastes such as urea diffuse into the bath and are thus removed from the blood. The cellophane is impermeable to protein so there is no loss of plasma proteins.

## OBJECTIVES FOR THE STUDY OF THE URINARY SYSTEM

At the end of this unit you should be able to:

1. State the two major functions of the urinary system.
2. State the major differences between the composition of intracellular fluid, interstitial fluid, and plasma.
3. State the location of the kidneys.
4. Describe the structure of the kidney in terms of the location of the renal cortex, renal medulla, renal pelvis, and hilus.
5. Name the different parts of a renal tubule.
6. Describe the vascular structures associated with a tubule.
7. State the function of the ureters.
8. Describe the location, structure, and function of the bladder.
9. Describe the differences between the function of the urethra in males and in females.
10. Describe the processes of glomerular filtration, tubular reabsorption, and tubular secretion.
11. State the relationship between the amount of any substance excreted and the amount of that substance filtered, reabsorbed, and secreted.
12. Name the two factors that determine the amount of any substance filtered.
13. Name the gland that secretes aldosterone and state the effect of aldosterone on the kidney.
14. State the stimulus for renin release by the kidney.
15. Explain the relationship between renin, angiotensin, and aldosterone.
16. Describe how the kidney would conserve sodium if the amount of sodium in the body decreased below normal.
17. Name the two stimuli that activate the thirst center and stimulate the posterior pituitary to secrete ADH.
18. Describe the effects of ADH on the kidney.
19. Name the two factors that determine the amount of $H_2O$ reabsorbed by the kidney.
20. Describe how the kidney conserves water if the blood volume drops or if the osmolarity of the blood increases.
21. State the relationship between aldosterone and potassium secretion.
22. Describe how the kidney would respond to an increase in plasma potassium.
23. Describe the changes that take place in the body in an attempt to compensate for acidosis and for alkalosis in terms of buffers, respiratory changes, and changes in the composition of the urine.
24. Distinguish between respiratory acidosis, metabolic acidosis, respiratory alkalosis, and metabolic alkalosis in terms of causes and the means by which the body compensates.
25. Describe the process of urination.

# 18
# The Reproductive System

MALE REPRODUCTIVE TRACT
MALE ACCESSORY GLANDS
MALE REPRODUCTIVE HORMONES
SPERMATOGENESIS
SEMINAL FLUID
DELIVERY OF SPERM TO THE FEMALE
MALE PUBERTY
FEMALE REPRODUCTIVE TRACT
MENSTRUAL CYCLE
PREGNANCY
LACTATION
FEMALE PUBERTY
FEMALE MENOPAUSE
SEX DETERMINATION

The function of the male and female reproductive systems is to produce new individuals for the continuation of the life of the species. In order to do this, egg cells produced by the female must be fertilized by sperm cells produced by the male. Once an egg has been fertilized, the female reproductive system must provide for the protection and nourishment of the developing fetus until it is capable of independent life.

## MALE REPRODUCTIVE TRACT

### Testes

Shortly before the birth of a male child the testes descend from the abdominal cavity through the inguinal canal into the scrotal sac. Because sperm can only be formed at a temperature that is below the normal body temperature, failure of the testes to descend will lead to sterility. Failure of the testes to descend is

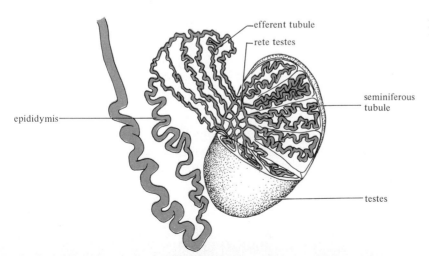

**Figure 18-1** Lobule of the testes.

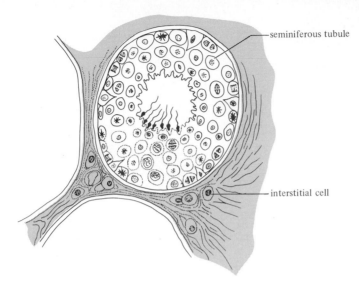

**Figure 18-2** Anatomical relationship between seminiferous tubules and interstitial cells of Leydig.

known as **cryptorchidism.** This condition can be corrected by surgery.

The testes are oval structures about 1½ inches long. They are covered by visceral peritoneum. Below the visceral peritoneum there is a layer of fibrous connective tissue. This tissue extends inward dividing the testes into lobules. As can be seen in Figure 18-1, each lobule contains a number of coiled **seminiferous tubules.** It is within these tubules that **spermatogenesis,** the formation of sperm, takes place. Once sperm have been formed in the seminiferous tubules they drain into a plexus called the **rete testes.** From the rete testes, sperm move through the **efferent ductules** into the epididymis, a tube lying on the outside of the testes.

The cells between the seminiferous tubules are called the **interstitial cells of Leydig.** These cells produce the male reproductive hormone **testosterone.** Figure 18-2 illustrates the relationship between the seminiferous tubules and the interstitial cells of Leydig.

### Epididymis

The epididymis is a coiled tube that lies on the superior and posterior surface of the testes. It receives, via the efferent ductules, the sperm formed in the seminiferous tubules. The sperm which enter the epididymis are not fully mature and it is within this tube that the final stages of maturation take place. At the base of the scrotum the epididymis joins with the ductus deferens.

### Ductus Deferens (Vas Deferens)

As can be seen in Figure 18-3, the ductus deferens (vas deferens) arises as a continuation of the epididymis at the base of the

# Male reproductive tract

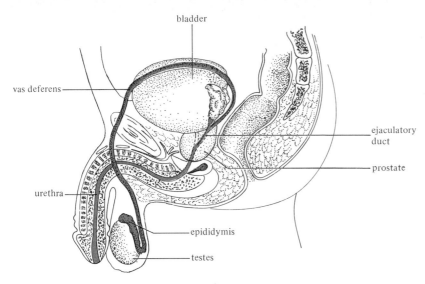

**Figure 18-3** Male reproductive system (sagittal section).

scrotum. It passes through the inguinal canal into the abdomen as part of the spermatic cord, which also includes the testicular arteries, veins, lymphatics, and nerves. Within the abdomen the ductus deferens passes over the lateral surface of the bladder from front to back. It then passes down the back of the bladder and enters the posterior surface of the prostate gland, as can be seen in Figure 18-4. At the point at which the ductus deferens enters the prostate

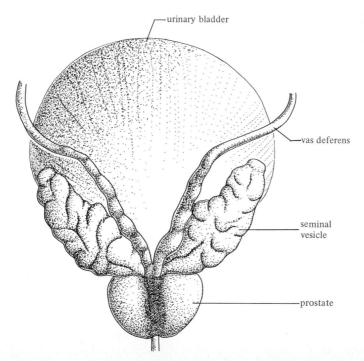

**Figure 18-4** Vas deferens and seminal vesicles (posterior view).

gland it joins with the duct from the seminal vesicle on that side to form the ejaculatory duct.

### Ejaculatory Duct

The two ejaculatory ducts are formed within the prostate by the union of the ductus deferens and the duct from the seminal vesicle on each side. Within the prostate the ejaculatory ducts join with the urethra slightly below the point it leaves the bladder.

### Urethra

In the male the urethra is part of both the urinary system and the reproductive system. Originating at the base of the bladder, it passes through the prostate gland where it is joined by the reproductive ejaculatory ducts. From the prostate the urethra passes through the penis and opens to the outside.

### Penis

The interior of the penis contains three bands of erectile tissue, as illustrated in Figure 18-5. These bands are the two dorsal **corpora cavernosa** and the ventral **corpus spongiosum.** The urethra runs through the substance of the corpus spongiosum. Erectile tissue contains large venous sinusoids that fill with blood during an erection.

At the end of the penis the corpus spongiosum enlarges to form the **glans penis.** This enlarged end of the penis is covered by the **prepuce,** or foreskin, which is often removed by circumcision.

## MALE ACCESSORY GLANDS

### Prostate

The prostate gland is located below the bladder and in front of the rectum, as can be seen in Figure 18-3. It surrounds both the ejaculatory ducts and the urethra.

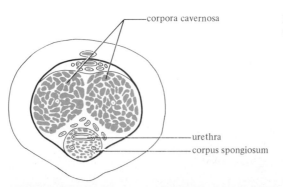

**Figure 18-5** Penis (cross section).

The prostate gland is covered by a fibrous capsule. Beneath this capsule lies a thick layer of smooth muscle which extends into the interior forming a network in which the glandular cells are embedded. Ducts from the glandular cells open into the urethra.

For unknown reasons it is very common for the prostate to enlarge as a man gets older. Because the enlarged prostate can compress and close off the urethra, prostate enlargement can cause serious urinary problems. Failure of the bladder to empty completely can lead to a bladder infection which can spread up the ureters to the kidneys.

### Seminal Vesicles

The two seminal vesicles are located behind the bladder and in front of the rectum. Like the prostate they contain glandular cells embedded in a network of smooth muscle. The main duct draining these cells joins the ductus deferens to form the ejaculatory duct.

### Bulbourethral Glands

These glands are located posterior and lateral to the urethra just below the prostate. Ducts from the glandular cells of the bulbourethral gland join the urethra as it enters the corpus spongiosum.

## MALE REPRODUCTIVE HORMONES

**Testosterone.** Testosterone is a steroid hormone produced in the testes by the interstitial cells of Leydig. It is synthesized by these cells from cholesterol and is quite similar in structure to the female steroids, estrogen and progesterone. The production and secretion of testosterone by the testes are dependent on the presence of luteinizing hormone (LH), one of the hormones secreted by the anterior pituitary gland.

Testosterone carries out a large number of functions. It is essential for the development and maintenance of the male reproductive structures. Spermatogenesis in the seminiferous tubules requires the presence of testosterone, although the major control over spermatogenesis is exerted by the pituitary hormone FSH. Testosterone is responsible for the development of the male secondary sexual characteristics. These include the male pattern of body hair distribution; the enlargement of the larynx, leading to a deeper voice; the enlargement of skeletal muscles; an increased production of red blood cells; and an increase in the basal metabolic rate. Testosterone has a growth-promoting effect. However, testosterone also acts to close the epiphyseal cartilages in long bones which limits the extent to which a male can grow in height. This is illustrated by the fact that eunuchs, castrated males, tend to be taller than normal males. Testosterone has an inhibitory effect on the secretion of LH by the pituitary. This acts as a negative feedback which serves to maintain a stable blood level of testosterone.

**FSH and LH.** The pituitary hormone FSH acts to stimulate spermatogenesis in the seminiferous tubules. Once puberty is reached, the blood level of FSH is maintained at a stable concentration so that spermatogenesis is a continuous, rather than an intermittent activity.

LH, or ICSH (interstitial cell stimulating hormone) as it is sometimes called in the male, stimulates the testes to secrete testosterone. As is the case for FSH, the blood level of LH is maintained constant after puberty. As a result the rate of testosterone secretion by the testes is also kept constant. Figure 18-6 is a schematic illustration of the interactions of FSH, LH, and testosterone.

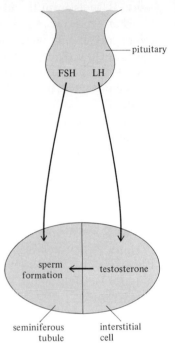

**Figure 18-6** Interactions of follicle-stimulating hormone (FSH), luteinizing hormone (LH), and testosterone.

## SPERMATOGENESIS

Spermatogenesis, the formation of sperm cells, takes place in the seminiferous tubules of the testes. In order for this process to take place, the hormones FSH and testosterone must be present. LH is

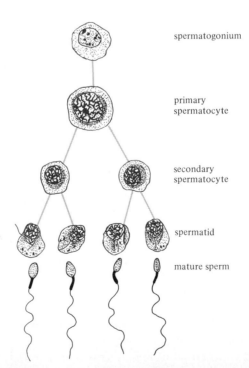

**Figure 18-7** Spermatogenesis.

required indirectly in that it is necessary for the production of testosterone.

The entire process of spermatogenesis takes approximately 75 days. It begins with the **spermatogonia,** the primitive cells adjacent to the walls of the seminiferous tubules, and proceeds in stages toward the center of the tubules. Figure 18-7 illustrates some of the important stages in the development of sperm.

The first stage involves the growth of a spermatogonium into a **primary spermatocyte.** This primary spermatocyte then undergoes a reduction division, a process known as meiosis, to form two **secondary spermatocytes.** In the reduction division the number of chromosomes is reduced from 46 to 23. Each secondary spermatocyte then undergoes mitosis and forms two **spermatids.** The spermatids then develop into sperm without further cell division.

Figure 18-8 illustrates the structure of a fully developed sperm cell, known as a **spermatozoan.** Approximately 0.05 mm long, a spermatozoan has three main parts: a head, a midpiece, and a tail. The **head** of the spermatozoan is almost completely filled with the nucleus, which contains the chromosomes and genes. On top of the nucleus there is a small amount of cytoplasm known as the **acrosome.** The acrosome contains enzymes which can break down the membrane of the egg during fertilization. Mitochondria and contractile filaments are found within the **midpiece** of the spermatozoan. These filaments continue into the relatively long **tail** which is similar in structure to a cilia. Contraction of the filaments cause the tail to move in such a way as to propel the entire cell.

Although most of the development of a sperm cell takes place in the seminiferous tubules the final stages take place in the epididymis. Once formed the sperm are stored in the ductus deferens prior to ejaculation. At the low pH of the male genital tract, the sperm are quite immobile and inactive. Sperm can stay fertile in the male tract for as long as 40 days.

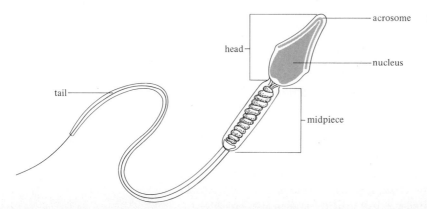

**Figure 18-8** Structure of a spermatozoan.

## SEMINAL FLUID

At the time of ejaculation, seminal fluid is added to the sperm by the seminal vesicles, the prostate gland, and the bulbourethral gland. This fluid is alkaline and serves to neutralize the acid fluid surrounding the sperm in the male ducts and the acid secretions of the vagina. The ejaculate, or semen, has a pH of about 7.5. This is essential because sperm are not motile at a low pH. The glandular secretions, particularly those of the seminal vesicles, also contain fructose which serves as a source of energy for the sperm.

## DELIVERY OF SPERM TO THE FEMALE

### Erection

Delivery of the sperm from the male reproductive tract into the female reproductive tract involves two basic processes: erection and ejaculation. An erection involves vasodilation of the arterioles of the penis and the filling of the erectile tissue in the penis with blood. Impulses arising from stimulation of the penis (particularly the glans penis), other erogenous zones, or from higher brain centers activate parasympathetic efferents in the sacral region of the spinal cord. These parasympathetic efferents stimulate vasodilation in the penile arterioles. When the penile arterioles dilate, blood flows rapidly into the venous sinusoids of the corpora cavernosa and corpus spongiosum. As these areas fill with blood and expand, they compress the veins which drain blood from the penis. Thus, a large quantity of blood is trapped in the penis causing it to swell and become erect. Sympathetic stimulation of the penile arterioles causes them to constrict. This leads to the penis becoming flaccid.

### Ejaculation

Ejaculation, the expulsion of semen from the male, involves two stages: emission and ejaculation proper. During emission stimulation by sympathetic efferents causes peristaltic contractions of the ductus deferens and epididymis. These contractions propel sperm into the urethra. At the same time contraction of the seminal vesicles, prostate, and bulbourethral glands adds seminal fluid to the urethra. Ejaculation proper involves contraction of the bulbocavernosus muscle, a skeletal muscle that surrounds the erectile tissue at the base of the penis. As this muscle rhythmically contracts semen is expelled from the urethra.

After two days of continence about $2\frac{1}{2}$–$3\frac{1}{2}$ ml of semen are ejaculated. More frequent ejaculation leads to the formation of a smaller volume of ejaculate. Although the number can vary greatly the average sperm count is about 100 million sperm per milliliter of ejaculate. If the sperm count is less than 20 million per milliliter there is usually infertility. However, it should be kept in mind that it is the quality of the sperm as well as the number of sperm that determines whether or not a male is fertile.

## MALE PUBERTY

Prior to puberty the secretion of FSH, LH, and testosterone is quite low and the testes do not produce sperm. In a boy of about 14 the hypothalamus begins to secrete FSH and LH releasing factors at a vastly increased rate. The pituitary responds to these releasing factors by increasing its secretion of FSH and LH. Once FSH and LH are secreted the testes respond by secreting testosterone and producing sperm. As testosterone secretion is increased, the reproductive organs enlarge and the male secondary sexual characteristics develop. It is not known what causes the increased secretion of FSH and LH releasing factors at puberty by the hypothalamus. However, it is known that the pituitary and testes are ready to respond to stimulation at any age. Thus it is the change in the rate of secretion of releasing factors by the hypothalamus which brings about puberty.

Although the rates of testosterone secretion and spermatogenesis decrease as a man gets older, there is no abrupt end to male reproductive function. There are documented cases of men being fertile in their eighties.

## FEMALE REPRODUCTIVE TRACT

### Ovaries

As can be seen in Figure 18-9, the two ovaries are located in the pelvis on either side of the uterus and just below the fallopian tubes. They are embedded in the posterior surface of the broad ligament that attaches the uterus to the walls of the pelvic cavity.

The ovaries are composed of an outer cortex and an inner medulla. Below the covering epithelium of the cortex there is a connective tissue framework which contains the **ovarian follicles.** At birth there are approximately 400,000 of these follicles present in the ovaries. Throughout the reproductive years one follicle will grow, develop a mature egg, and secrete the female reproductive hormones during each menstrual cycle. By menopause those follicles which did not mature have degenerated so that there are no longer any follicles left in the ovaries.

The inner medulla of the ovaries does not contain follicles. It is composed of loose connective tissue and contains the blood vessels which supply the ovaries.

## 420 THE REPRODUCTIVE SYSTEM

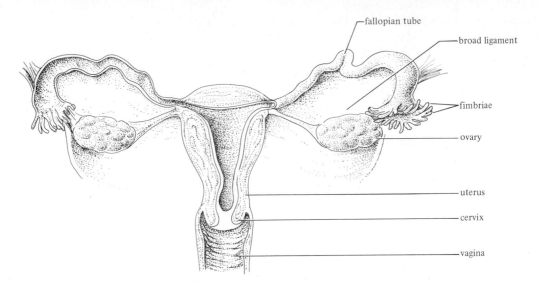

**Figure 18-9** Female reproductive system (posterior view).

### Fallopian Tubes

The fallopian tubes, or oviducts, extend through the upper portion of the broad ligament from the ovaries to the uterus. At the distal, or ovarian end, they are not attached directly to the ovaries, as can be seen in Figure 18-9. There is a small gap between the ovaries and the fallopian tubes. This means that eggs released by the ovaries must pass for a short distance through the abdominal cavity before they can enter the fallopian tubes. Fingerlike extensions of the fallopian tube, the **fimbriae,** act to pull the released egg into the tube.

The inner surface of the fallopian tubes is lined with a mucous membrane that contains both ciliated cells and secretory cells. Movement of the cilia along with peristaltic contraction of smooth muscle in the tubular wall act to move eggs down the tube. The mucous lining of the fallopian tube is continuous with the lining of the vagina and uterus. It is also continuous with the visceral peritoneum which lines the abdominal cavity. Unfortunately, this provides a route whereby infectious organisms that gain entry into the vagina can move into the abdomen and cause peritonitis.

### Uterus

The uterus is located in the pelvis posterior and superior to the urinary bladder, as seen in Figure 18-10. It is attached to the walls of the pelvic cavity by four ligaments: the broad ligament, the cardinal ligament, the round ligament, and the uterosacral ligament. The largest and most important of these is the **broad ligament** which extends from the lateral surface of the uterus to the lateral wall and floor of the pelvic cavity. Both the ovaries and the fallopian tubes are attached to the broad ligaments.

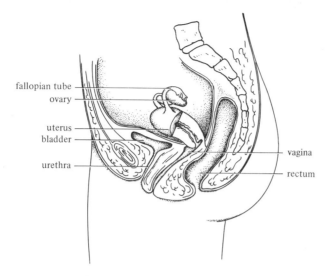

Figure 18-10 Female reproductive system (sagittal section).

The uterus is divided into two major areas: the cervix and the body. The **cervix,** which is the lower portion of the uterus, projects a small distance into the vagina (Figure 18-10). A transverse slit, the **os uteri,** is the opening between the uterus and vagina.

The **body** of the uterus is the upper, triangular portion above the bladder. When empty it lies flat on top of the bladder with its cavity extending in an anterior-posterior direction. It moves up and down as the bladder fills and empties.

The inner mucous membrane of the uterus is called the **endometrium.** It is composed of two layers: a **superficial layer** which expands during each menstrual cycle and is then shed during menstruation, and a **deep layer** which serves to nourish and regenerate the superficial layer. The bulk of the uterine wall is composed of smooth muscle which provides the force necessary for the delivery of the developed fetus. A serous membrane which is part of the visceral peritoneum covers the outer surface of the uterus.

### Vagina

The vagina passes obliquely upward and backward from its external opening to the cervix of the uterus. It lies behind the urethra and in front of the rectum. The external opening of the vagina is partially covered by a membrane, the **hymen.** At the uterine end of the vagina there are anterior, posterior, and lateral recesses where the vagina extends past the opening into the uterus.

A mucous membrane lines the inner surface of the vagina. This membrane contains glands which secrete the mucus that lubricates the vagina during intercourse. The bulk of the vaginal wall is composed of smooth muscle. This muscle, along with the smooth muscle of the uterine wall, contracts during the female orgasm. It is possible, although not established, that these contractions may in some way help transport the sperm to the egg.

### External Genitalia

Figure 18-11 illustrates the female external genitalia. The **mons pubis,** or mons veneris, is a pad of fat that lies in front of the pubic symphysis. Two longitudinal folds of skin, the **labia majora,** extend from the mons to the anus. These folds of skin contain hair follicles, sweat glands, and sebaceous glands. They cover a thick layer of underlying fat. Two thinner folds of skin, the **labia minora,** are located between the labia majora. Anteriorly, the labia minora meet to form the **prepuce,** a fold of skin that serves as a type of hood over the clitoris. The **clitoris** contains erectile tissue and is very similar in structure to the penis. It contains many sensory endings which play an important role in female sexual arousal. During arousal the clitoris fills with blood and becomes swollen.

The space posterior to the clitoris and between the labia minora is known as the **vestibule.** It contains the urethral opening in front and the vaginal opening in back. The ducts of the mucus secreting **Bartholin's glands** open into the vestibule at the vaginal opening. Mucus secreted by these glands, which are located beneath the labia minora, serves as a lubricant during intercourse along with the mucus secreted by the vagina.

The area from the posterior wall of the vaginal opening to the anus is known as the **perineum.** This area is sometimes cut, a procedure known as an **episiotomy,** to widen the birth canal during delivery.

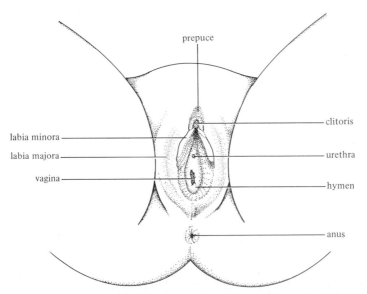

**Figure 18-11** Female external genitalia.

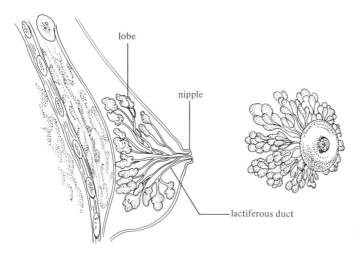

**Figure 18-12** Structure of the breast.

### Breasts

The breasts are located on the ventral surface of the thorax, extending from the 2nd to the 6th rib and from the sternum to the axilla. Figure 18-12 illustrates the internal structure of the breasts. Each breast is divided into about 20 lobes. Within each lobe there are smaller lobules which consist of glandular tissue surrounded by fat. The glandular tissue is composed of glandular cells that surround a duct. Interlobular ducts from each lobule drain into the main lactiferous duct of each lobe. The lactiferous ducts extend to the nipple where they open to the outside. A pigmented area, the **areola,** surrounds each nipple.

## MENSTRUAL CYCLE

The pattern by which the reproductive cells of the female, the eggs, are produced and released is very different from the pattern by which the male produces and releases sperm. Over a period of slightly less than a month one follicle in the ovary grows, produces and releases a mature egg, and secretes the reproductive hormones estrogen and progesterone. During this time period these hormones cause changes in the lining of the uterus so that it will be prepared to receive the egg if it is fertilized. If the egg is not fertilized the uterine lining is eventually shed and a new cycle begins. The pituitary hormones, FSH, LH, and LTH serve to regulate the growth and development of the ovarian follicle, its release of the egg, and its secretion of estrogen and progesterone. A menstrual cycle refers to the various events which take place in the ovary, uterus, and pituitary from the time one follicle starts to develop until the uterine lining is shed and a new cycle begins.

For convenience, the first day of the cycle is considered to be the day on which menstruation, the shedding of the uterine lining, begins. Thus in the following discussion, when we say that something happens on day 14, we mean 14 days after the onset of menstruation.

**Ovarian Events**

Starting on about day 3, one ovarian follicle, known also as a Graafian follicle, begins to develop under the influence of pituitary FSH and LH. The stages in the development of a follicle are illustrated in Figure 18-13. From about day 3 to day 14 the follicle undergoes a period of growth. During this period the egg undergoes a series of changes in which a primitive oogonium develops into a mature ovum. These changes include growth and a meiotic reduction division in which the number of chromosomes is reduced from 46 to 23. As the egg is developing, a fluid-filled space, the **antrum,** is formed in the middle of the follicle. This period of egg development, from day 3 to day 14, is known as the **follicular phase.** During the follicular phase cells located in the follicular membrane, which surrounds the antrum, secrete estrogen. The estrogen stimulates the endometrium of the uterus to thicken and acts on the pituitary to inhibit the secretion of FSH. This inhibition of FSH prevents the development of a second follicle once the first has begun its development.

On approximately day 14 the follicle, which has moved to the surface of the ovary, ruptures and releases its egg into the abdominal cavity. Movement of the fimbriae then acts to pull the released egg into the fallopian tube. Just prior to ovulation the pituitary greatly increases its secretion of LH. It is believed that this increased LH acts on the follicle to cause it to rupture. The release of the egg from the follicle is known as **ovulation.**

After the egg has been released the cells of the follicle fill with

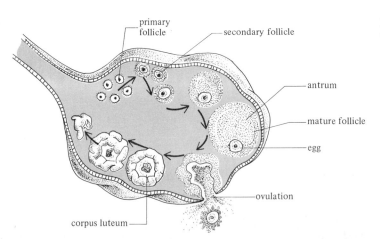

**Figure 18-13** Stages in follicle development.

fat to form the **corpus luteum.** These cells now secrete both estrogen and progesterone, which stimulate further development of the endometrium. Estrogen continues to inhibit the secretion of pituitary FSH and progesterone inhibits the secretion of pituitary LH. If a fertilized egg does not reach the endometrium of the uterus, on about day 26 the corpus luteum begins to degenerate. It then becomes filled with scar tissue and forms the corpus albicans. When estrogen and progesterone secretion cease, the endometrium undergoes degenerative changes, and after day 28 it begins to shed. This is the start of the next cycle.

### Uterine Events

From about day 5 to day 15 the endometrium forms a new epithelium which thickens as a result of estrogen stimulation. During this period new blood vessels grow into the superficial layer of the endometrium and the uterine glands grow. This period is known as the proliferative phase of the uterus.

After the corpus luteum is formed in the ovary, the endometrium continues to grow in response to estrogen and progesterone. The endometrial glands become even larger and begin to secrete the endometrial fluid. There is a continuation of the development of blood vessels in the superficial layer, and the stromal cells of the deep layer form deposits of fat and glycogen.

If a fertilized egg does not reach the uterus, the endometrium will begin to degenerate on about day 26 when the corpus luteum stops secreting estrogen and progesterone. As the estrogen and progesterone levels drop, the endometrial arteries supplying the superficial layer constrict. In the absence of a blood supply, the cells of the superficial layer degenerate and die. After a few days, on day 1 of the next cycle, the arteries dilate and blood flows into the weakened capillaries of the superficial layer. These capillaries then burst, causing a minor hemorrhage.

As blood flows out of the capillaries it washes away the dead epithelial cells of the superficial layer. These hemorrhages occur in one area of the endometrium at a time so that the entire period of menstruation lasts about five days. The areas which are shed at the beginning of menstruation are already being regenerated by the time the final area are shed. Menstruation involves the loss of about 30 ml of blood.

### Pituitary Events

**FSH.** FSH secretion by the pituitary is high at the beginning of the cycle. As the developing follicle begins to secrete estrogen, the level of FSH secretion begins to fall. It continues to fall until day 26 when the corpus luteum degenerates. At this point FSH secretion is no longer inhibited by estrogen and the rate of secretion increases.

**426** THE REPRODUCTIVE SYSTEM

The main action of FSH is to stimulate a new follicle to begin its development. Once this occurs FSH secretion is inhibited so that a second follicle will not develop at the same time.

**LH.** At the beginning of the menstrual cycle the pituitary secretes only a small amount of LH. It is thought that there must be a small amount of LH present along with the FSH for a follicle to begin its development. On about day 13 the pituitary begins to secrete LH at a greatly increased rate. The LH then causes ovulation and the formation of the corpus luteum in the ovary on day 14. As the corpus luteum secretes progesterone, the secretion of LH by the pituitary is inhibited. The blood level of LH falls until the 26th day when the corpus luteum degenerates. At this point, when the inhibitory effect of progesterone is removed, the rate of LH secretion begins to increase.

**LTH.** The secretion of luteotropic hormone (LTH), or prolactin as it is sometimes called, by the pituitary is increased once the corpus luteum is formed in the second half of the menstrual cycle. LTH seems to act on the corpus luteum to stimulate it to secrete estrogen and progesterone once it has been formed under the influence of LH. The feedback control of LTH secretion is not very well understood at the present time.

### Summary of the Menstrual Cycle

Figure 18-14 summarizes the changes which occur in the ovary, pituitary, and uterus during a menstrual cycle. Under the influence of FSH and a small amount of LH, a follicle begins to

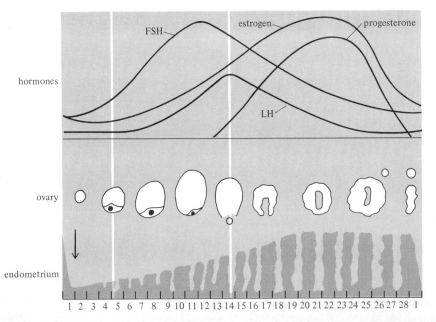

**Figure 18-14** Summary of menstrual cycle.

develop in the ovary on about day 3. From day 3 until day 14 the follicle grows, forms a mature ovum, and secretes estrogen. The estrogen stimulates the growth of the endometrium and inhibits the secretion of FSH.

On day 14 the follicle ruptures and releases its egg under the influence of LH. The follicle cells then fill with fat and begin to secrete progesterone as well as estrogen. Estrogen and progesterone cause a further increase in endometrial growth and the beginning of secretion by the endometrial glands. The secretion of hormones by the corpus luteum seems to be influenced by pituitary LTH.

In addition to stimulating the development of the endometrium, estrogen and progesterone inhibit the secretion of FSH and LH by the pituitary. If pregnancy does not occur the corpus luteum begins to degenerate on about day 26. As the corpus luteum degenerates and stops secreting estrogen and progesterone, the endometrium begins to degenerate.

After the 28th day the superficial layer of the endometrium is shed. Menstruation lasts from the 1st to 5th day of the next cycle. The degeneration of the corpus luteum on day 26 frees the pituitary from estrogen and progesterone inhibition. It responds by secreting FSH and LH which then reach high enough levels to cause a new follicle to develop by day 3 of the next cycle.

## PREGNANCY

Once ejaculated, sperm can remain fertile in the female reproductive tract for somewhere between 24 and 72 hours. The exact mechanism by which sperm move from the upper end of the vagina, where they are deposited, into and through the uterus and then up the fallopian tube to the egg is not well understood. Certainly the propulsive movements of the sperm's tail play a role. However, the sperm are probably also aided by peristaltic contractions of the female tract.

After being released from the ovarian follicle the egg is moved down the fallopian tube by peristaltic contractions and the beating of cilia. It is fertile for about the 18 to 24 hours it takes to reach the midpoint of the tube. Thus fertilization usually takes place somewhere in the upper half of the fallopian tube. If the egg is not fertilized it undergoes degenerative changes and is eventually expelled.

Although only one sperm actually enters and fertilizes the egg, many sperm must be present in the vicinity of the egg for fertilization to occur. It is believed that a large number of sperm are needed for the secretion of sufficient enzyme to break down the membrane of the egg. Once this enzyme, secreted by the acrosomal caps of the

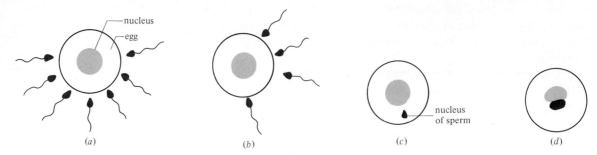

Figure 18-15 Fertilization. (a) Sperm surround egg. (b) Sperm comes in contact with egg. (c) Nucleus of one sperm enters egg. (d) Nucleus of sperm and egg join.

sperm, breaks down a portion of the egg membrane, the head of a single sperm enters the egg. The entry of one sperm then produces changes in the egg membrane so that no additional sperm can enter. Within the egg the nucleus of the sperm joins with the nucleus of the egg to form a fertilized egg, or **zygote.** The zygote almost immediately undergoes a mitotic division. Figure 18-15 illustrates the steps in the fertilization of an egg.

For the next three or four days after fertilization, the egg slowly moves down the fallopian tube to the uterus. This movement is brought about by both peristaltic contractions and the beating of cilia. As the egg moves down the tube it undergoes numerous cell divisions so that by the time it reaches the end of the tube it has formed a solid ball of cells called a **morula.**

Within the uterus the egg spends three or four days floating in the uterine fluid prior to embedding in the endometrium. During this time nourishment is received from the secretions of the uterine glands, and the egg continues dividing. By the time it is ready to implant in the endometrium the egg is formed into the shape of a hollow ball of cells called a **blastocyst.** Figure 18-16 illustrates the structure of a blastocyst. The outer layer of cells is called the **trophoblast.** It takes part in forming the placenta and the membranes which later surround the fetus. The inner group of cells will develop into the embryo and then the fetus.

Approximately six to eight days after the egg has been fertilized it begins to burrow its way into the endometrium, usually on

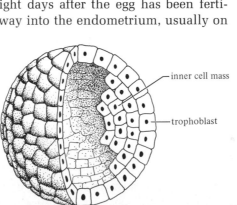

Figure 18-16 Blastocyst.

the dorsal surface of the uterus. This is accomplished through the breakdown of the endometrium at the site of implantation by enzymes secreted by the trophoblast. Materials released by the digested cells of the endometrium serve as nourishment for the embryo. Eventually the blastocyst becomes completely buried in the endometrium. The portion of the endometrium that covers the blastocyst is called the **decidua capsularis.** The rest of the endometrium is called the **decidua parietalis.**

### Placenta

The placenta serves as an organ for the exchange of material between the fetal and maternal circulation. Figure 18-17 illustrates the structure of the placenta. When the blastocyst implants in the endometrium, extensions of the trophoblast, the **villi,** grow into the endometrium and break down endometrial tissue. As endometrial capillaries are broken down, sinuses are formed around the villi. These sinuses contain maternal blood supplied by the uterine arteries and drained by the uterine veins. At the same time blood vessels grow into the villi of the trophoblast from the developing embryo. These vessels are branches of the umbilical arteries and veins. Thus fetal blood flows from the umbilical arteries into the capillaries of the villi and is then drained back to the fetus by the umbilical vein. Maternal blood flows from the uterine arteries into the sinuses surrounding the villi and is then drained into the

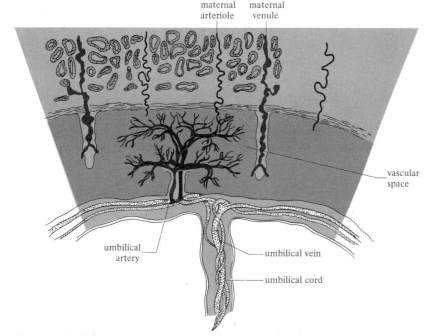

**Figure 18-17** Placenta.

uterine veins. As the fetal and maternal blood flow in close proximity through the placenta, substances can diffuse between the two bloods. Oxygen and nutrients diffuse from the maternal blood to the fetal blood whereas carbon dioxide, urea and other wastes diffuse from the fetal blood to the maternal blood.

**Embryonic Membranes**

As can be seen in Figure 18-18, the developing fetus is surrounded by two membranes: an outer **chorion** and an inner **amnion.** Both of these membranes surround a cavity filled with fluid.

**Hormones of Pregnancy**

As the blastocyst implants in the endometrium, cells of the trophoblast secrete a hormone called **chorionic gonadotropin.** This hormone prevents the degeneration of the corpus luteum so that it continues to secrete estrogen and progesterone. The estrogen and progesterone are needed to maintain the endometrium and the developing fetus. After about three months the corpus luteum does degenerate and the placenta itself takes over the secretion of estrogen and progesterone. In addition to chorionic gonadotropin, estrogen, and progesterone, the placenta seems to secrete a number of other hormones. A hormone called chorionic growth hormone—prolactin, CGP—acts in a manner similar to both prolactin and growth hormone. Another hormone secreted by the placenta has thyroid-stimulating hormone activity.

**Development of the Fetus**

During the first three months after conception all of the major

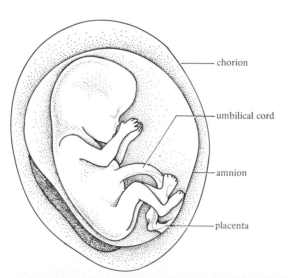

**Figure 18-18** Embryonic membranes.

organ systems of the fetus develop. At the end of this period the fetus is approximately 10 centimeters long. During the next six months the fetus grows. After five months the fetus weighs about 1 pound and at birth it usually weighs somewhere between 6 and 9 pounds.

### Fetal Circulation

Due to the fact that the fetus exchanges gases, nutrients, and wastes at the placenta its circulatory system has special adaptations. Figure 18-19 illustrates the fetal circulation. Blood flows to the placenta through the two **umbilical arteries** which branch off the internal iliac arteries in the pelvis. After passing through the capillaries of the placenta the blood returns to the fetus through the umbilical vein. Some of the returning blood flows through the fetal liver and then through the hepatic veins into the inferior vena cava. However, the bulk of the returning blood bypasses the liver by flowing from the umbilical vein into the **ductus venosus,** and then directly into the inferior vena cava.

Because there is no exchange of gases at the fetal lungs, the fetal circulation is designed in such a way that blood returning to

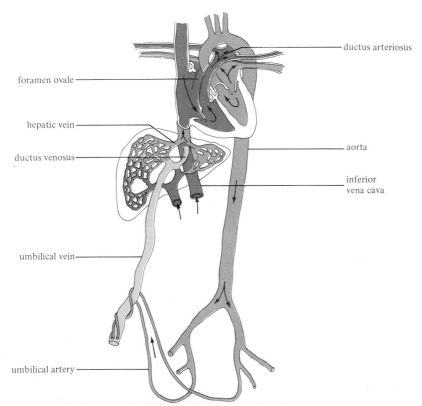

**Figure 18-19** Fetal circulation.

the fetal heart can bypass the lungs and be pumped directly back into the systemic circulation. The two structures which allow the blood to bypass the lungs are the foramen ovale and the ductus arteriosus. The **foramen ovale** is a hole in the wall between the right atrium and the left atrium. It allows blood returning to the right side of the heart to pass directly into the left side of the heart and be pumped into the aorta without ever flowing to the lungs. The **ductus arteriosus** is a short vessel connecting the pulmonary artery and the aorta. Blood that flows from the right atrium into the right ventricle and is then pumped into the pulmonary artery can flow through the ductus arteriosus into the aorta. This enables the blood pumped by the right side of the heart to enter the systemic circulation instead of flowing to the lungs. In general, blood returning to the right atrium from the inferior vena cava flows through the foramen ovale, whereas blood returning from the superior vena cava flows down into the right ventricle and then enters the aorta through the pulmonary artery and ductus arteriosus.

At birth the ductus venosus, ductus arteriosus, and foramen ovale close off. If the ductus arteriosus and foramen ovale do not close, the newborn will not have enough blood flowing to the lungs for the adequate exchange of oxygen and carbon dioxide. In this situation the structures must be closed by surgery.

### Parturition (Delivery)

Intermittent, mild uterine contractions usually begin about 30 weeks after conception. By the fortieth week these contractions develop into the strong contractions of labor. It is not clear at this point what physiological changes initiate labor; however, once labor begins the posterior pituitary gland secretes large amounts of oxytocin. This hormone acts on the uterus to stimulate strong muscular contractions.

The process of parturition, or labor, can be divided into two stages. The contraction of the first stage acts to dilate the cervix so that the fetus will be able to pass through. Once the cervix is sufficiently dilated the second stage of contractions acts to expel the fetus through the cervix and vagina. As the fetus passes through the birth canal the head usually acts as a wedge to open the passageway for the rest of the body. Delivery of the fetus during the second stage can be aided by contractions of the skeletal muscles of the abdomen.

Once the fetus passes out of the mother's body the umbilical blood vessels of the fetus and uterine blood vessels of the mother constrict. This causes the placenta to separate from the endometrium. It is then expelled by further contractions of the uterine smooth muscle.

## LACTATION

Lactation, the formation and secretion of milk, depends on the action of a number of hormones. Both glucocorticoids and growth hormones must be present, although it is not clear what role they play in lactation. Growth and development of the breasts after puberty is a result of the combined action of estrogen and progesterone. During pregnancy the rate at which these hormones are secreted is vastly increased. In response the breasts enlarge. It is felt that estrogen primarily causes development of the ducts and that progesterone primarily causes development of the secretory alveoli.

After delivery the anterior pituitary begins to secrete a large amount of prolactin, a hormone which causes the secretory cells of the breast to form milk and release it into the ducts. When the newborn infant begins to suck on its mother's nipples a reflex is initiated which leads to the secretion of oxytocin by the posterior portion of the pituitary. Oxytocin causes contraction of the ducts within the breast and the expulsion of milk. For the first two or three days after delivery the secretion of the breasts has protein and lactose, but very little water. This initial secretion is known as **colostrum.**


## FEMALE PUBERTY

Female puberty, like male puberty, depends on hypothalamic changes which lead to the increased secretion of FSH and LH releasing factors. These changes usually occur somewhere between the ages of 12 and 16. Once the releasing factors are secreted, the pituitary will secrete FSH and LH, and follicle development will begin in the ovary. Although the first cycles tend to be anovulatory and irregular, a normal pattern of menstrual cycles is soon established. The secretion of estrogen and progesterone during the menstrual cycles leads to the development of female secondary sexual characteristics, such as widening of the hips, enlargement of the breasts, and the growth of pubic hair.


## FEMALE MENOPAUSE

By the time a woman gets to be somewhere between 45 and 55 all of the ovarian follicles have either already developed or degenerated. At this time there are no longer any follicles left in the ovaries and menstrual cycles cease. The cessation of the menstrual cycles and the consequent end of estrogen and progesterone secretion is known as **menopause.** Women can go through a number of secondary physical and emotional changes during menopause. Symptoms often include headaches, hot flashes, muscular pains, and emotional upsets.

## SEX DETERMINATION

The sex of an individual depends on two chromosomes: the X chromosome and the Y chromosome. A female has two X chromosomes, whereas a male has an X chromosome and a Y chromosome.

Because a female has two X chromosomes, all of the eggs she forms by meiosis will have X chromosomes. On the other hand half of the sperm cells formed by a male will have X chromosomes and half will have Y chromosomes.

If a sperm with an X chromosome fertilizes the egg the child will be XX and therefore a girl. If a sperm with a Y chromosome fertilizes the egg the child will be XY and therefore a boy.

Within the fetus both male and female structures develop from the same primitive cells. In a genetic XY male these primitive cells begin secreting a testosterone-like substance about six weeks after conception and the male reproductive structures begin to develop.

A genetic XX female does not secrete this chemical and begins to develop female reproductive structures about nine weeks after conception. Thus it seems that the presence of this substance causes the formation of the male system and its absence causes the development of the female system. If testosterone is injected into a genetic female the male system will develop.

## OBJECTIVES FOR THE STUDY OF THE REPRODUCTIVE SYSTEM

At the end of this unit you should be able to:

1. Describe the internal structure of the testes.
2. State the two functions of the testes.
3. Describe the duct system that leads from the testes to the outside of the body.
4. Describe the internal structure of the penis.
5. State the location of the prostate gland, the seminal vesicles, and the bulbourethral glands.
6. Name the three male reproductive hormones and state the function of each.
7. Describe the structure of a mature sperm.
8. Describe the important components of the seminal fluid secreted by the prostate, seminal vesicles, and bulbourethral glands.
9. Explain the mechanism of erection and ejaculation.
10. Describe the changes that take place in a male at puberty.

11. State the location of the ovaries.
12. State the location of the fallopian tubes, uterus, and urethra.
13. Distinguish between the location of the cervix and the body of the uterus.
14. Describe the structure of the uterine wall.
15. Describe the location and structure of the various parts of the female external genitalia.
16. Describe the internal structure of the breasts.
17. Describe the changes that take place in an ovarian follicle during the menstrual cycle.
18. Name the structure formed by the follicle once an egg has been released.
19. Describe the changes that take place in the uterus during the menstrual cycle.
20. Describe the changes in the blood concentrations of the following hormones during a menstrual cycle: FSH, LH, estrogen, and progesterone.
21. State the approximate lengths of time sperm and eggs remain fertile in the female reproductive tract.
22. Describe the process of fertilization.
23. State approximately how long the fertilized egg remains in the fallopian tube.
24. State the approximate length of time between fertilization and implantation.
25. Describe the structure of the placenta.
26. Name the hormone released by the placenta immediately after implantation and state the function of this hormone.
27. State the function and location of the following structures characteristic of the fetal circulation: ductus arteriosus, foramen ovale, and ductus venosus.
28. Describe the process of parturition.
29. Describe the process of milk formation and secretion.
30. Describe the changes that take place in a female both at puberty and at menopause.
31. Explain the chromosomal difference between a male and a female.

# Glossary

If you cannot find the word you are looking for in this Glossary, check the Index for a text reference.

**absorption** to take into the blood
**acid** a molecule that releases hydrogen ion
**acromegaly** a disease involving the excessive secretion of growth hormone after closure of the epiphyseal cartilages; characterized by thick hands, feet, and jaws
**action potential** an electrical impulse in a nerve or muscle cell; a wave of depolarization and repolarization that passes along a cell membrane
**active transport** the use of cellular energy to move a substance from an area of lower concentration to an area of higher concentration
**adenoids** enlarged pharyngeal tonsils
**adenosine triphosphate** the form of energy that cells use to do work
**adipose** having to do with fat
**ADP** adenosine diphosphate
**afferent** toward
**agglutinate** to clump together
**alveoli** tiny sacs at the end of the respiratory tract; site of gas exchange with the blood
**amino acid** fundamental combining unit of a protein; contains amino group, acid group, and characteristic side chain
**androgen** male reproductive hormone
**anemia** a condition in which the red blood cell count or amount of hemoglobin is deficient, resulting in a reduction in the amount of oxygen transported by the blood
**anterior** in front
**arth-** having to do with joints
**-ase** indicates an enzyme
**atom** the basic unit of which molecules are formed
**atomic number** the number of protons in the nucleus of a particular type of atom
**ATP** adenosine triphosphate
**auditory** having to do with hearing
**autonomic nervous system** the portion of the nervous system that controls the internal environment
**axon** the portion of a neuron that conducts impulses away from the cell body

**baro-** having to do with pressure
**base** a substance that combines with hydrogen ion

**brachial** having to do with the upper arm
**buffer** an acid-base pair that plays a role in the regulation of pH

**carbohydrates** sugars; combinations of carbon, hydrogen, and water
**cardiac** having to do with the heart
**central nervous system** the brain and spinal cord
**cervical** having to do with the neck
**cervix** the lower end of the uterus, opening into the vagina
**cilia** tiny hair-like process extending from the external surface of a cell
**CNS** central nervous system
**cochlea** snail-shaped structure in the inner ear; contains the receptor cells for hearing
**coenzyme** a substance necessary for an enzyme to function
**collagen** a protein found in the space between cells
**colon** the large intestine, excluding the cecum and rectum
**concentration** amount per unit volume
**conjunctiva** lining of the interior surface of the eyelid
**coronal** indicating a front view of the body or a body part
**corpus** body
**cortex** the outer layer
**costal** having to do with ribs
**covalent bond** a chemical bond formed by the sharing of electrons
**cranial** having to do with the head
**cretinism** a disease caused by the lack of thyroxine in the fetus or young child; characterized by underdevelopment of all body systems
**cutaneous** having to do with the skin
**cyanosis** bluish color in the skin
**-cyte** a cell

**decussate** crossing as in an X
**defecation** the elimination of feces through the anus
**dendrite** the portion of a neuron that conducts impulses toward the cell body
**depolarization** a reduction in charge separation across a cell membrane
**diabetes insipidus** a disease caused by a lack of antidiuretic hormone; characterized by excessive urination
**diaphysis** the central shaft of a long bone
**diastole** cardiac relaxation
**diffusion** the movement of a substance from an area of higher concentration to an area of lower concentration
**disaccharide** combination of two simple sugar molecules
**distal** away from the origin
**DNA** deoxyribonucleic acid; the chemical of which genes are composed
**dorsal** in back

**edema** an accumulation of intercellular fluid
**effector** a structure that can bring about a change in the internal environment
**efferent** away from
**electrolyte** a particle with an electrical charge
**element** a substance made up of identical atoms
**-emia** in the blood
**emulsify** to break down large fat drops into smaller, water-soluble droplets
**endo-** inner
**endocrine gland** a structure that secretes hormones into the blood
**endometrium** the inner lining of the uterus
**energy** the ability to do work
**enzyme** a biological catalyst; a protein that speeds up a chemical reaction
**epi-** outside
**epiphysis** the end of a long bone
**epithelium** tissue in which the cells are close together, with little intercellular substance
**exocrine gland** a structure that secretes a chemical into a body tube (other than a blood vessel or onto the body surface)
**external** toward the outside
**extracellular** outside of cells

**fallopian tube** tube connecting the ovary and uterus
**femoral** in the region of the thigh
**filtration** the screening of particles below a certain size
**foramen** an opening
**frontal** in the region of the forehead; a front view of the body

**ganglion** nervous tissue composed primarily of nerve cell bodies located outside the central nervous system
**gastro-** having to do with the stomach
**genesis** formation
**glaucoma** a disease characterized by an increase in the pressure within the eyeball
**gluconeogenesis** the synthesis of glucose from noncarbohydrate sources
**glyco-** having to do with sugar
**glycogen** a polysaccharide; a storage form of energy
**glycolysis** the conversion of glycogen or glucose to pyruvic acid
**goiter** an enlargement of the thyroid
**gonad** reproductive structures; testes and ovaries
**gray matter** cell bodies and synapses in the central nervous system

**hemat-** having to do with the blood
**hemolysis** the destruction of red blood cells
**hepat-** having to do with the liver
**hilus** a depression in an organ where blood vessels and nerves enter or leave
**hist-** having to do with tissues
**homeostasis** the maintenance of a stable internal environment
**hormone** a regulating chemical secreted into the blood by an endocrine gland
**hydro-** having to do with water
**hyper-** above; above normal
**hyperopia** farsightedness; an inability to focus on close objects
**hypertonic solution** having a larger solute concentration than normal red blood cells
**hypertrophy** an increase in the size of an organ that does not involve tumor formation
**hypo-** below; below normal
**hypophysis** the pituitary gland
**hypotonic solution** having a lower solute concentration than normal red blood cells

**inferior** below
**inter-** between
**intercellular** between cells
**internal** toward the inside
**interstitial** the space between cells
**intracellular** within a cell
**-iole** small
**ion** an electrically charged atom or molecule
**ionic bond** a chemical bond between a positive ion and a negative ion
**iris** a pigmented muscle that regulates the amount of light reaching the retina
**isometric contractions** muscle contractions that do not move bones
**isotonic solution** having the same solute concentration as normal red blood cells
**isotonic contractions** muscle contractions that move bones
**-itis** inflammation

**kali-** having to do with potassium
**kilo-** one thousand

**lacrimal** having to do with tears
**lactation** the secretion of milk from the breast
**larynx** the voicebox, located at the top of the trachea
**lateral** toward the side
**ligament** connective tissue that connects bone to bone
**lingual** having to do with the tongue
**lipid** one of a group of carbon compounds that do not dissolve in water
**lumbar** having to do with the lower back
**luteal** having to do with the corpus luteum
**-lysis** denoting dissolving or breaking down

**macrophage** a cell capable of removing large particles by phagocytosis

**mammary** having to do with the breast
**meatus** an opening
**medial** toward the midline
**medulla** inside or center of an organ
**meiosis** cell division in which the new cells have half the number of chromosomes as the original cell
**meninges** the coverings of the brain and spinal cord
**metabolism** the sum of all chemical reactions that take place in the body
**micro-** small
**micturition** urination
**milli-** one-thousandth of
**mitosis** cell division in which the new cells are genetically identical to the original cell and each other
**mitochondria** major site of ATP formation in cells
**molecule** the combination of two or more atoms held together by electrical forces
**monosaccharide** simple sugar
**motor** having to do with movement
**myo-** having to do with muscle
**myopia** nearsightedness; an inability to focus on distant objects
**myxedema** extreme hypothyroidism in an adult

**nephron** a kidney tubule
**neuron** a nerve cell
**nucleus** center; in a cell, the structure that contains the genes

**occipital** having to do with the back of the head
**olfactory** having to do with smell
**oral** having to do with the mouth
**orbital** in the region of the eye
**os** opening
**osmolarity** the total solute concentration of a solution
**osmosis** the diffusion of water through a semipermeable membrane
**osteo-** having to do with bone
**-ostomy** forming an opening

**-otomy** cutting into
**ovulation** the release of an egg from an ovarian follicle

**parietal** having to do with the body walls; having to do with the top of the head
**parturition** giving birth
**patellar** having to do with the knee
**pelvic** having to do with the hips
**peptide bond** the chemical bond between two amino acids
**peripheral** toward the outside
**peristalsis** a wave of muscular contraction
**peritoneum** the membrane lining the inner surface of the abdominal cavity and the outer surface of most abdominal organs
**pH** a measurement of the amount of hydrogen ion in a solution
**phagocytosis** the engulfing and digesting of material by a cell
**pharynx** throat; junction of the nasal and oral cavities
**plantar** having to do with the foot
**plexus** an interconnecting network of nerves or blood vessels
**polar** a separation of electrical charge
**poly-** many
**polyuria** excessive urination
**posterior** in back
**pressure** force per unit area
**proprioceptors** receptors sensitive to the location and state of contraction of skeletal muscles
**proximal** toward the origin
**puberty** the onset of sexual maturity
**pulmo-** having to do with the lungs
**pupil** the opening in the middle of the iris
**pyrogen** a substance that acts to raise body temperature

**receptor cells** cells that respond to stimuli by initiating electrical impulses
**releasing factor** a hypothalamic hormone that stimulates the secretion of a pituitary hormone
**renal** having to do with the kidney

**repolarization** a return to the normal charge separation across a cell membrane
**reticular** intertwined; like a net
**retina** the layer of receptor cells in the eyeball
**ribosome** a structure in the cell involved in the synthesis of proteins
**RNA** ribonucleic acid; a substance used in the synthesis of protein

**sagittal** a side view of the body or a body part
**sarcomere** the basic contractile unit of a muscle
**sclerosis** a hardening of a tissue or organ
**scrotal sac** external sac that contains the testes
**septum** a partition
**sinus** space
**solute** a dissolved substance
**solvent** a substance in which other substances can be dissolved
**sphincter** a muscular ring that can close a passageway
**stenosis** constriction or narrowing of a passageway
**steroid** a lipid containing four interconnected carbon rings
**sty** an inflammation of the sebaceous glands associated with the hair follicles of the eyelid
**superior** above
**suture** a joining together
**synapse** the junction between two neurons or between a neuron and a muscle cell
**synovial** having to do with the fluid in joints
**systole** cardiac contraction

**temporal** having to do with the side of the head
**tendon** connective tissue that connects muscles to bones
**thoracic** having to do with the chest
**thrombo-** having to do with blood clots
**tissue** a group of similar cells and the material between these cells
**tonsil** a mass of lymphatic tissue located in the pharynx
**transverse section** a cross section of a body part
**tropic** something that stimulates
**tympanic membrane** eardrum; membrane separating the external and middle ears

**ureter** tube through which urine flows from the kidney to the bladder
**urethra** tube through which urine flows from the bladder out of the body; in men semen also leaves body through this tube

**vas deferens** tube through which sperm move from the testes to the urethra
**vascular** having to do with blood vessels
**ventral** in front
**ventricle** a cavity; the lower chambers of the heart
**visceral** having to do with the internal organs
**viscosity** the resistance of a liquid to flowing

**white matter** parts of neurons, called fibers, that carry information up and down the central nervous system

**zygote** fertilized egg cell

# Index

A band, 122
Abdominal aorta, 269, 273–274
Abdominal wall, 145–146
Abdominopelvic portion of ventral cavity, 16
Abduction, 116
ABO system, 247–249
Absorption, 331
Absorptive state, 367–368
Accommodation, 196
Acetabulum, 110
Acetylcholine, 123–124, 126–128, 129, 162–164, 206
Achilles tendon, 113, 148
Achlasia, 343
Acids, 28–29, 376–377, 401–407
Acinar cells, 233
Acne, 83
Acromegaly, 219
Acromion process, 105
Acrosome, 417
Actin, 122, 126–128, 253–254
Action potential, 126
Active transport, 49–51
Addison's disease, 230
Adduction, 116
Adenine, 61–66
Adenoids, 311
Adenosine diphosphate (ADP), 129
Adenosine triphosphate (ATP), 53–60, 128–131, 364–368, 370–371
Adipose cells, 76–77
Adipose tissue, 77
Adrenal glands, 229–232
Adrenocorticotropic hormone (ACTH), 82, 217, 221

Afferent arteriole, 389
Afferent neurons, 162–164
Afferent (sensory) pathways, 156, 297, 327–328
Agglutinogen, 247–249
Agranular leukocytes, 243
Agranulocytosis, 245–247
Air, composition of, 322–323
Albumin, 239
Aldosterone, 300, 384, 399, 400
Alimentary canal, 331
Alkalosis, 406, 407
Alpha globulins, 239
Alveolar air, 322–323
Alveoli, 7, 315–316, 323
Amino acids, 34–35, 49, 52, 57–66, 335, 356, 367, 368, 369, 392–393
Aminopeptidase, 356
Amnion, 430
Ampulla, 206
Ampulla of Vater, 349
Anabolic reactions, 35
Anal sphincter, 358
Anatomical location, terms describing, 14
Anatomy, 2
Androgens, 229, 230
Anemias, 241
Anemic hypoxia, 325
Aneurysm, 269
Ankle bones, 113
Antagonists, 135
Anterior chamber, 195
Anterior column, 164
Antibody formation, 244–245
Antidiuretic hormone (ADH), 221–222, 398–400

Antigens, 244–245, 247–249
Antihemophilic factor, 246
Antrum, 424
Aorta, 269, 273–274
Aortic arch, 269
Appendicitis, 357
Appendicular skeleton, 104–115
Appendix, 357
Aqueous humor, 195
Arachnoid layer, 179
Arbor vitae, 174
Areola, 423
Arm, 105–107, 139–141
Arrector pili muscles, 83
Arterial pressure, 295–300
Arteries, 237, 267–276, 431
Arterioles, 389, 418
Arteriosclerosis, 268, 296
Arteriovenous capillaries, 277
Arthritis, 114–115
Articular cartilage, 89
Ascending aorta, 269
Ascending colon, 358
Association areas, 178
Astigmatism, 197
Atherosclerosis, 268–269
Atlas, 101–102
Atmospheric air, 322
Atmospheric pressure, 317
Atomic number, 21–23
Atomic weight, 21
Atoms, 20–25
Atria, 260–262
Atrioventricular node, 258–259
Atrioventricular valves, 256–257
Auditory nerve, 205–206
Auricle, 201

Autoimmune diseases, 115, 245
Autonomic nervous system, 156–157, 207–209
Axial skeleton, 93–104
Axillary artery, 273
Axillary vein, 281
Axis, 101, 102
Azygos veins, 282

Balance, sense of, 206
Baroreceptors, 297
Bartholin's glands, 422
Basal ganglia, 176–177
Basal metabolic rate, 378
Base pairing, 62
Bases, 28–29, 401–407
Basilar membrane, 204
Basilic vein, 280–281
Basophils, 243
Beriberi, 337
Beta globulins, 239
Biceps brachii, 142
Bicuspid valve, 256
Bile, 350–351
Bilirubin, 242
Bladder, 8, 16, 390
Blastocyst, 428
Blood, 7, 74, 237–249, 323–326, 431–432
Blood cells, 87
Blood flow, 301–302
Blood plasma, 238–239
Blood platelets, 245–246
Blood pressure, 294–300
Blood sugar, 31, 232, 233–234
Blood vessels, 7, 80–81, 237, 267–268
Body cavities, 16
Body planes, 14–15
Body regions, 12–13
Body temperature regulation, 329–381
Bolus, 341–343
Bone marrow, 90–91
Bone matrix, 87
Bones, 74, 86–87, 89–116
Bone tissue, 87–88
Bowman's capsule, 388
Brachial artery, 273
Brachialis, 142
Brachial plexus, 168
Brachial veins, 281
Brachiocephalic artery, 271
Brachiocephalic veins, 281–282
Brain, 16, 156, 170–178, 199–201, 205–206
Brainstem, 170
Breasts, 423, 433
Broad ligament, 420
Broca's area, 178
Bronchi, 7, 314
Bronchial tree, 315

Bronchioles, 315
Buccinator, 136
Bulbourethral glands, 415, 418

Calcaneus, 113
Calcium, 86–87, 92–93, 246, 401
Calyces, 387
Canaliculi, 88
Canal of Schlemm, 195
Cancellous bone, 88
Canines, 339
Capillaries, 76, 237, 276–277, 287–291, 301–304
Carbohydrates, 30–32, 335, 355
Carbon, 24, 29–30, 32, 59
Carbon dioxide, 6, 7, 8, 57, 309, 325–326
Carbonic anhydrase, 325
Carboxypolypeptidase, 352, 355
Cardiac cycle, 260–262
Cardiac muscle, 78, 253–254
Cardiac orifice, 343
Cardiac output, 262–264
Cardiac region, 343
Cardiac sphincter, 342
Carotid arteries, 271–272
Carpal bones, 108
Carriers, 49
Cartilage, 74, 88–92
Catabolic reactions, 35
Cataract, 197
Catecholamines, 232
Cauda equina, 167
Caudate nucleus, 177
Cecum, 357
Celiac artery, 274–275
Celiac ganglia, 208
Cell membrane, 42–51
Cells, 2–4, 40–68, 121–127, 157–159
Cellular antibodies, 244
Cellulose, 31, 32, 335
Central fissure of Rolando, 176
Central nervous system, 12, 155–156, 179–182
Centriole, 42
Centromere, 68
Cephalic phase, 345
Cephalic vein, 280
Cerebellar cortex, 174
Cerebellar hemispheres, 174
Cerebellum, 170, 174–175
Cerebral aqueduct, 180
Cerebral arteries, 272
Cerebral cortex, 176, 177–178, 201
Cerebral hemispheres, 175–176
Cerebrospinal fluid, 180, 181–182
Cerebrum, 170, 175–178
Cervical plexus, 168
Cervical vertebrae, 100
Cervix, 421
Chemical bonds, 24, 52

Chemical buffers, 403–404
Chemical reactions, 35–37
Chemical transmitter, 160
Chemistry, 20–37
Chemoreceptors, 300
Chief cells, 344
Cholecytokinin, 350–351
Cholesterol, 350, 357
Cholinesterase, 123
Chondrocytes, 89
Chorion, 430
Choroid membrane, 195–196
Choroid plexuses, 181
Chromosomes, 40, 68, 434
Chylomicrons, 357
Chyme, 346–347
Chymotrypsin, 355
Chymotrypsinogen, 351–352
Ciliary body, 195
Ciliary muscles, 195
Circle of Willis, 272–273
Circulatory system, 7–8. See also Blood; Blood vessels; Heart; Lymphatic system
Circumduction, 116
Citric acid, 56
Clavicle, 104
Cleft palate, 97
Clitoris, 422
Coccygeal vertebrae, 100
Cochlea, 203
Cochlear duct, 204
Coenzymes, 37
Collagen fibers, 74–75
Collarbone, 104
Collecting duct, 388
Colliculi, 172
Colloid osmotic pressure, 239
Colon, 357–358
Colostrum, 433
Columnar epithelium, 73
Common bile duct, 349
Common hepatic duct, 349
Compact bone, 88, 89
Concentration, 28
Conchae, 98
Conduction, 379–380
Condyles, 111, 112
Conjunctiva, 193
Conjunctivitis, 193
Connective tissue, 73–78, 120–121. See also Blood; Bone; Cartilage; Lymph
Constrictor muscles, 341
Convection, 380
Cornea, 194, 197
Coronal plane, 15
Coronal suture, 99
Coronary arteries, 269–270
Coronary sinus, 255, 278
Corpus callosum, 175

# Index

Corpus luteum, 425
Cortisol, 229
Costal cartilages, 104
Covalent bond, 25
Cranial nerves, 167, 178–179
Cranial portion, 16
Cranium, 93–97
Creatine phosphate, 129–130
Creatinine, 384
Cretinism, 254
Cristae, 42
Crown, 339
Cryptorchidism, 412
Curare, 129
Cushing's disease, 230–231
Cyanosis, 82
Cystic duct, 349
Cytoplasm, 41
Cytosine, 61–66

Dead space, 322
Decidua capsularis, 429
Decidua parietalis, 429
Deciduous teeth, 338–339
Decussate, 172
Deep layer of endometrium, 421
Deep veins, 281
Defecation, 360–361
Deglutition, 341–342
Delivery, 432
Deltoid, 141
Dendrites, 158–159
Dense connective tissue, 78
Dentin, 339
Deoxyribonuclease, 352
Deoxyribonucleic acid (DNA), 61–68
Dermis, 82
Descending colon, 358
Detrusor muscle, 390
Diabetes insipidus, 222
Diabetes mellitus, 233, 376–377
Diaphragm, 144–145
Diaphysis, 89–90, 92
Diarthroses, 114
Diastole, 260–262, 294–295
Diencephalon, 172
Diffusion, 45–46, 49
Digestive system, 5–6, 331–361
Digestive tract, 6, 73, 331, 338–347
Diglyceride, 33
Dilator muscle, 195
Dipeptidase, 356
Disaccharide, 31
Distal tubule, 388
Dorsal cavity, 16
Dorsiflexion, 116
Double bond, 30
Ductus arteriosus, 432
Ductus deferens, 412–414
Ductus venosus, 431
Duodenocolic reflex, 360

Duodenum, 346–347, 352
Dura mater, 179

Ear, 201–206
Edema, 303–304
Effectors, 156
Efferent arteriole, 389
Efferent ductules, 412
Efferent neurons, 162–164, 207
Efferent (motor) pathways, 156, 298–300, 328
Ejaculation, 418–419
Ejaculatory duct, 10, 414, 418–419
Elastic cartilage, 89
Elastin fibers, 75
Electrolytes, 27, 357, 395–397
Electrons, 20–21, 24–25
Elements, 21–24
Embolus, 247
Embryonic membranes, 430
Enamel, 339
Endocardium, 254
Endochondral bone formation, 91
Endocrine disease, 215
Endocrine glands, 11–12, 73, 212
Endocrine system, 11–12, 212–234
Endolymph, 204
Endometrium, 421
Endomysium, 120
Endoneureum, 166
Endoplasmic reticulum, 41–42
Energy metabolism, 35, 51–60, 364–381
Enterogastrone, 347
Enterokinase, 351–352
Enzymes, 36–37, 344–345, 351–352, 355, 356, 367
Eosinophils, 243
Epicardium, 253
Epicranius, 135–136
Epidermis, 81–82
Epididymis, 412
Epiglottis, 312
Epimysium, 120
Epinephrine, 229, 231, 232, 302, 372–373
Epineureum, 166
Epiphyseal cartilage, 92
Epiphysis, 89, 92
Episiotomy, 422
Epithelial tissue, 71–73
Erection, 418
Erythroblast, 240
Erythroblastosis fetalis, 249
Erythrocytes, 238, 239–242
Esophagus, 6–7, 342–343
Essential amino acids, 335
Estrogen, 384, 425, 433
Ethmoid bone, 96
Eustachian tube, 201–202
Evaporation, 380

Eversion, 116
Excitatory postsynaptic potential (ESP), 161–162
Excitatory transmitters, 161–162
Excretion, 393–395
Exocrine glands, 213
Exopthalamus, 226
Expiration, 321
Expiratory reserve, 322
Expired air, 323
Extension, 116
External auditory meatus, 96, 201
External environment, 3, 4
External nares, 310
External obliques, 146
Extracellular fluid, 385
Extrinsic factor, 241
Eye, 192–201
Eyelids, 192–193
Eye movement, 137–138

Face, 93, 97–98, 135–137
Facial nerve, 191
Facilitated diffusion, 49
Falciform ligament, 347
Fallopian tubes, 10, 420
False pelvis, 111
False ribs, 104
Fats, 32–33, 52, 57–60, 356–357, 365–366, 368, 369–370
Fat-soluble vitamins, 336–337
Fatty acids, 32–33, 59, 357, 365–366, 367, 370–371, 377
Feet, 113–114
Femoral artery, 276
Femoral nerve, 168–169
Femoral vein, 283–284
Femur, 111
Fetal skull, 91, 98–99
Fetus, 430–432
Fever, 381
Fibrin, 246
Fibrinogen, 239, 244, 246
Fibroblasts, 75
Fibrocartilage, 89
Fibula, 112
Filiform papillae, 338
Fimbriae, 420
Flat bones, 91
Flexion, 116
Floating ribs, 104
Follicle-stimulating hormone (FSH), 217, 416, 419, 425–426, 433
Follicular phase, 424
Fontanels, 91, 99
Foramen magnum, 97
Foramen of Magendie, 180–181
Foramen ovale, 432
Foramina of Luschka, 181
4th ventricle, 180
Frenulum linguae, 338

Frontal bone, 94, 96
Frontalis, 136
Frontal lobe, 176, 178
Frontal plane, 15
Fructose, 30, 31
Fundus, 343
Fungiform papillae, 338

Galactose, 30, 31
Gallbladder, 349
Gamma globulins, 239
Gases, 26, 323–326
Gastric arteries, 274–275
Gastric phase, 345
Gastric pits, 343
Gastrin, 345
Gastrocnemius, 148
Gastrocolic reflex, 360
Gastroduodenal artery, 275
Gastroesophageal sphincter, 342–343
Gastroileac reflex, 360
Genes, 40–41, 61–68
Glans penis, 414
Glaucoma, 195
Glenoid cavity, 105
Glomerular filtration, 371–372
Glomerulonephritis, 407
Glomerulus, 388–389
Glossopalatine, 311
Glossopharyngeal cranial nerve, 191
Glucagon, 233, 234, 372
Glucocorticoids, 229, 374
Gluconeogenesis, 229
Glucose, 30–31, 52–54, 364, 365, 369–378, 392–393
Gluteus maximus, 146
Glycerol, 33, 59, 357, 366
Glycogen, 31, 129–131, 335, 368, 369
Glycolysis, 53–54, 56–57
Goblet cells, 79
Goiter, 226–227
Golgi apparatus, 42
Gonadotropin, 217
Gout, 115
Granular leukocytes, 242–243
Graves' disease, 225–226, 227
Gray matter, 164, 170
Growth hormone, 218–219, 373–374
Guanine, 61–66
Gyrus, 177

Hair, 82–83
Hamstring muscles, 147
Hands, 108, 143
Hard palate, 97
Haustra, 359
Haversian canals, 88
Haversian system, 88
Head movement, 138–139
Hearing, 204–206
Heart, 7, 16, 237, 252–264

Heart rate, 262, 263
Hematocrit, 238
Hemiazygos veins, 283
Hemocytoblast, 240
Hemoglobin dissociation curve, 324
Hemorrhage, 304–307
Hemorrhoids, 359
Heparin, 76
Hepatic artery, 274
Hepatic flexure, 358
Hepatic portal vein, 286
Hepatic sinuses, 348
Hepatic veins, 285
Hilus, 386
Hip bones, 108–111
Histamine, 76, 243
Histoxic hypoxia, 325
Homeostasis, 3–4
Hormones, 4, 11–12, 93, 212–224, 227–228, 233–234, 371–374, 384–385, 415–416, 425–427, 430
Horns, 164
H region, 122
Humerus, 105–106
Humoral antibodies, 244
Hyaline cartilage, 89
Hydrochloric acid, 344–345, 355
Hydrogen, 24, 29, 30, 32, 59
Hymen, 421
Hyoid bone, 100
Hyperopia, 198
Hyperparathyroidism, 228
Hypertonic solution, 48
Hypertrophy, 133
Hypoadrenalism, 230
Hypogastric arteries, 275–276
Hypogastric vein, 284
Hypoparathyroidism, 228
Hypopharyngeal sphincter, 341, 342
Hypophyseal fossa, 216
Hypophyseal portal veins, 221
Hypophyseal stalk, 216
Hypothalamic releasing factors, 220
Hypothalamus, 172–174, 381, 397
Hypotonic solution, 48–49
Hypoxia, 325
Hypoxic hypoxia, 325

I bands, 122
Ileocecal valve, 357
Ileum, 352
Iliac arteries, 275–276
Iliac crest, 110
Iliac veins, 284
Ilium, 109–110
Impulse, 126
Incisors, 339
Incus, 202
Inferior vena cava, 255, 284
Inhibitory postsynaptic potential (IPSP), 161–162

Inhibitory transmitters, 161–162
Inner ear, 201, 203–204
Insertion, 135
Inspiration, 319–320
Inspiratory reserve, 322
Insulin, 233, 371–372, 374–376
Interatrial septum, 254
Intercostals, 145
Internal capsule, 177
Internal environment, 3, 4
Interneurons, 162–164
Interphase, 68
Interstitial cell stimulating hormone (ICSH), 416
Interventricular septum, 255
Intervertebral disc, 102–103
Intestinal phase, 345
Intracellular fluid, 385
Intramembraneous bone formation, 91
Intrapleural pressure, 317, 318, 320
Intrapleural space, 315
Intrapulmonary pressure, 317–318, 320
Intrinsic factor, 241
Inversion, 116
Ionic bonds, 24–25
Ions, 25
Iris, 195
Irregular bones, 91
Ischial tuberosity, 110
Islets of Langerhans, 233
Isometric contractions, 119
Isotonic contractions, 119
Isotonic solution, 48

Jaundice, 242
Jejunum, 352
Joints, 114–116
Jugular veins, 279–280
Juvenile diabetes, 376

Keratin, 81
Keratinizing cells, 81–82
Ketones, 370, 377, 385
Kidneys, 8, 16, 300, 385, 386–389, 391–401, 404–409
Kinetic energy, 51
Knee, 111
Krebs (citric acid) cycle, 53–60, 364, 365, 366, 367
Kupffer's cells, 348
Kyphosis, 103

Labia majora, 422
Labia minora, 422
Lacrimal bones, 97
Lacrimal canals, 97, 193
Lacrimal glands, 193–194
Lacrimal sac, 193

Lactase, 355
Lactation, 433
Lacteals, 357
Lactic acid, 130–131, 369
Lactose, 31, 335
Lacunae, 88
Lamellae, 87
Large intestine, 6, 357–360
Laryngopharynx, 312
Larynx, 7, 312–313
Late-onset diabetes, 376
Lateral column, 164
Lateral corticospinal, 206–207
Lateral fissure of Sylvius, 176
Lateral ganglia, 208
Lateral lobes, 222
Latissimus dorsi, 140
Latogenic hormone (LTH or prolactin), 218
Legs, 111–114, 146–148
Lens, 195, 196–198
Lentiform nucleus, 177
Leukemia, 245
Leukocytes, 238
Levator palpebrae superioris, 136
Ligaments, 78, 102
Lingual tonsils, 312
Lipids, 32–34, 335–336
Liquids, 26
Liver, 11, 16, 347–351, 367, 369, 370
Long bones, 89–91, 92
Longitudinal fissure, 175
Loop of Henle, 387–388
Loose connective tissue, 77–78
Lordosis, 103
Lumbar arteries, 275
Lumbar puncture, 182
Lumbar veins, 284
Lumbar vertebrae, 100
Lungs, 7, 16, 314–328, 431
Luteinizing hormone (LH), 217–218, 221, 415–416, 419, 425, 426, 433
Luteinizing hormone-releasing factor, 220
Luteotropic hormone (LTH), 425 426–427
Lymph, 74
Lymphatic system, 287–292
Lymphocytes, 243, 292
Lysosomes, 42

Macrophages, 75–76
Macula lutea, 198–199
Malleolus, 112
Malleus, 202
Maltase, 355
Maltose, 335
Mandible, 97
Marrow, 87
Masseter, 138
Mass movements, 360

Mast cells, 76
Mastication, muscles of, 138
Mastoiditis, 96
Mastoid process, 94–95
Matter, states of, 25–26
Maxilla bones, 97
Medial epicondyle, 106
Median cubital vein, 281
Median fissures, 164
Mediastinum, 314
Medulla oblongata, 170–172
Medullar cavity, 90
Melanin, 82
Melanocytes, 82
Melanocyte-stimulating hormone (MSH), 82, 220
Membranes, 79–80
Meninges, 179–180
Menopause, 433–434
Menstrual cycle, 423–427
Mesenteric arteries, 275
Mesenteric ganglia, 208
Mesenteric vein, 286–287
Mesentery, 334
Messenger RNA, 63–66
Metabolic acidosis, 406–407
Metabolic alkalosis, 407
Metabolic rate, 378
Metabolism, 35, 51–60, 364–381
Metacarpals, 108
Metatarsals, 113
Microvilli, 353–354
Micturition, 407
Midbrain, 172
Middle ear, 201–203
Mineralcorticoids, 229
Minute volume, 322
Mitochondria, 42
Mitosis, 66–68
Mitral valve, 256–257
Molars, 339
Molecules, 24–25, 27, 36
Monocytes, 243
Monoglycerides, 33, 357
Monosaccharides, 30–31
Mons pubis, 422
Morula, 428
Motor neurons, 123–128, 129, 131–133, 206
Motor unit, 132
Mouth, 338
Mucin, 79
Mucosa, 331
Mucous membranes, 79
Mucous neck cells, 344
Mucus, 79, 340
Muscle fibers, 120
Muscle tissue, 78
Muscular dystrophy, 129
Muscularis externa, 332
Muscularis mucosae, 332

Muscular system, 8, 78, 86, 119–152, 162–164, 175, 196, 197
Myasthenia gravis, 129
Myelin sheath, 159
Myenteric plexus, 332–333
Myocardium, 253
Myopia, 197–198
Myosin, 122, 126–128, 253–254
Myxedema, 224–225

Nails, 83
Nasal bones, 97
Nasal cavities, 310–311
Nasal septum, 309
Nasolacrimal duct, 193
Nasopharynx, 311
Negative feedback, 221
Negative pressure, 317
Nerves, 4, 166–170, 178–179
Nerve tissue, 79
Nervous system, 11, 12, 155–182, 206–209, 374–376
Neuroglial cells, 159
Neurolemma, 159
Neuromuscular junction, 123, 124–125
Neurons, 123–129, 131–133, 157–164, 166, 206–207
Neutrons, 20–21
Neutrophils, 243
Nitrogen, 24, 29, 30, 58
Nodes of Ranvier, 159
Nonessential amino acids, 335
Norepinephrine, 209, 231–232, 302, 372–373
Normoblast, 240
Nuclear membrane, 41
Nucleolus, 41
Nucleotides, 62
Nucleus of atom, 20, 21
Nucleus of cell, 40
Nutrients, 334–338

Obliques, 138
Occipital bone, 96
Occipitalis, 136
Occipital lobe, 176
Odontoid process, 102
Olecramon process, 107
Olfactory cranial nerve, 192
Omentum, 334
Ophthalmic artery, 272
Optic chiasma, 201
Optic disc, 199
Optic nerve, 199
Oral cavity, 6
Orbicularis oculi, 136
Orbicularis oris, 136
Organ of Corti, 204
Organ systems, 5–12
Oropharynx, 311

Osmolar concentration, 46–47, 385–386
Osmoreceptors, 397
Osmosis, 46–49
Osmotic pressure, 47–48
Osteoarthritis, 115
Osteoblasts, 89
Osteoclasts, 92
Osteocytes, 87–88
Os uteri, 421
Oval window, 202
Ovarian arteries, 275
Ovarian follicles, 419
Ovarian veins, 284
Ovaries, 10, 419, 424
Oviducts, 10
Ovulation, 424
Oxidative phosphorylation, 56, 131, 364, 367
Oxygen, 6, 7, 24, 29, 30, 32, 59, 323–325
Oxygen debt, 131
Oxyntic cells, 344
Oxytocin, 222, 433

Pacemaker, 258
Palate, 97
Palatine bones, 98
Pallidum, 177
Pancreas, 11, 16, 233–234, 351–352
Pancreatic amylase, 352
Pancreatic duct, 349
Pancreatic lipase, 352, 356–357
Pancreozymin, 352
Papillary muscles, 257
Paralysis, 128–129
Parasympathetic nervous system, 209
Parathormone, 93, 227–228
Parathyroids, 227–228
Parietal bones, 94
Parietal cells, 344
Parietal layer, 79–80
Parietal lobe, 176
Parietal peritoneum, 334
Parietal pleura, 315
Parieto-occipital fissure, 176
Parotid glands, 340
Parturition, 432
Patella, 111
Pectoralis major, 141
Pedicles, 101
Peduncles, 174–175
Pellagra, 337
Pelvic brim, 111
Pelvic girdle, 108–111
Penis, 414, 418
Pepsin, 344–345, 355
Pepsinogen, 344, 345–346
Peptide bond, 34–35
Pericardium, 79–80
Perilymph, 204

Perimysium, 120
Perineum, 422
Perineureum, 166
Periodic table, 21–23
Periodontal membrane, 340
Periosteum, 89
Peripheral nervous system, 12, 156, 159
Peristalsis, 333, 342
Peritoneum, 79, 333–334
Peritonitis, 360
Pernicious anemia, 241
Peroneal nerve, 170
pH, 29, 404
Phagocytosis, 51, 159
Phalanges, 108, 114
Pharyngeal tonsil, 311
Pharyngopalatine arches, 311–312
Pharynx, 6, 311, 341
Phospholipids, 33, 357
Phrenic arteries, 275
Phrenic nerve, 168
Physiology, 2
Pia mater, 179–180
Pinocytosis, 51
Pituitary gland, 215–222
Pituitary hormones, 216–222, 373–374, 425–427
Placenta, 429–430
Plantar flexion, 116
Platysma, 137
Pleura, 79, 315, 318
Pleurisy, 315
Plexuses, 167–168, 332–333
Polar molecule, 27
Poliomyelitis, 129
Polymorphonuclear leukocytes, 242–243
Polysaccharides, 31, 335
Polyunsaturated fats, 33
Polyuria, 408
Pons, 172
Popliteal artery, 276
Popliteal vein, 283
Portal hypertension, 287
Portal system of veins, 285–287
Post-absorptive state, 369
Posterior chamber, 195
Posterior column, 164
Posterior communicating artery, 272
Potassium, 392–393, 400–401
Potential energy, 51
Potential osmotic pressure, 47–48
Pregnancy, 427–432
Premolars, 339
Prepuce, 422
Presbyopia, 198
Prime movers, 135
Procarboxypolypeptidase, 351–352
Progesterone, 384, 425, 433
Prolactin, 425, 426–427, 433

Prolactin inhibitory factor, 220
Pronation, 116
Prostate gland, 413–415, 418
Protein, 34–35, 60–66, 335, 355–356, 368
Prothrombin, 239, 246
Protons, 20–21, 24–25
Protraction, 116
Proximal tubule, 387–388
Pterygoid, 138
Ptyalin, 340
Puberty, 419, 433
Pubic arch, 110
Pubis, 109–110
Pudendal nerve, 170
Pudendal plexus, 170
Pulmonary arteries, 269
Pulmonary circulation, 256
Pulmonary embolism, 247
Pulmonary valve, 257
Pulmonary veins, 255
Pulse pressure, 295
Pupil, 195
Putamen, 177
Pyelonephritis, 407
Pyloric sphincter, 343
Pylorus, 343
Pyrogens, 381
Pyruvic acid, 53–56, 59, 130, 131, 366, 369

Quadriceps femoris, 147

Radial artery, 273
Radial veins, 281
Radiation, 379
Radius, 106–107
Ramus, 167
Reabsorption, control of, 394
Receptors, 12, 155, 297
Rectum, 6, 358
Rectus abdominis, 146
Rectus muscles, 137–138
Red blood cells, 238, 239–242
Red marrow, 90–91
Refraction, 196
Renal arteries, 275
Renal cortex, 387
Renal medulla, 387
Renal papillae, 387
Renal pelvis, 387
Renal pyramids, 387
Renal tubules, 387–388
Renal veins, 284–285
Renin, 396
Reproductive systems, 8–11, 411–434
Residual air, 322
Respiration, muscles of, 144–145
Respiratory acidosis, 406
Respiratory alkalosis, 407
Respiratory control center, 328

Respiratory receptors, 327
Respiratory system, 6–7, 170, 309–328
Rete testes, 412
Reticular activating system, 172
Reticular fibers, 75
Reticular formation, 172
Reticulocyte, 240
Reticuloendothelial system, 242
Retina, 195, 196, 198–201
Retraction, 116
Rheumatoid arthritis, 115
Rhodopsin, 199
Rh system, 249
Riboflavin, 337
Ribonuclease, 352
Ribonucleic acid (RNA), 63–66
Ribosomes, 41–42, 63–66
Ribs, 104
Rickets, 93
Rima glottis, 313
Root canal, 340
Roots of hair, 82–83
Roots of spinal cord, 164
Roots of teeth, 339–340
Rotation, 116
Rough endoplasmic reticulum, 42
Round window, 204

Sacculus, 206
Sacral artery, 275
Sacral vertebrae, 100
Sacroiliac joint, 109
Sagittal plane, 15
Salivary glands, 340–341
Saphenous vein, 283
Sarcolemma, 121, 123, 128
Sarcomere, 122
Sarcoplasm, 121
Sarcoplasmic reticulum, 122–123
Saturated fats, 33
Scapula, 104–105
Scar tissue, 92
Schwann cells, 159
Sciatic nerve, 169–170
Sclera, 194
Scoliosis, 103–104
Scrotal sac, 9
Sebaceous glands, 83
Sebum, 83
Secretin, 352
Sella turcica, 97
Semicircular canals, 202–203
Semilunar notch, 107
Semilunar valves, 257
Seminal fluid, 418
Seminal vesicles, 415, 418
Seminiferous tubules, 412
Semispinalis capitis, 138, 139
Senses, 177–178, 185–206
Serosa, 333

Serous membranes, 79–80
Sex determination, 434
Shaft, 82
Short bones, 91
Shoulders, 104–105, 139–141
Sigmoid colon, 358
Sinoatrial node, 258
Sinusoids, 277
Skeletal bones, 8, 93–115
Skeletal muscles, 8, 78, 86, 133–152, 206–207
Skin, 72–73, 80–83
Skull, 93–99
Small intestine, 6, 346–347, 352–357
Smell, sense of, 191–192
Smooth endoplasmic reticulum, 42
Smooth muscle, 78
Sodium, 392–393, 395–397, 399
Soleus, 148
Solids, 26
Solvent, 27
Sperm, 8–10, 411, 416–419, 427–428, 434
Spermatic arteries, 275
Spermatic veins, 284
Spermatids, 417
Spermatogenesis, 412, 416–417
Spermatogonia, 417
Spermatozoan, 417
Sphenoid bone, 96
Sphincter muscle, 195
Sphincter of Boyden, 349
Spina bifida, 101
Spinal cord, 16, 156, 162–166
Spinal nerves, 166–170
Spinal portion, 16
Spinal reflex, 162–164
Spinocerebellar, 190
Spinous process, 101
Spleen, 291
Splenic flexure, 358
Splenic vein, 287
Stagnant hypoxia, 325
Stapes, 202
Starch, 31–32, 335
Stenosis, 257
Sternocleidomastoids, 138–139
Sternum, 104
Steroids, 33–34, 336, 384–385
Stomach, 6, 16, 343–347
Striated muscle, 121
Stroke volume, 262, 263–264
Sty, 193
Subclavian arteries, 273
Subclavian vein, 281
Sublingual glands, 340
Submandibular glands, 340
Substantia nigra, 177
Sucrase, 355
Sucrose, 31, 335
Sugars, 30–32, 49

Sulcus, 177
Summation, 132
Superficial layer, 421
Superficial veins, 280–281, 283–284
Superior vena cava, 255
Supination, 116
Suprarenal arteries, 275
Suprarenal veins, 285
Surfactant, 316
Survival, 2
Suspensory ligaments, 195
Sutures, 99
Swallowing, 341–342
Sweat glands, 81, 83
Sympathetic nervous system, 208–209
Symphysis pubis, 109
Synapses, 160–162, 164, 166, 170
Synaptic knob, 160
Synarthrose, 114
Synergists, 135
Synovial fluid, 114
Synovial joints, 114
Synovial membranes, 80, 114
Systemic circulation, 256
Systole, 260–262, 294, 295

Talus, 113
Target cells, 213
Tarsae, 113
Taste, sense of, 190–191
Taste buds, 190–191
Tectoral membrane, 204
Teeth, 338–340
Temperature, 36
Temporal bones, 94, 96–97
Temporal lobe, 176
Temporal muscle, 138
Tendons, 78, 86
Teniae coli, 359
Terstitial cells of Leydig, 412
Testes, 9–10, 411–412
Testosterone, 10, 34, 384–385, 412, 415–416, 419
Tetanus, 132–133
Thalamus, 172, 174, 201
Thiamine, 337
3rd ventricle, 180
Thoracic arteries, 273
Thoracic portion, 16
Thoracic vertebrae, 100
Thorax, 103–104
Thrombocytes, 238, 245–246
Thrombus, 246–247, 269
Thymine, 61–66
Thymus, 291–292
Thyrocalcitonin, 227
Thyrocervical trunk, 273
Thyroid gland, 222–227
Thyroid-stimulating hormone (TSH), 217, 221, 224, 226

Thyroid-stimulating hormone-releasing factor, 220, 221, 224
Thyroxine, 223, 224
Tibia, 112
Tibial arteries, 276
Tibialis anterior muscle, 147–148
Tibial veins, 283
Tidal volume, 322
Tissues, 4, 5, 71–83
Toes, 114
Total lung capacity, 322
Trachea, 7
Tracheostomy, 313
Tracts, 164
Transfer RNA, 64–66
Transverse abdominis, 146
Transverse colon, 358
Transverse plane, 15
Trapezius, 139–140
Triceps brachii, 142
Tricuspid valve, 256–257
Triglycerides, 33, 335–336
Trigone, 390
Trophoblast, 428
Troponin, 126–128
True capillaries, 277
True pelvis, 111
True ribs, 104
Trypsin, 355
Trypsin inhibitor, 352
Trypsinogen, 351–352
t system, 122, 123
Tubular reabsorption, 392–393
Tubular secretion, 393
Tumor, 66
Twitch, 132
Tympanic canal, 204
Tympanic membrane, 201

Ulna, 106–107
Ulnar artery, 273
Ulnar veins, 281
Umbilical arteries, 431
Unsaturated fats, 33
Urea, 59, 384
Uremia, 408
Ureters, 8, 389–390
Urethra, 8, 10, 390–391, 414
Uric acid, 384
Urinary system, 8, 384–409
Urine, 8, 385, 386, 389, 391–393, 400
Uterus, 420–421, 425
Utriculus, 206

Vagina, 10, 421–422
Vallate papillae, 338
Valves, heart, 256–257
Vas deferens, 10, 412–414
Vasoconstriction, 268
Vasodilation, 76, 268
Vasomotor center, 297
Veins, 237, 277–287, 431
Ventral cavity, 16
Ventricles of brain, 180
Ventricles of heart, 260–262
Ventricular folds, 312
Vertebral artery, 273
Vertebral column, 100–103
Vertebral foramen, 101
Vestibular branch, 206
Vestibular canal, 203–204
Vestibule, 202–203, 422
Villi, 353, 429
Visceral layer, 79, 80
Visceral peritoneum, 334
Visceral pleura, 315
Viscosity, 296
Vision, 196–201
Vital capacity, 322
Vitamins, 37, 241, 246, 336–338
Vitreous chamber, 195
Vocal cords, 7
Volkmann's canals, 88
Voluntary nervous system, 156
Vomer, 97

Waldeyer's ring, 312
Water, 26–28, 47–48, 397–400
Water-soluble vitamins, 337–338
White blood cells, 77, 238, 245
White matter, 164, 177
Wrist bones, 108

Yellow marrow, 90, 91

Z lines, 122, 123
Zygomatic arch, 95
Zygomatic bones, 97
Zygomatic cells, 344
Zygomatic muscles, 136
Zygomatic process, 94–95
Zygote, 428